COLLECTED STUDIES SERIES

Astronomical Instruments and Their Users

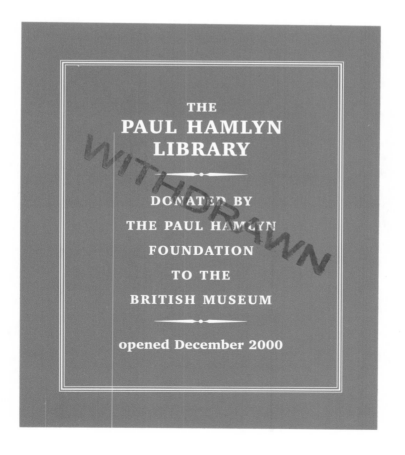

To Rachel

Allan Chapman

Astronomical Instruments and Their Users

Tycho Brahe to
William Lassell

VARIORUM
1996

This edition copyright © 1996 by Alan Chapman.

Published by VARIORUM
        Ashgate Publishing Limited
        Gower House, Croft Road,
        Aldershot, Hampshire GU11 3HR
        Great Britain

        Ashgate Publishing Company
        Old Post Road,
        Brookfield, Vermont 05036–9704
        USA

ISBN 0–86078–584–X

British Library CIP Data
        Chapman, Allan
        Astronomical Instruments and Their Users: Tycho Brahe to William
        Lassell. — (Variorum Collected Studies Series: 530)
        1. Astronomical Instruments. I. Title
        522.2

US Library of Congress CIP Data
        Chapman, Allan
        Astronomical Instruments and Their Users: Tycho Brahe to William
        Lassell / Allan Chapman.
           p. cm. — (Variorum Collected Studies Series: CS530)
        Includes index (cloth: alk. paper)
        1. Astronomical Instruments—History. I. Title. II. Series: Collected
        Studies: CS530.
        QB86.C56   1996   96–15909
        522'.2'09–dc20   CIP

The paper used in this publication meets the minimum requirements of the
        American National Standard for Information Sciences - Permanence
        of Paper for Printed Library Materials, ANSI Z39.48-1984. ∞ ™

Printed by Galliard (Printers) Ltd
        Great Yarmouth, Norfolk, Great Britain

COLLECTED STUDIES SERIES CS530

CONTENTS

This volume contains xii + 320 pages

PUBLISHER'S NOTE

The articles in this volume, as in all others in the Collected Studies Series, have not been given a new, continuous pagination. In order to avoid confusion, and to facilitate their use where these same studies have been referred to elsewhere, the original pagination has been maintained wherever possible.

Each article has been given a Roman number in order of appearance, as listed in the Contents. This number is repeated on each page and quoted in the index entries.

Some of the articles published in this collection, numbered III, VIII, X, XVI and XVII, were originally published in page formats that were not suitable for reproduction in this volume. In consequence, these articles have been re-set. The original pagination has been clearly indicated on the running-head line, and the specific page-divides by a square bracket, e.g. [321], in the text.

PREFACE

The ability to measure the natural world lies at the heart of all of the sciences, for without precise quantification, it is not possible to develop theoretical models or frame coherent laws of nature. Astronomy undoubtedly provided the role model in this respect, for even in antiquity its embodiment of geometrical principles into a collection of instruments made it possible to quantify and predict the heavens in a way that could not be emulated in the life sciences until the twentieth century.

Considering the centrality of instrumentation to science, it is unfortunate that such a relatively small part of the contemporary output of scholarship in the history of science is devoted to it. And even less attention is given to a scholarly understanding of those craftsmen and scientists who devised and used instruments.

Yet a study of instrumentation within a given science need not be restricted to the 'antiquarian', nor a study of the lives of its practitioners to the 'heroic', any more than an analysis of the social context of both has to be trammelled by currently fashionable sociological or deconstructionist theories.

What has always fascinated me about the past, and the reason why I became a historian and not a working scientist, is the individual peculiarity of human beings, and its expression in the creation of specific and ingenious objects. One might say that my interest in history stems from an interest in the quantification of differences: differences in nature, in the human eyes and minds that behold nature, and in the tools that beholding eyes have devised to help them, for man is the supreme toolmaker. This fascination with human diversity on the one hand, and the didactic power of objects on the other, is what first turned, and continues to attract, my attention to the history of science, and in particular, to the history of astronomy.

With the exception of items I, II and VII, all of the articles in this volume are concerned with a particular personage and his engagement with a problem in astronomy which could only be solved through that piece of physical ingenuity which we call an instrument: a form of human-cum-instrument biography, in fact. Articles II and VII are exceptions, however, for they deal with instruments that still survive in modern museum collections and have been made to yield up physical data about their own construction and accuracy, while leaving us in whole or partial ignorance of the individuals who made or used them. These two articles, on the late medieval astrolabe and on a collection of seventeenth-century astronomical quadrants, represent what might be seen as a form of 'unburied archaeology'. They go back to a time when astronomers were becoming increasingly aware of the relationship that exists between astronomical ideas and precise measurement, yet still lacked a clear concept of instrumental errors. My intention, in both of these pieces of research, was to subject surviving instruments to modern physical analysis, to see if one could establish a ceiling of accuracy for a given technology and epoch.

Our understanding of how astronomers in the past made and used their instruments can also be amplified by the practical procedures of reconstruction. Over the years, I have reconstructed a variety of instruments, and article IX recounts my work on the quadrant and astronomical radius of Pierre Gassendi. None of Gassendi's original instruments survive, although it was possible, from the accounts which he left of them, to reconstruct their scales and wooden working parts. While reconstructions can never tell us how accurate individual instruments from former centuries actually were, they can tell us a great deal about the practical constructional and handling problems that were inherent in a particular design. The factor of skill involved in making a scientific observation was often very large in the past, and I have always been concerned with trying to quantify and understand it. The articles already mentioned, along with article V on Jeremiah Horrocks, article XVII on William Lassell, and many more of the pieces in this book, deal in one way or another with the business of making physical objects produce significant data.

But the instruments themselves cannot be separated from the wider astronomical and social culture of a given period. How astronomers earned their livings and acquired the resources necessary to undertake lengthy and technically demanding pieces of research is part of this wider understanding, and many of the articles are concerned with this subject. It is true that much astronomy was funded from a clearly-defined public source, as the articles on Tycho Brahe (III), George Airy (XIV-XVI), and, to some extent, the Astronomers Royal (VIII), make clear, although during its first century of existence, the Greenwich Royal Observatory was so cash-starved that its first five Directors required some form of private means to keep going. These men, in fact, were in a sense cousins to that particularly English class of astronomers, the 'Grand Amateurs'.

Because the public funding of original research was virtually restricted to work that was of use to the Navy until after 1860, a great deal of British astronomy lay in the hands of wealthy amateurs. This self-funded tradition, and the economic factors that occasioned it, have been neglected by the majority of academic historians of science, who have generally preferred to look for scientific innovation in institutional settings, in the context of salaried research. Perhaps this is one reason why the history of British astronomy has been relatively neglected. Yet it is a field of immense richness, technically, socially, and from the standpoint of original archival material. It invites us to ask why a diverse collection of individuals who had made or inherited substantial fortunes should choose to spend their money in resolving nebulae, measuring the elements of double stars, or searching for new satellites in the outer solar system. It is this personal diversity, and the sheer unpredictability of individual motivation, which make this aspect of the history of science so important in understanding the social and intellectual priorities of former centuries.

On the other hand, equally significant historical insight can be gained from the study of a consciously professional career, if one approaches it in the right way. John Herschel, in article XIII, clearly straddles the professional and Grand Amateur divide, for while he was a Cambridge-trained mathematician, he held no formal scientific post, apart from voluntary ones in learned societies, and paid for forty years' worth of original research out of an inherited fortune. But without doubt, the first British 'professional' astronomer was George Biddell Airy, who negotiated hard, and publicly, for his salaries and who regarded his time as accountable to the State, as he made clear in his defence following the discovery of Neptune. The actions, social approaches, and sense of 'place' of these two men, and others like them, interest me deeply, for they show us much about the complexion of the early Victorian scientific world. And these men and their colleagues were great devisers and users of instruments.

Herschel tells us a good deal about how he, and his married-into-money Grand Amateur father (article XII), went about making and using large reflecting telescopes for cosmological research. Airy, a natural engineer, was not only concerned with the development of precision technology for the service of the nation, but with using it to minimise human error when quantifying the constants of nature, as is discussed in articles XV and XVI.

I hope that some of the articles in this collection will provide fresh assessments of certain astronomers and astronomical instruments. For instance, article V should dispel the long-standing, but primarily Victorian, myth that Jeremiah Horrocks was an Anglican clergyman; and my three articles on Airy will show that he was much more than a rigid disciplinarian whose concerns with regularity lost the discovery of Neptune for Britain. But most of all, I hope that this collection will demonstrate the value of specific case-studies, in both technology and biography, to forming an understanding of the history of astronomy.

ALLAN CHAPMAN

Wadham College, Oxford
1996

ACKNOWLEDGEMENTS

I wish to express my warmest thanks to Professor Gerard L'E. Turner, who first directed my attention towards the academic study of historical scientific instruments, and who encouraged me to produce this volume. I also thank Dr. John Smedley and his editorial staff at Variorum for guidance in its production. Debts of gratitude are, in addition, owed to the staffs of the Bodleian and Radcliffe Science Libraries, Oxford; and to the Warden and Fellows of Wadham College, Oxford, for convivial academic sanctuary over the years. The staff of the Royal Society Library have been unfailingly helpful, and particular thanks are due to Peter Hingley and Mary Chibnall of the Royal Astronomical Society Library for their ready assistance. An especial debt of gratitude is owed to the staff of the Museum of the History of Science, Oxford, and in particular to Tony Simcock, whose unrivalled knowledge of archival and artefact sources has been invaluable in the researching of many of the papers included in this collection.

Permission to print individual articles has been most generously granted by Messrs. Taylor and Francis (I, IV, VII); Science History Publications Ltd. (II, XIV); The British Astronomical Association (III); Blackwells Science Ltd. (V, XII); The Historic Society of Lancashire and Cheshire (VI); Elsevier Science Ltd. (VIII, XIII, XVI, XVII); Variorum, Aldershot (IX, XI); The Scientific Instrument Society (X); The Antiquarian Horological Society (XV).

But most of all, I thank my wife Rachel, who has reset articles III, VIII, XVI and XVII to conform to the required page size, and has given me her assistance in numerous other ways. This volume is dedicated to her.

I

A Study of the Accuracy of Scale Graduations on a Group of European Astrolabes

Summary

Precision measurements have been made of the scales on a group of European astrolabes manufactured between $c.$ 1450 and 1659. Little is known from documentary sources of the construction and scale-dividing methods used by late-medieval craftsmen. The measurements of the present group of twenty-four scales have been analysed statistically, so that the parameters of accuracy expected of them can be deduced. Scribing marks and other features give clear indications of how the scales were constructed.

Contents

1. Introduction

Historians possess relatively little reliable knowledge about the construction and accuracy of early astronomical instruments. This is a problem of some importance when one realizes the extent to which the ability to make accurate observations constituted the basis of calculation and creative research in the past. At a time when astronomy was concerned primarily with celestial cartography, the quality of research was frequently regulated by the homogeneity of the degree graduations of the instruments against which the stellar and planetary angles were taken.

Perhaps the main drawback in understanding the practical accuracy of instruments in any given period has been the lack of documentary material available to the researcher. Generally speaking, the only contemporary sources of information are the astronomer's own accounts of how accurate he believed his observations to be, and as scales were often read at face value, after applying a few standard corrections, the historian is left with little indication of the intrinsic accuracy of a particular instrument or observation.

Despite the lack of documentary evidence, however, a large number of early instruments survive in museum collections throughout the world, and the purpose of the present paper is to consider how the physical characteristics of their graduated scales may be deduced, when subjected to modern analytical techniques. This investigation has involved the examination of two specific types of instrument; the small diameter circles of the medieval astrolabe, and the large radius quadrants and sextants surviving from the seventeenth century. Each class of instrument required a different technique of measurement, suitable to its dimensions and characteristics, and both provided illuminating results. In the present paper, discussion will be confined to the astrolabe, and similar small circular instruments. The results derived from the seventeenth-century pieces are published separately.[1]

The literature relating to the astrolabe, both in terms of primary and secondary sources, is substantial. A succession of Islamic and European authors wrote about the instrument, and in his *Treatise on the Astrolabe*, Chaucer produced the first scientific text in the English language.[2] In recent times, the history and significance of the astrolabe has received considerable treatment from R. T. Gunter, Henri Michel, and others.[3]

Most modern studies, however, tend to be taxonomic in character. They sometimes stress the social and cultural importance of the instrument, but with the exception of some scholars who have examined almucantar projections, little attention has been paid to the astrolabe's practical accuracy.[4] Since many medieval astrolabes survive in good condition in museum collections, it was decided to examine a selection of them as specimens of graduated scales, in the hope of learning something of the parameters of accuracy within which the medieval scale divider operated.

2. The measurement of early scales

One of the few examinations of an early graduated scale to have been published to date is Derek Price's attempt to ascertain the accuracy of a 46-degree fragment of the divided scale of the Antikythera instrument, dating from *c.* 80 B.C.[5] From an enlarged photograph of the scale, Price made linear measurements of the size of each degree, and for a scale diameter of three inches, found the resulting degrees to be remarkably regular in their disposition. From a photographic enlargement of $2·6 \times$, Price concluded that the scale possessed an error of ± 6 minutes of arc.

Illuminating as these values certainly are, the technique of measurement was unlikely to achieve critical results, as they were obtained from a paper photograph, and made in linear, rather than direct angular measurements. With medieval astrolabes, however, there is a wider data base than that available with the Antikythera fragment.

The basic apparatus used to make the measurements described in this paper consisted of an engineer's dividing engine, with an $11\frac{3}{4}$-inch plate, and verniers reading down to 1 second of arc. Above the plate was placed a low power microscope to facilitate accurate alignment. An astrolabe was then placed upon the plate, centred, and the microscope crosswires aligned with the magnified graduations on the medieval scale. The engine plate was next advanced one degree at a time by the micrometer

[1] Allan Chapman, 'The Design and Accuracy of some Observatory Instruments of the Seventeenth Century', *Annals of Science*, 40 (1983), 457–71.

[2] Geoffrey Chaucer, *A Treatise on the Astrolabe*, in *The Works of Geoffrey Chaucer*, edited by F. N. Robinson (Cambridge, Mass., 1957), pp. 544–63; see p. 545, sections 7–8. D. J. de Solla Price, *The Equatorie of the Planetis* (Cambridge, 1955), which is a study of Peterhouse MS 75.1.

[3] R. T. Gunther, *Astrolabes of the World*, 2 vols (Oxford, 1932); Henri Michel, *Traité de l'Astrolabe* (Paris, 1947). For a bibliography, see F. R. Maddison, 'Early Astronomical and Mathematical Instruments. A Brief Survey of Sources and Modern Studies', *History of Science*, 2 (1963), 17–50.

[4] Chaucer (footnote 2), p. 545, speaks of quadrant dividing. Similar vague instructions appear in the Latin treatise, 'Un traité de l'astrolabe du XV siècle', edited by Henri Michel, *Homenaje a Millas-Vallicrosa*, II, (Barcelona, 1956), Cansejo Superior de Investigaciones Cientificas; it is recommended '...quatorum divedetur in 90 equales partes per 5ᵃˢ et 5ᵃˢ distinctas', p. 64, section 131ʳ.

[5] Derek J. de Solla Price, *Gears from the Greeks; the Antikythera Mechanism*, (New York, 1975), also published in *Transactions of the American Philosophical Society*, new series 64, part 7 (November, 1974), pp. 18–22.

screw, so that the precise angular extant of each 'degree' mark on the astrolabe could be measured by the engine. With practice, about fifty minutes of time were required to measure and tabulate the errors of each degree on an astrolabe scale.

Twenty-two graduated scales from twelve separate instruments of European manufacture, covering the period *c.* 1450 to 1659, were examined, taken from the collection in the Museum of the History of Science, Oxford. Twenty of the scales were 'paired' or taken from the front and back scales of the same astrolabe, the remaining two being from separate instruments.

Although the twelve instruments were constructed over two centuries, nine of them were dated between 1550 and 1600, the outsiders being two astrolabes of *c.* 1450 and 1521, and a divided plate dated 1659. This reduced the chronological range that I had originally intended to cover, but was occasioned by the quality of the instruments available. A cursory examination of the thirteenth- and fourteenth-century instruments in the Museum's collection revealed graduations of such manifest irregularity as not to warrant critical study, from which one might conclude that Chaucer's astrolabe was probably not accurate much beyond half a degree at the most, for any single observation.

After the scales had been measured, the results were analysed with the hope of eliciting three types of information. Firstly, there was the purely visual examination of the scales through the microscope, to detect guide dots or special scratches that may have been laid off by the maker as preliminary reference points to the full graduation of the circle. Some circles had many such dots, while on others they were entirely absent. Secondly, the errors were plotted on to graphs with the first to fourth quadrants of the circle arranged in descending order, in the hope of revealing repetitions, or error curves between facing quadrants. Thirdly, all the data was examined by computer to produce an error distribution map, and obtain mean and standard deviation errors for each astrolabe as a whole, and for specified groups of degrees.

In a short treatise on dialling published in 1593, Thomas Fale advised the aspiring scale graduator to divide each quadrant of the circle into three equal parts, using the radius.[6] The three 30-degree arcs were next trisected into 10-degree spaces, bisected to 5 degrees, 'and eche of these into five, if you can'. In the absence of the other evidence, one assumes that a similar process was used to divide the astrolabes examined. It is significant to notice that in Fale's description, divisions down to 5 degrees could be obtained by a conventional geometrical procedure, but no clues are given as to how single degrees should be obtained. As the subsequent examination indicates, most of the astrolabes in this study bear graduation characteristics that approximate closely to instruments divided by this method.

The tools required for this operation are not specified, but probably consisted of the compasses, ruler, and scriber. When visiting the workshops of Persian astrolabists in 1674, John Chardin was informed by his European escort that the Persian practice was fundamentally the same as that currently employed in the West, and that they used similar tools. A heavy pair of iron screw set dividers with burin points, a beam compass, and a straight edge were used to scribe the almucantar projections on the astrolabes,

---

[6] Thomas Fale, *Horolographia* (London, 1593), p. 1. Though drawings and instructions for the construction of various dials are known to exist from at least *c.* 1450, i.e. Bodley MS. 68, reproduced by R. T. Gunther, *Early Science in Oxford*, (Oxford, 1923), II, 38–9, they do not describe the drawing of astronomical angular graduations. For early dialling accounts, see D. J. de Solla Price, 'The Little Ship of Venice—a Middle English Instrument Tract', *Journal of the History of Medicine*, 15 (1960), 399–407. Also A. W. Fuller, 'Universal rectilinear dials', *Mathematical Gazette*, 41 (1957), 2–24.

and while Chardin does not describe the process of original degree graduation, the same tools were probably used to produce the circumference divisions.[7] Little time is required to divide a circle with such tools using the techniques described above, and I was personally able to produce a set of 360 regular divisions in a little under one hour.

Because the instruments chosen for this study varied in radial size between 5·78 and 16·25 inches, an attempt was made to reduce them to one standard, or common linear unit equivalent, so that the skills of various craftsmen could be compared directly, irrespective of the size of instrument upon which they worked. From the diameter of each circle, the circumference length could easily be calculated, and from these, the linear extent, in fractions of an inch, of each degree and minute of arc. It was thus possible to compare the mean errors, or average accuracy of one circle against another in the following manner:

Instrument no. 9
diameter   = 10·69 inches
circumference = 33·58 inches
therefore
    1 degree   = 0·0933 inches
    1 minute   = 0·00155 inches.

The size of the one minute space was next multiplied by the mean error in minutes of the instrument, to ascertain the smallest overall tolerance to which the graduator could work:

mean error of scale = 1·49 minutes of arc.
0·00155 × 1·49 = 0·00232 inches.

No.	Maker and Date	Museum number	Diameter (inches)	Mean error (minutes)	Engraving error (inches)
(1)	*Fr. Morillard Lugdunen* [i.e. Lyons] *faciebat Narb.* [i.e. Narbonne] *Anno M. VI*^{c} [1600]'	57–84/178	5·78	3·14	0·00264
(2)	*Io: Dom: Feciolus Trident: faciebat* ... *1558*'	73–11/2	5·81	7·06	0·00597
(3)	Anon.; '1521'. Possibly German	IC–252	5·97	8·72	0·00755
(4)	'Erasmus Habermel'; c. 1585	IC–278	6·66	2·29	0·00222
(5)	'AEgidius cuiniet antuerpianus facieb. A°. 1560'	IC–224	7·75	9·15	0·01031
(6)	Anon.; c. 1570. Possibly French	57–84/21	8·28	1·25	0·00151
(7)	*Regnerus Arsenius Nepos Gemmae Frisij fecit Louanij anno 1565*'	IC–229	10·46	1·36	0·00207
(8)	Anon.; after 1582. Possibly Flemish	52–1	10·59	1·50	0·00231
(9)	'Jeh Charla'; c. 1450	IC–163	10·69	1·49	0·00232
(10)	'Thomas II 1559' [i.e. Thomas Gemini] (The Queen Elizabeth astrolabe)	IC–575	13·16	2·03	0·00389
(11)	'Erasmus Habermel fe:'; c. 1590	57–84/19	14·31	2·37	0·00493
(12)	'Henricus Sutton Londini fecit. 1659'	IC–313	16·25	0·48	0·00113

Table 1.
The instruments examined all belong to the Museum of the History of Science, Oxford.

[7] J. Chardin, *Voyages* (London, 1686). Chardin's description of the Persian astrolabists was reprinted by H. Michel, 'Méthodes de tracé et d'exécution des Astrolabes persans', *Ciel et Terre*, 57th year, part 12 (December, 1941), 481–96.

When assessing the accuracy of each instrument, the best, or most accurate scale was taken, as representing the maker's uppermost limit of skill. The Table lists the instruments examined, with their characteristics and accuracy values.

When translated into thousandths of an inch in the figures shown in the sixth column, it is possible to appreciate the remarkable tolerances to which these craftsmen worked. Their divisions averaged out to within a few thousandths of an inch of each other, which is no mean feat when working with dividers and scriber. Although the bigger instruments tend to reveal lower engraving errors than the smaller ones— compare no. 2 (5·81-inch diameter, 0·00597 error) with no. 12 (16·25-inch diameter, 0·00113 error)—this is by no means a uniform occurrence, as no. 1 (5·78-inch, 0·00264 error) shows. It is interesting to note that Erasmus Habermel's 6·66-inch astrolabe, no. 4, shows a better 'absolute' accuracy than his princely 14·31-inch instrument, no. 11. The greatest overall error is displayed in the scales of astrolabe no. 5, with an error of 0·01031 for its 7·75-inch scale.

3. Interpretation of the measurements

No particularly noticeable improvement in accuracy seems to have taken place with time, for the *c.* 1450 astrolabe, no. 9, is quite equal to any of its sixteenth-century counterparts. The exception is the seventeenth-century divided plate by Henry Sutton, no. 12, which reveals its maker as a graduator of superlative skill. Circle dividers, it seems, were accomplished masters of a conservative craft.

Accuracies varied surprisingly little with instrument diameters. One would indeed have expected greater variations, especially between instruments by different makers, for there must have been an inevitable 'clumsiness' factor operating between the graduation of large and small scales. Ideally, of course, comparisons should have been made between the large and small instruments of one single maker, but the logistics of bringing all of one man's work together for measurement, from museums around the world, has made this out of the question for the present time.

Several instruments, especially nos. 6, 10, 7, and 9 were seen to bear systematic circumference marks when examined microscopically, suggesting their use as construction points in the drawing of the scale. This is further substantiated by their falling at the 10-degree points, in accordance with Thomas Fale's description, cited above. It has been concluded from these preliminary scratch marks that the instruments were originally divided on the brass, unlike pieces lacking such marks which might have been copied from a workshop dividing plate or protractor.

Reference has already been made to the use of graphs to plot the scale errors for each of the quadrants on an astrolabe circle. In this way, it was not only possible to construct a visual pattern for the errors of each astrolabe, but to compare the pattern for each individual quadrant with the others, and thereby notice if error patterns repeated in different parts of the same circle. In addition to plotting graphically the place of each single degree, separate distribution graphs were also drawn for the 5- and 10-degree points for each quadrant on the twenty-two scales, and their resulting curves were found to display a much greater regularity than the curves drawn for the single unit degrees. On certain instruments, especially nos. 11, 5 and 1, it was even possible to detect superimposed 30-degree curves, though this characteristic was not so pronounced as expected. (Figure 1.)

No written evidence is available to suggest how the single unit degrees were drawn to fill the 10-degree spaces, and in the hope of eliciting this information the computer was made to print a continuous analytical graph for each scale, to highlight the error

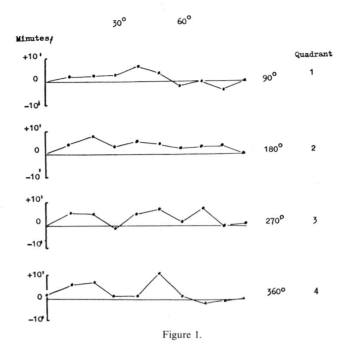

Figure 1.

Astrolabe Number 1. Graph showing the error distribution in minutes of arc, for the four quadrants of the astrolabe's circular scale. The horizontal axis of the graph indicates zero minutes error, with up to + 10 minutes above, and − 10 minutes below. Each dot designates one ten degree division on the astrolabe scale. Note the appearance of 30, 60 and further sexagesimal degree curves, the accuracies of which are higher and more uniform than their integers.

patterns within the 10-degree spaces. It was hoped that a regular sequence might emerge, indicating the mode of subdivision, and in conjunction with the graph, tables of degree widths as they were adjusted to fill strategic intervals.

Tables of degrees for each succeeding integer within the 5- and 10-degree groups were listed next, i.e. the 2nd, 12th, 22nd; 3rd, 13th and 23rd degrees, in the hope of finding a pattern such as might have occurred had the workman divided them from a standard 10-degree template, so that each space would have shown the same error. (Figure 2.)

Only on instruments nos. 3 and 4, however, could anything like a recurring sequence be found, with error flows that plunged rather too regularly to seem random. But whether this was the product of template subdivision, or chance occurrence, it is impossible to be certain.

On many of the instruments periodic patterns occurred within the intervals of a given circle, but the overall accuracy of the divisions was not sufficiently regular to substantiate the use of a template. Yet, for template errors to become apparent on each 10-degree space, one would have to assume a perfect workman who, with no more than simple hand-held tools, could adjust and copy his template perfectly through all the thirty-six courses of the circle. Within routine circular division, the use of a template

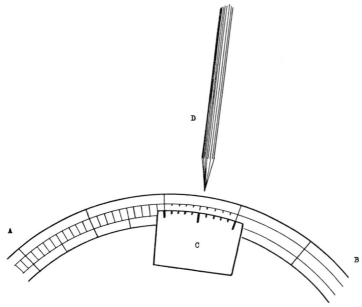

Figure 2.

Use of the 10-degree template, being used to subdivide to single degree digits. A–B is part of the
circle to be divided, C is the template, and D the scribing tool.

would appear an obvious accessory, but beyond its possible application to two
instruments, nothing more substantial may be asserted.

Bearing in mind the ease with which individual scribing errors could occur when
using the tools of 1550, attention was confined not to the detail, but rather the *drift* of
the numerical values obtained for a given instrument. Single values were always
suspect, and only through averages and standard deviations taken over many degrees
could worthwhile conclusions be reached.

When the graphs for the four quadrants of each circle were compared with each
other, several instruments were found to exhibit repetition errors in quadrants that
faced each other across the circle. This was particularly noticeable in no. 11 (back scale)
and in both the scales of no. 9. The quadrants of these instruments were found to
display a regular 1–3, 2–4 repetition, especially noticeable between quadrants 2 and 4
on the back scale of no. 9. It may be suggested that these repetitions were occasioned by
the instrument maker scribing across the circle with a straight edge so that after
graduating one half of the circle, he could copy divide the other (Figure 3). Such
practices would naturally produce a twofold rotational symmetry error between
opposite quadrants. (Figures 4 and 5.)

What is more difficult to explain is the occurrence of repetition in contiguous or
adjacent quadrants, as seems to occur in no. 3, where the quadrants of the front scale
show a 1–4; 2–3 repetition, as shown in Figure. 6. The most likely explanation is that the

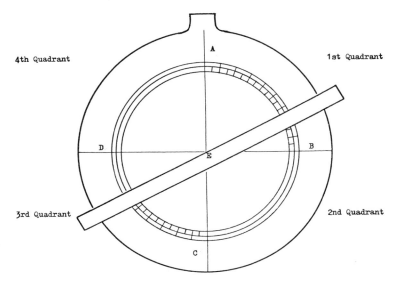

4th Quadrant   1st Quadrant

3rd Quadrant   2nd Quadrant

Figure 3.

Scribing across the circle to produce two-fold rotational symmetry in facing quadrants. The quadrant A–B is graduated in the manner described. A ruler is then layed across the centre E, and the quadrant C–D copied from it. An identical procedure would be followed to fill in the quadrants B–C and D–A.

maker used a pre-graduated template or protractor to strike off two contiguous quadrants as he worked his way around the circle. But instrument no. 3 is especially interesting because its errors not only repeat within the *same* circle, but also between the back and front circles. Although the errors from zero vary considerably between the two circles, the drift patterns display a remarkable similarity between the two fourth quadrants and the second and third quadrants of both circles.

These error patterns emerge quite distinctly, in spite of the inaccuracies of some individual degrees on this 5·97-inch astrolabe, and one wonders whether the anonymous maker used a protractor of 90 or 180 degrees rather than a 360-degree dividing plate, to scribe all eight quadrants. The use of a protractor would also agree well with the absence of construction marks on the face of the scale, as such marks would be unnecessary on a copy instrument. Similarly, the mean errors of + 8·72 minutes (front) and + 10·53 minutes (back), along with the standard deviations of 15·03 and 12·89, also exhibit the same closeness which one would expect for two scales divided from the same template.

Dividing plates and templates were often used for the routine division of mathematical instruments. Chardin describes the use of a 'basin'-shaped astrolabe dividing plate, which he saw in Persia in 1674,[8] and as we know from Robert Hooke, graduating plates were commonplace in contemporary English workshops. The

[8] Michel (footnote 7), 495.

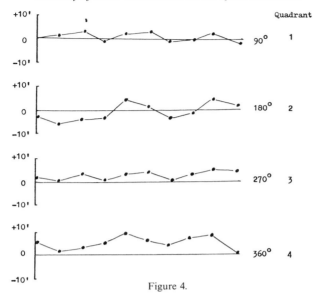

Figure 4.

Astrolabe Number 11. Back scale, showing two-fold rotational symmetry between the 10 degree dots in quadrants facing each other across the circle. The error patterns of the quadrants 1–3 and 2–4 produce closely related curves.

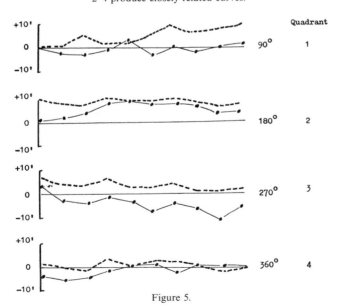

Figure 5.

Astrolabe Number 9. Front scale —●—●—●—. Back scale - - - -. Note the rotation between quadrants 1–3 and 2–4 on both the front and back scales.

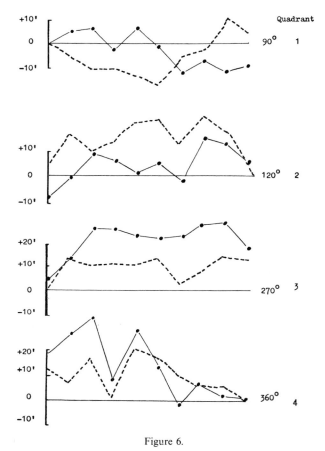

Figure 6.

Astrolabe Number 3. Front —●—●—●—. Back----. In spite of the wide fluctuations, note the close similarity in error patterns, between quadrants 1–4, 2–3 (front), and 2–3 (back).

Museum of the History of Science, Oxford, possesses an Islamic *dastur*, or dividing plate, probably dating from the nineteenth century, which belongs to a traditional genre of instruments.[9] (Figure 7.)

    Examination for correlations between the front and back scales of the astrolabes was next extended to the eleven remaining pieces, and their error graphs compared. Close correlations were discovered between the two scales of astrolabe no. 8, an unsigned instrument of *c.* 1582 that bore a strong resemblance to the astrolabe no. 7, which carries the inscription 'Nepos Gemma Frisius'. (Figure 8.)

[9] Museum of the History of Science, Oxford, Accession no. 69–51. No attempt has been made to measure the accuracy of this instrument because of the obvious crudity of its graduations. I am not aware of any European *dasturs* that have survived, although in Robert Hooke's *Animadversions ... on Hevelius* (London, 1674), p. 14, the use of a 10-foot dividing plate is mentioned.

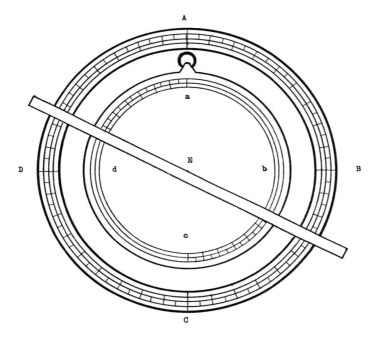

Figure 7.

Principle of the dividing plate. A, B, C, D, is the graduated perimeter of the dividing plate, and a, b, c, d, the blank astrolabe to be divided. After centring within the plate, the divisions are transferred to the astrolabe by the ruler and scriber across the pivot E.

Though the mean accuracy between the two scales of no. 8 varied by several minutes, −5·72 minutes (front) and +1·50 minutes (back), their drift patterns once again followed very similar curves, with almost identical standard deviations of 1·83 and 1·81 respectively. I consider that both scales of this astrolabe were graduated from the same template or dividing plate, and that in measuring the instrument's error pattern, one is really measuring at second hand the errors of this original plate. The total 7·22-minute mean error disparity from true zero between the scales is easily explained in terms of the preliminary adjustments made to the astrolabe blank, to fit it to the dividing plate. A starting mark could have been made at the crown of the astrolabe from which the copied divisions were intended to run and any slight incongruity between this mark and the zero point on the dividing plate would have occasioned an overall displacement in the rest of the scale. Though the zero readings on the two scales would thus vary slightly, their internal consistencies, as represented in their mutual standard deviations, would remain unaffected, in accordance with the observed results. A similar displacement was also detected in the half degree spaces of this instrument, with mean errors of +25·30 minutes (front) and +31·77 minutes (back), with standard deviations of 2·54 and 1·81 respectively. Further evidence for the copy division of this astrolabe is also adduced from the absence of regular construction

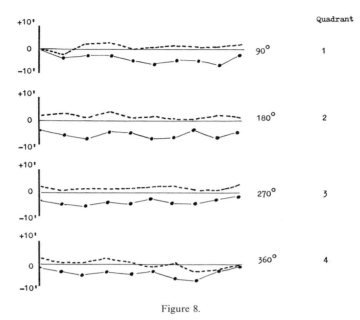

Figure 8.

Astrolabe Number 8. Front —●—●—●—. Back ----. Correlations may be discerned in the error patterns of the corresponding quadrants of both front and back scales.

marks on its surface, the only marks present being those falling at 0 and 180 degrees, which would have been used to mark the preliminary bisection for adjustment to the dividing plate. Of all the sixteenth-century astrolabes examined, no. 8 presented the best appearance under the microscope, its incised divisions being extremely uniform and easy to measure. When viewed microscopically, indeed, the divisions of some of the other instruments examined in this study appeared very rough and uneven.[10]

Mention has already been made of the presence of construction dots upon the scale of no. 9, which encouraged me to study more closely the dot sequences of the other instruments, to see what might be learned by visual examination. On the back scale of no. 9, the dots were seen at their most elaborate, being located at 5-degree intervals all around the scale. Their size, shape and disposition varied, moreover, with the angle encompassed. The zero dot at the crown appeared through the microscope as a clear, circular punch hole, while its opposite, at 180 degrees, was a well incised square. The visual markings further indicate that a compass opened to the same angle as the radius was used to strike off the 60-degree points, for each 60-degree dot (i.e., 60, 120 and so on), consisted of a well incised rectangular indentation slanting diagonally from left to right across the circumference line. As the slope of all 60-degree dots ran in the same direction it is likely that they were struck off in a continuous clockwise sequence around the circle from the same scribing compass.

[10] Because of their alternate hatchings, the degrees on the Queen Elizabeth astrolabe (no. 10 in this study), appear very irregular through the microscope.

The 5- and 10-degree dots also appeared as rectangular indentations, but in almost every case they slanted from right to left. No special marks designated the 30-degree spaces, but as the distribution graph tends to indicate wave patterns around these points, they were probably the first dots to be laid off with the right-to-left facing scribing tool. None of these marks presents the appearance of scratches, but rather distinct indentations, resembling punched holes. The different slants and sizes of dots suggest that at least two distinct processes were used in the graduation, and possibly, two separate compasses each with its characteristic point, indented by finger pressure. This practice agrees well with Chardin's description of the Islamic astrolabists, who also used two compasses in their work. Astrolabe no. 6, a French instrument of *c.* 1570, also displays a curious configuration of scale dots, most of which fall around the 5- and 10-degree marks. In addition, there is a broken sequence of fourteen contiguous dots on the back scale that may have been test marks, used to establish the ratio of the single degrees.

Most of the astrolabes examined contain eccentricity errors, for it was rare to find an instrument where the central rivet hole which carried the alidade coincided exactly with the geometrical centre of the circle, although such errors would not have been of much practical consequence on scales that were only read to a half degree by their original users.[11]

4. Examination of an equatorium, c.1600

Some time after completing the examination of the instruments in the Museum of the History of Science, Oxford, I succeeded in obtaining a set of photographic plates of the equatorium of *c.* 1600 in the Merseyside County Museum's collection, Liverpool.[12] It was possible to place these $4\frac{1}{2} \times 5\frac{1}{2}$-inch negatives on to the dividing engine in the normal way, after a simple mount had been devised to secure background illumination of two of its 360-degree scales. Shortly after measuring the negatives, the 10·5-inch diameter equatorium itself was brought to Oxford, so that it was possible to examine the same scales by direct measurement. As there were no significant discrepancies between these and the photographic measurements, it is hoped that the examination of photographic plates may prove a way by which instruments in overseas collections may be studied and compared.

The analysis of the Liverpool equatorium measurements were taken from the inner, or second 360-degree scale around the perimeter of the Planetary Plate, and from the extreme outer scale on the Lunar Plate. When the graphs were plotted from the scale errors, it became clear that the third and fourth quadrants of the Lunar Plate scale exhibited a very similar flow pattern in their 10-degree divisions, suggesting the use of some sort of protractor or template to divide them. (Figure 9.) When the errors for the Planetary Plate scale were plotted, a repeating pattern again became obvious. Indeed, when the four Planetary Plate graphs were turned right to left and read *backwards*, one found that quadrants three and four displayed error patterns that were congruent to their equivalent third and fourth quadrants on the Lunar Plate. In short, the degree errors 180 to 270, and 270 to 360 on the Lunar Plate, were almost identical with the degrees 270 to 180, and 360 to 270 on the Planetary Plate, forming a mirror inversion.

---

[11] Some of the eccentricities were due to pivot wear, others to inaccurate drilling of the centre. They may suggest a division of labour in the making of astrolabes, as the drilling was handed over to a less skilled workman than the one who divided the scale.

[12] For an exhaustive study on the nature of this instrument, see John North, 'A Post-Copernican Equatorium', *Physis. Rivista internazionale di Storia della Scienza*, 11 (1969), 418–57.

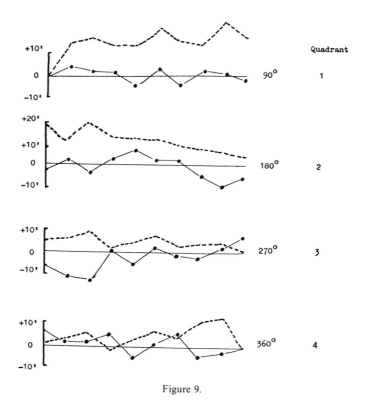

Figure 9.

Liverpool Equatorium. Front, or Planetary Plate: second scale —●—●—●— from outer edge.
Back, or Lunar Plate: outer scale ---- read clockwise from '*' mark on crown.

Although the error graphs of the third and fourth quadrants of each scale show clear congruency, producing an unmistakable pattern, it should be emphasized that other quadrants on the instrument possess related patterns. The first quadrant of the Lunar Plate is a good case in point, for while it has a large graduation error of around +20 minutes, the pattern is virtually identical with those of the third and fourth quadrants. (Figure 10).

It must be borne in mind how easy it is, when working with hand-held tools, to mis-align a single degree mark, or even a whole quadrant, if the protractor is not exactly in place. When the degree lines themselves subtend an angle of four or five minutes wide, it must have been impossible for a craftsman to make perfect alignments between adjacent courses of divisions. Once again, this stresses the importance of broad error patterns rather than single degrees in interpreting the measurements of scale accuracies.

At first sight, there is no obvious reason why the errors and scribing sequences were inverted between the Planetary and Lunar scales of the instrument, for if the graduator was laying off two adjacent quadrants of the same scale, he must have worked with his

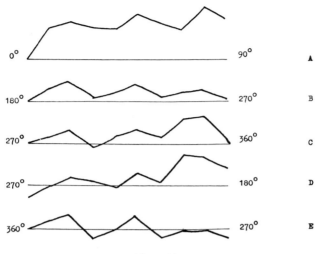

Figure 10.

Liverpool Equatorium. Comparison of error distribution curves for five of the eight quadrants. A. Lunar scale, 1st quadrant, 0°–90°. B. Lunar scale, 3rd quadrant, 180°–270°. C. Lunar Scale, 4th quadrant, 270°–360°. D. Planetary, 3rd quadrant, 270°–180° (inverted). E. Planetary, 4th quadrant, 360°–270° (inverted).

protractor or template facing upwards, thereby producing errors that all flowed in one direction. But as all known sixteenth-century protractors are engraved on opaque materials, such as brass, it is difficult to now how they could have been read in the inverted position, scale downwards, as would have been necessary when drawing the respective Planetary and Lunar scales. It is likely that the graduations were struck off from a protractor with a denticulated edge, so that the bottom of each tooth guided the scribing tool to give the exact position of each degree, without the need for alignments by eye alone. Such a denticulated template, cut into a thin sheet of brass, could have been used equally well either way up, to produce inadvertently the same error pattern, running either clockwise or anticlockwise, depending upon its position when in use.

5. Conclusions

From the astrolabes, circles, and equatorium examined in this study, it has been possible to draw some important conclusions regarding the construction and accuracy of early instruments. Evidence of certain geometrical construction procedures showed that some instruments were divided directly, while others were copy graduated. The type and shape of the construction dots tell us something about the tools used in manufacture, together with the use of 10-degree templates to subdivide to single degrees. Twofold rotational symmetry patterns between opposite quadrants of an astrolabe suggest mixed methods of half original, and half copy, graduation work, while my own attempts to divide circles using these techniques indicate something of the speed with which they could be made to work.

Quite apart from suggesting the possible use of denticulated protractors to lay off scales, the measurements of the Liverpool equatorium also show that reliable results

I

488

can be deduced from photographic plates, as well as from actual instruments. This opens up the prospect of being able to collate and examine all the extant work of one specific workshop—such as Arsenius of Louvain—to see if related errors are to be found in the whole of its output. The extension of such 'archaeological' techniques might thus provide a basis of information about what has hitherto been an area relegated to guesswork and conjecture; namely, how were scientific instruments made in the past, and how accurately could they be made to perform?

Acknowledgments

I wish to acknowledge the assistance and advice given to me by many persons and institutions during the course of this research. In particular, I wish to thank Mr F. R. Maddison, Curator, and the technical staff of the Museum of the History of Science, University of Oxford, for granting access to the astrolabes. Special thanks go to Dr Gerard L'E. Turner, for both technical and editorial advice on this project. I also thank Mr Frank Petit of the Computer Studies Department, University of Oxford, for advice on data analysis, and Mr Brian Busby and his staff in the University of Oxford Department of Engineering Science for providing an abundance of technical help and loan of equipment. Outside Oxford, I am very grateful to Mr Julian Ravest and Dr Robert Smith of the Merseyside County Museum, Liverpool, for photographing the equatorium, and for bringing it to Oxford so that I was able to examine it.

II

THE ACCURACY OF ANGULAR MEASURING INSTRUMENTS USED IN ASTRONOMY BETWEEN 1500 AND 1850

The principal task of astronomical research, until well into the nineteenth century, was the measuring and cataloguing of stellar and planetary positions. Only after the painstaking accumulation of positional data could the theoretician hope to find workable solutions to such astronomical problems as the Earth's motion in space, the universal nature of gravitation, and the details of planetary motion.

The accumulation of such data required the use of large radius sextants and quadrants, and astronomers fully realized by the early seventeenth century, that the accuracy with which craftsmen could graduate the scales of their instruments constituted a barrier to observational research. Several important elements were present in the construction of superior measuring instruments. There was the artisan's skill in executing the elaborate compass geometry required to draw consistent divisions. Following from this was the problem of actually sighting the instruments, so that the advantages of accurate scales were not lost by poor alignment. Lastly, there were the necessary improvements to be made to the astronomer's own techniques of observation, with their demands on eye and hand, if he was to take full advantage of the new tools placed at his disposal.

The interaction between theoretician, practical astronomer and craftsman forms an interesting process that is too complex to examine closely in the present study, although it does pose major questions to our understanding of the growth of the physical sciences, and has been alluded to elsewhere.[1] In the history of astronomy in particular, one encounters several incidents in which the theorist required finer working data from the observational astronomer, yet could not obtain it because there were problems at the workshop level, as the ingenuity of the craftsman was unable to produce an instrument capable of reading reliably to the 'next decimal'. The best example of this kind concerns the stellar parallax, which came to be demanded as a proof of Copernican astronomy, unsuccessfully investigated for three centuries, and eventually measured in the 1830s; by which time craftsmen were capable of producing instruments showing the fractions of a single second of arc, as were necessary to detect even the largest parallaxes.

But demand does not always move from the theoretician downwards. When John Flamsteed became the pioneer of applying telescopic sights to graduated instruments in the 1670s, it became possible to supply Newton with data of such accuracy that he was able to go a long way towards solving the ancient problem of the lunar theory. It is more than a coincidence that Newton's main work on the lunar theory was performed in the 1690s, when one remembers that in 1689 the Royal Observatory brought into use a mural arc of revolutionary design, built by Abraham Sharp, which was used in conjunction with a new observational technique devised by Flamsteed. Indeed, it was Newton's failure to acknowledge his "obligation to the Royal Observatory", as Flamsteed put it, that played such a major part in the hostility which later grew between Flamsteed and Newton.[2]

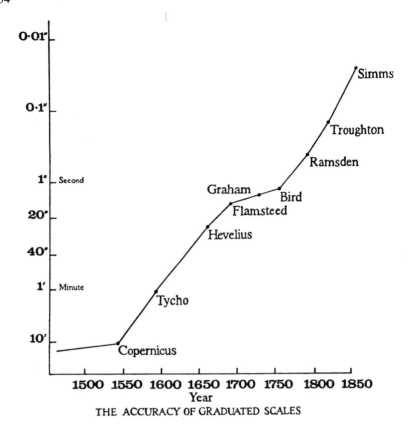

THE ACCURACY OF GRADUATED SCALES

Fig. 1.

In the history of astronomy between 1500 and 1850, one can discern three crucial developments in instrument making, which enabled the science to move ahead with a new certainty. The first of these came with Tycho Brahe. By the adaptation of diagonal scales to the limbs of his instruments, he could read small fractions of a degree without extending the radii, thus obtaining compact and stable instruments. Similarly, Tycho's invention of improved sights for his sextants and quadrants made it possible to eliminate the alignment errors which he claimed were as large as one-eighth of a degree on the instruments of Copernicus.

The second breakthrough came in the third quarter of the seventeenth century, following the invention of the telescopic sight and the micrometer. Though both of these devices had been invented by William Gascoigne sometime around 1640, their systematic application to instrument design did not come until the period after 1660.[3] The significance of the telescopic sight and micrometer came from their ability to increase the astronomer's own powers of resolution, and hence his ability to utilize his instruments more fully. As Robert Hooke had demonstrated, the average human eye could only resolve angles of about one minute of arc, so that the prospect of making measurements was limited physiologically.[4] By 1660, the geometrical art of quadrant dividing had already passed beyond this one-minute barrier in the production of accurate scales, so without an improvement to the old naked-eye sights, further research would have been futile. The application of telescopes to the sighting arms of quadrants allowed this barrier to be broken through, as the seeing power of the eye came to be artificially extended.

The micrometer, likewise, extended angular resolving power, either in the form of the filar micrometer, set in the field of the telescope, or else adapted as a fine adjusting screw onto the limbs of quadrants. In consequence, the one-minute-of-arc ceiling of 1660 was extended to around 15″ by 1700, and 8″ by 1725, thus making it possible to detect such small quantities as were necessary to discover the aberration of light, and the nutation.

The third great breakthrough came in the late eighteenth century, when Jesse Ramsden and Edward Troughton replaced the quadrant with the full circle as the most reliable design for a graduated instrument. Because a circle can be rotated, it was possible to graduate it and cross-check its graduations with micrometer microscopes in a way that was impossible with a quadrant. Thus, the accuracy of 1″ or 2″ obtained by Bird on a quadrant, was reduced to ½″ by 1800. It is interesting to note that by this period, the leading instrument makers had merged into the scientific community, for both Ramsden and Troughton were Fellows of the Royal Society, and moved on terms of equality with the astronomers whom they served.

The refinement of late eighteenth-century circle-dividing techniques continued into the nineteenth century with such makers as Simms and Jones, though by the end of the nineteenth century, the course of astronomical research had shifted to areas outside the scope of this paper. The development of astronomical photography and the measurement of astronomical positions from photographic plates after 1880 rendered the old circle-dividing tradition redundant, whilst by the end of the century, the predominant emphasis in research had shifted away from positional, into physical astronomy.

The consequences of the three crucial developments in instrument technology discussed in this paper may be traced in Figure 1. Particularly significant is the degree to which the quadrant, by the mid-eighteenth century, was exhibiting a condition of diminishing returns in research, as it became increasingly difficult to incorporate new developments into the old shape.

Bird's efforts to make the quadrant measure a single second were greater in both magnitude and kind than those which had been necessary for Hevelius to achieve 20″. The slowing-down in the rate of quadrant accuracy is demonstrated in the graph.

With the application of the full astronomical circle, the limiting factors that applied to the quadrant were overcome. The use of micrometer microscopes, multiple cross-checking and improved engineering standards, allowed the single second to be divided first into tenths and then into hundredths, as indicated on the graph.

NOTE ON DIVIDING ENGINES

The instrument accuracies examined in this study were the work of "original division", and not the work of dividing engines. Though Ramsden and Troughton produced important dividing engines in the late eighteenth century, these machines were used entirely in the graduation of sextants, theodolites and similar small radius instruments, in which the master graduations on the engine were copied onto the work. Engines were not used to graduate major astronomical instruments until the development of Simms's self-correcting engine in the mid-nineteenth century, and all observatory instruments before 1850 were hand-graduated. Simms's engine *did* graduate the circle of Airy's transit circle installed at Greenwich in 1850, but this was the only instrument in the present study to be so divided.

SOURCES FOR FIGURE 1

The angular quantities used to compile the graph were drawn from a variety of sources. The majority consist of contemporary estimates made by astronomers and practitioners such as Tycho Brahe and John Bird. Others are historical assessments, made by one astronomer (such as Sir George Shuckburgh) about the capacities of his predecessors, whilst several are the products of modern reductions. It should be noted, moreover, that when reductions of observations have been performed, they often give values that come very close to the original estimated values.

Medieval Europe. No reliable figures exist for the medieval period, although it was possible to ascertain the division accuracy of a series of astrolabes in the Museum of the History of Science, Oxford, by direct measurement on a dividing engine. Though the scales were often found to be divided to less than $\frac{1}{2}°$, the simple sights, and centre errors, would have increased the actual observing accuracy to about 1°. See Allan Chapman, "The archaeology of the graduated scale", *Annals of science* (forthcoming).

Copernicus. The only sixteenth-century estimate of Copernicus's accuracy comes from Tycho Brahe, who possessed the Polish astronomer's "Ptolemaic rulers". He estimated that they could not be more accurate than $\frac{1}{8}°$ $(7\frac{1}{2}')$ or $\frac{1}{10}°$ (6'). See Tycho Brahe, *Astronomiae instauratae mechanica* (1598), translation by H. Raeder and E. and B. Strömgren (Copenhagen, 1946), 46. These values seem to have been adopted by subsequent commentators, *viz.* Sir George Shuckburgh, "An account of an equatorial instrument", *Philosophical transactions*, lxxxiii (1793), 67–128, p. 75, where the figures 5', 8' and 10' are cited. William Pearson, in the article "Circle" in Abraham Rees's *Cyclopaedia*, viii (London, 1819), Sig. Ee 2r, gives 8' or 10'. (Pearson's tables of historical accuracies seem to have been based on Shuckburgh's original of 1793. In following references, they will be designated "Shuckburgh" and "Pearson", the pagination remaining the same throughout.)

Tycho Brahe. In *Astronomiae instauratae progymnasmata*, 224, Tycho claimed that he could make his results agree to 1'. See J. L. E. Dreyer, *Tycho Brahe* (1890, repr. New York, 1963), 353. The *Progymnasmata* value was also accepted by Shuckburgh and Pearson in their accuracy tables. Under favourable conditions, reductions have shown that Tycho could go down as low as 20"; see G. L. Tupman, "A comparison of Tycho Brahe's meridian observations", *The observatory*, xxiii (1900), 132–5, 165–71, and Walter G. Wesley, "The accuracy of Tycho Brahe's instruments", *Journal of the history of astronomy*, ix (1978), 42–53. Wesley has calculated that the Tychonic instruments could work to an average accuracy of 30" to 50".

Johannes Hevelius. Shuckburgh and Pearson give the same values, 15" to 20".

John Flamsteed. In 1721–22, James Pound claimed to have detected a 15" error in Flamsteed's observations, that were occasioned by a graduation error in the mural arc; letter, J. Crosthwait to A. Sharp, 27 January 1721/22 (Royal Society manuscript, referenced "Sharp Letters", *FlSh xxiv.d*; letters bound in date order). Both Shuckburgh and Pearson cite Flamsteed's error as 10" to 12" for the sextant, and were probably unaware of Pound's estimated error for the arc.

George Graham. Halley is said to have read the scales of the 8 ft Greenwich quadrant to $7\frac{1}{2}"$; see James Bradley, *Miscellaneous works and correspondence of the Rev. James Bradley*, ed. by S. P. Rigaud (Oxford, 1832), lv. In 1746, Bradley claimed to have detected a $15\frac{3}{4}"$ error in the Graham quadrant at Greenwich, although this was after the instrument is known to have changed shape. Shuckburgh and Pearson both give the same value; 7" to 8".

John Bird. Bird claimed that his 8 ft quadrants were accurate to within 2"; John Bird, *A method of dividing astronomical instruments* (London, 1767), 13–14. In 1753, Bradley found a $1\frac{1}{2}"$ error in the Greenwich quadrant; Bradley, *Works*, lxxxvii. In the reductions he made with the south quadrant at the Oxford Observatory, Knox-Shaw claimed to have found an error of $1\frac{1}{2}"$, although this later cancelled out; see, Thomas Hornsby, *The observations of the Rev. Thomas Hornsby*, ed. by H. Knox-Shaw et al. (London, 1932), 79.

Jesse Ramsden. From his exhaustive examination of the circles on the "Shuckburgh Equatorial", Sir George Shuckburgh believed that the dividing error amounted to less than $0"·5$;

Shuckburgh, "Description", 102. From a comparison of observations made with Ramsden's "Palermo" and "Armagh" circles, John Pond calculated that these instruments showed an error of about 1″; see, J. Pond, ". . . Description of an astronomical circle", *Philosophical transactions*, xcvi (1806), 420–54, pp. 421–2.

Edward Troughton. Pond, "Description", claimed that by cancelling out errors with opposite microscopes, the errors of his "Westbury Circle" could be reduced to 0″·25. Edward Troughton's Greenwich Circle of 1812 carried six microscopes capable of measuring down to 0″·1; see Derek Howse, *Greenwich Observatory*, iii (London, 1975), 27.

William Simms. The Great Transit Circle built for the Greenwich Observatory in 1850 could read to 0″·06, by means of six cross-checking microscopes capable of isolating scale errors; see Howse, *Greenwich Observatory*, iii, 44. The consistency of the transit's errors is indicated in G. B. Airy's *Astronomical Observations at the Royal Observatory, 1852* (London, 1854), Appendix II, 17–19.

Some years ago, an angular accuracy graph was compiled by H. Mineur, and included in H. T. Pledge, *Science since 1500* (London, 1939), 291. Mineur did not cite his sources, although he named the astronomers who attained the specific peaks in his graph. Because it is likely that Mineur and myself used the same sources for the overlapping parts of our respective graphs, there is a close similarity between them. It must be stressed, however, that the graph in this paper is drawn quite independently of Mineur's, and from the sources cited herein.

REFERENCES

1. Allan Chapman, *Dividing the circle* (Amersham, in press).
2. Of course, no one is claiming that Newton's work on the lunar theory was initiated by Flamsteed, although the important point remains that Newton required reliable observational 'fixes' if his theory was to be taken seriously as a true representation of nature. In 1690, the only instrument in Europe capable of providing such fixes to the requisite degree of accuracy, moreover, was the mural arc in the Royal Observatory.
3. Many references to the work of Gascoigne are extant; see Chapman, *Dividing the circle*. For primary sources, see William Derham, "Extracts from Mr Gascoigne's and Mr Crabtree's letters, proving Mr Gascoigne to have been the inventor of the telescopick sights of mathematical instruments, and not the French", *Philosophical transactions*, xxx (1717), 603–10.
4. Robert Hooke, *Some animadversions on the first part of Hevelius, his 'Machina Coelestis'* (London, 1674), 7.

III

Tycho Brahe — Instrument Designer, Observer and Mechanician

70   Accustomed as modern astronomers are to thinking of their science as progressive, and based on an ever-improving instrument technology, it is hard to envisage a time when astronomy was seen as a conservative discipline. Yet if one believed that the principal features of the heavens had already been catalogued by the ancients, then one could do little more than make adjustments for precession and other quantities. Medieval astronomers regarded the heavens as a great clock, the workings of which were generally understood, and which were read periodically with astrolabe and quadrant to keep the calendar in adjustment.[1]

Not until the fifteenth and sixteenth centuries did new scientific circumstances demand a change of approach. The Julian Calendar was running into error, the Earth's motion around the Sun was being seriously proposed, while the new star of 1572 demonstrated that change could take place in space, beyond the 'sphere of the Moon'. Astronomers were also coming to challenge the very existence of the crystal sphere of heaven, while comets were being considered as astronomical rather than meteorological bodies. None of these questions could be answered by simply adjusting existing observations, but demanded a fresh examination and measurement of all bodies in the sky. Though Tycho Brahe was not the first Renaissance astronomer to observe the heavens, he was the first to realise that long-term, systematic observation was required, and that the quality of these observations — and the cosmological conclusions which could be drawn from them — depended on the improving accuracy of instruments.

The main event which compelled this new way of thinking upon Tycho was a new astronomical phenomenon, for which there was no precedent in existing catalogues: the New Star of 1572.[2]

When the New Star appeared, in November 1572, it was confidently expected by astronomers to display a parallax. It was assumed, in accordance with prevailing cosmological beliefs, that no new phenomenon could occur beyond the Moon on the grounds that only the terrestrial regions were subject to change, so the star must be in the upper atmosphere. As the Moon, which was considered to occupy the extreme upper limit of the atmosphere, displayed an obvious parallax, the essentially meteorological New Star, being closer than the Moon, should show an even bigger one. Tycho set about measuring the New Star's position from its adjacent stars in Cassiopeia during the winter of 1572-3, observing as it transited the northern meridian.

The conclusion delivered a serious blow to classical cosmology, by demonstrating that there could be changes and new stars in deep space. Equally important was the avenue via which Tycho arrived at this conclusion: not by the rules of logic or philosophical abstraction, but by mathematical data yielded by a precise measuring instrument. In many ways, one might say that it was Tycho's observation of the New Star of 1572, even more than Copernicus's theory, which saw the birth of modern astronomy, for it identified the essential working ingredients of a scientific problem: [71] the need for accurate observation, exact instrumentation and conclusions based on careful measurements as opposed to purely theoretical criteria. Over the next twenty-five years, Tycho was to devise a new working method for astronomical research, grounded in the systematic use of instrumental evidence, and described in detail in his *Mechanica* and *Progymnasmata*.[3] These two books, along with Tycho's other published works, established a new way of doing science, searching for and utilising new knowledge, which was to be as important for the philosophy of science as Kepler's Laws or Newton's gravitation.

Tycho's observations of the New Star were performed with an instrument of his own devising, comprising a hinged pair of 6-foot beams, opening like a pair of compasses against an accurate scale of 60°, controlled by a fine screw. Unlike most of the instruments of his day, this 'sextant' was designed to fulfil one single function — the precise measurement of a vertical angle.[4] Most sixteenth-century instruments were devised to be multi-purpose, and could measure a variety of different types of phenomenon, such as the time, and vertical and horizontal angles. Tycho established a fundamental principle of instrumentation: the more functions you expected an instrument to perform, the less accurately would it do each one of them. Supreme accuracy depended on an instrument that was designed to fulfil only one specialised task, and which could be left in one critical adjustment (Fig. 1).

The New Star, and the major book in which he published his conclusions and observing methods, made Tycho's astronomical reputation by the time that he was twenty-seven. Following an invitation from King Frederick II of Denmark, he set up his famous observatory of Uraniborg, on the island of Hven, off Copenhagen, to commence a quarter-century research programme which was to change the course of astronomy. Fortunate insofar as he had ample Royal funds to back him during King Frederick's life, Tycho set about a complete revision of the northern heavens, to become the first European since antiquity to re-map the sky from scratch.

Tycho's work was original not only in the way in which it stressed the use of specialised instruments to attack specific astronomical problems, but because of his wider contributions to the process of technological research and

III

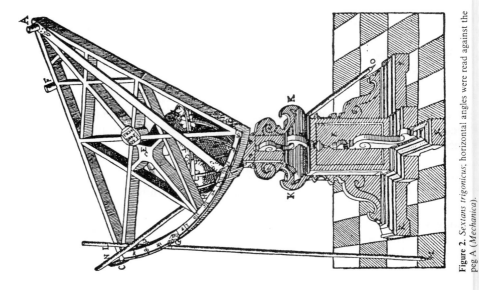

Figure 2. *Sextans trigonicus*; horizontal angles were read against the peg A (*Mechanica*).

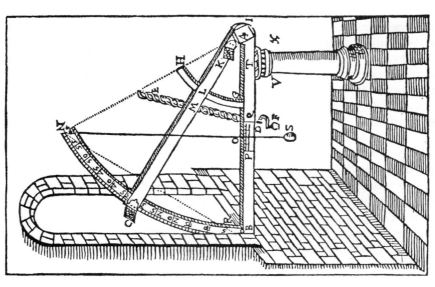

Figure 1. Vertical sextant. By turning the screw, the hinged arm AC moved against the 60° scale, its plumb line bisecting the isosceles triangle ABN at O to cross-check the vertical adjustment (*Mechanica*).

development. Uraniborg was more than just the first well-equipped observatory in Europe — it was also a workshop and testing ground for new ideas in astronomical hardware. Tycho's skills as an observer are well known, but we often forget his perhaps greater skills as a mechanician and engineer. Though he did not make equipment with his own hands, he did design it, and between 1573 and 1597 evolved several 'families' of instruments, individual members of which he improved, re-built and perfected until he was able to obtain the extreme limit of accuracy possible to the naked eye. The modern scholar, Victor Thoren, has worked out a detailed chronology for the invention, use and superseding of many of Tycho's instruments from the observing logs included in the astronomer's *Opera Omnia*.[5]

Tycho's main instruments fell into three groups, each devised to fulfil a particular purpose, comprising the sextants, armillaries and quadrants. The sextants represented Tycho's most original and structurally significant instruments. They were used for measuring either vertical or horizontal angles between pairs of objects in the sky, and comprised a 60° arc, which was produced naturally by striking off the radius of any circle against its own circumference, in accordance with Euclidean procedures.[6] Though not the first astronomer to use a 60° instrument, Tycho was certainly the first to see the wide possibilities which this natural geometrical shape, and its enclosed isosceles triangle, afforded. His first instrument of this type had been a portable 'half sextant' of 30° made when travelling in Germany in the late 1560s.[7] It had been with such an instrument, and with a full sextant of 60°, that he had observed the New Star of 1572, as mentioned above.

Once this hinged two-beam design had been realised, Tycho improved it, to produce a series of sextants with rigid 60° frames of wood and metal, equipped with pairs of sights to enable two observers to read off large horizontal angles against a central sighting peg. In his 6-foot-radius 'Sextans Trigonicus' of 1584, the body of the sextant could be locked by a ball and claw to any [72] plane in the sky to observe a series of great interlocking spherical triangles around the zodiac or celestial equator.[8] Though accurate to about one minute of arc for a single observation, Tycho greatly refined his values by regularly repeating the same observation over months or years, so that when his observations were reduced in the nineteenth century, his standard stars were found to be accurate to 24 seconds of arc.[9] These were the stars which formed the foundation positions from which everything else was measured in the construction of his catalogue (Fig. 2).

The armillaries were large circular instruments, based on ancient Greek prototypes, in which great rings around six feet in diameter were set up to form skeleton spheres, to demarcate the zodiac, equator, meridian and pole. By taking sightings across the accurately graduated rings, one could observe

III

Right Ascension angles, measure the Sun's daily coordinates and fix the First Point of Aries. Because classical and medieval astronomers used the zodiac as their fundamental celestial plane, Tycho found it awkward to integrate ecliptic coordinates with latitude measurements made from the pole. In consequence, he simplified the armillary sphere by reducing its plane to that of the equator, thereby improving its ease of operation by simulating the daily rotation of the heavens. By this act, we must also remember, he also 'invented' the equatorial mount.[10]

Tycho built several armillaries at Uraniborg, each one more specialised and exact in its function, until he reduced the design to a single ring ten feet in diameter, moving around a polar axis, within a semicircle set within the equatorial plane. This instrument was so large that the observer stood inside it to make an observation, while it was mounted on one of the earliest sets of self-centering bearings in the history of precision mechanics (Fig. 3).[11]

The traditional problem with armillary spheres derived from the fact that the central axis upon which the heavy rings were mounted moved in a simple plug socket, which had to be fairly loose to permit the rings to turn. Tycho realised that the instrument could never be exact so long as the inevitable 'play' in the sockets remained. On his 10-foot armillary, however, he terminated the lower end of the polar axis with a specially-machined conical point bearing, which rested inside a corresponding conical hole cut inside an adjustable steel block. At the same time, the upper end of the polar axis was secured by a matching bearing. This design successfully eliminated play, for the weight of the instrument inevitably forced the pointed end into the cone bearing, while the upper axis was also self-centering. The resulting instrument would have been in perfect balance, easily moved, yet centering itself dead in any position. After a succession of design experiments with armillaries over a ten-year period, Tycho had evolved the ideal combination of specificity of [73] function, lightness, accuracy and stability.

The third of Tycho's principal instrument types was the quadrant. While the quadrant, like the armillary, had a lineage extending back to antiquity, its construction had always presented problems, for a 90° scale was not easy to divide and had to be produced by a combination of elaborate geometrical techniques.[12] While nothing is known of the way in which Tycho obtained his individual degree divisions, he nonetheless made several innovations in the overall design of the quadrant as a mechanical structure. Most of this work was concerned with attempts to solve two basic problems: (a) how to make an instrument where the degree divisions were physically large enough to allow accurate subdivision into minutes, and (b) how to achieve the same on a structure which was not physically cumbersome and likely to distort under its own weight.

In the past these two requirements had been irreconcilable within the same instrument, for if one built a quadrant that was large enough to show single minute subdivisions, it must, perforce, be as large as a house, and hence be too crude for those delicate adjustments necessary to effectively use its small graduations. This had formed a barrier to improved accuracy before Tycho, and his own solution to the problem was a classic in geometrical and engineering innovation.

In the *Astronomiae Instauratae Mechanica* (1598), where Tycho described the functions of his major instruments, it is clear that more effort was devoted to devising the perfect quadrant than to any other instrument. He describes nine quadrant designs, as opposed to five sextants and four armillaries. The purpose of the quadrant was simple in theory, but exacting in practice: it had to measure vertical angles between horizon and zenith. Without a reliable quadrant, it was impossible to fix any of the cardinal points of the sky, such as the latitude, solstices and equinoxes, or derive the exact time. It provided the fundamental vertical angles of bodies, from which the sextants and armillaries could next measure the horizontals, for a good set of declination angles was an essential prerequisite for measuring RAs.

Tycho's first attempt at a scale showing small subdivisions had been made in Augsberg, Germany, in the early 1570s, when he designed a 19-foot-radius instrument for his astronomical friend Paul Hainzel, who was also Burgomaster of the city. This massive instrument was built of wooden beams, so that the 19-foot-radius quadrant could be made to move in the vertical plane, while rotating to face any part of the sky on its great timber support pole.[13] Though its brass scale was big enough to be divided into individual minutes, the sheer bulk of the 30-foot-high structure obliterated that delicacy of touch necessary to use them (Fig. 4).

Tycho experimented with various ways of dividing a degree space, including the 'Nonius' scale, which was a complicated ancestor of the Vernier, but it was not until he hit upon the use of transversal, or diagonal lines, that his desired solution was found. While Tycho did not invent the diagonal scale, he was the first to develop and apply it to major astronomical instruments. The principle of the scale lay in utilising not the circumference edge of a degree division, but a diagonal line drawn on the surface of the quadrant between a consecutive pair of degrees. In this way, it was possible to draw a line that was much longer than the circumference edge line, and divide it into a greater number of fractions.[14] When an observation was made, one obtained the eventual angle by first reading the nearest whole degree, and then by counting which diagonal dot had been cut by the sighting arm edge, and adding the same number of minutes to obtain the complete angle. The diagonal scale made it possible to have an instrument divided

III

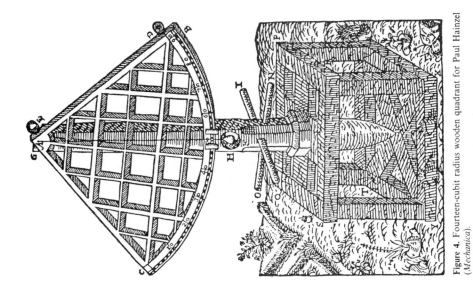

Figure 4. Fourteen-cubit radius wooden quadrant for Paul Hainzel (*Mechanica*).

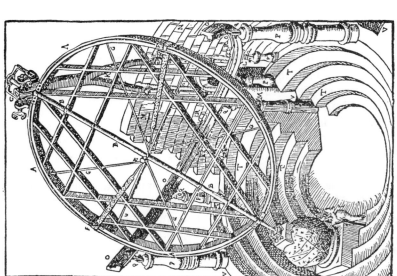

Figure 3. The Great Armillary of one ring, set upon a polar axis, for the measurement of declination angles across the central peg (*Mechanica*).

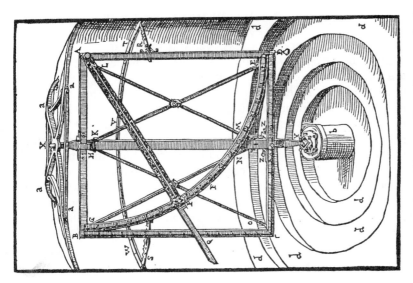

Figure 7. Steel quadrant within a square. The quadrant was equipped with both circular and linear sine scales, and rotated against a graduated horizon circle fixed to the surrounding wall (*Mechanica*).

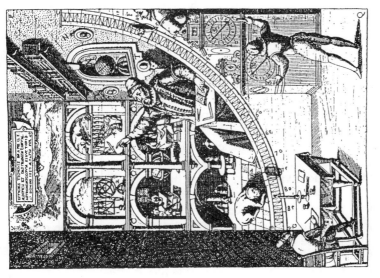

Figure 6. Tycho's Great Mural Quadrant. The portrait and view of the inside of Uraniborg were added, to fill up the blank wall space (*Mechanica*).

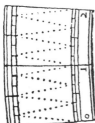

Figure 5.
Transversal dot scale,
or Diagonal, used on all of
Tycho's major instruments
(*Mechanica*).

down to single minute intervals or less, yet still have a quadrant that was small enough to be physically compact and easy to manage. It formed an elegant example of precision miniaturisation (Fig. 5).

From a variety of intermediary models, Tycho evolved two major quadrants, both of which incorporated his new diagonal divisions. Maximum rigidity in quadrant design was achieved in the mural quadrant of 1582, when he reduced the instrument to a heavy brass arc fastened to a masonry wall, to make the first scientifically significant mural quadrant set in the meridian.[15] With its radius of almost 7½ feet, it carried a diagonal scale divided down to 10-arc-second spaces to delineate the fundamental plane of the sky. Tycho was now able to take his principal observations as objects culminated in the meridian, to commence a procedure which would become standard in positional astronomy down to the twentieth century (Fig. 6).

His second major quadrant, of the same radius as the mural (so that their scales carried equal proportions), was his 'Steel quadrant in a square' of 1588.[16] While the mural quadrant provided an excellent absolute standard, he still needed an instrument which he could use outside the meridian, especially to observe the Sun's position in any part of its orbit. The steel quadrant obtained maximum rigidity by its metallic structure, braced as it was within a 6-foot square. The whole instrument stood on a strong vertical axis of steel, which was both self-centering and capable of rotation to face any part of the sky. In many ways, the steel quadrant in a square was Tycho's masterpiece, demonstrating as it did an understanding of light girder bracing, versatility and accuracy. Having experimented with wood and brass, and abandoned them where critical accuracy was necessary, he settled on the properties of steel as the ideal material for a light- [74] weight, rigid frame. It was made with the same radius as the mural quadrant, moreover, so that the paired 10-arc-second divisions on the two instruments could be interchanged (Fig. 7).

Much of Tycho's significance in the history of science comes from his quest to find the perfect shape, material and design for each type of observation. He also realised the importance of having several different instruments utilising different geometrical principles, such as those used in the sextants, armillaries and quadrants, to make the same observation. Crucial constants, such as the solar altitude or First Point of Aries, would be observed with a battery of instruments, both to obtain the best average value, and to cross-check the strengths and weaknesses of different designs.

Cross-checking and error analysis formed a major component in Tycho's working method, and he was a leading pioneer of its development. Almost every instrument which Tycho designed possessed an internal cross-check against itself, along with divisions of similar proportions to those on other instruments. Thus, his earliest vertical wooden sextant of 1572 exploited its sixth part of a circle shape to produce an isosceles triangle which was bisectable by a plumb line in accordance with a Euclidean proposition.[17] His armillaries could read the same angles in either clockwise or anti-clockwise directions around the sky to see if they always closed to zero at the First Point of Aries, while the steel quadrant was enclosed inside a square, the straight edges of which were engraved with a sine table, to cross-check the circular degrees.[18] Armillaries could also be tried against sextants when reading the same horizontal angle, and circular degrees against a linear sine scale when reading a vertical one.

Early in his career, about 1573, he came to realise that otherwise well-made instruments could be spoiled if their naked-eye sights were not precise. Up to Tycho's time, astronomers had used 'pinhole' or 'pinnule' sights on their instruments with which to sight the stars, but he came to realise that as an observer would never be certain of having the star at the centre of the sight hole, it was possible to introduce random errors of up to eight minutes in an observation.[19] It was absurd, therefore, having an instrument with scales graduated to single minutes if the sights contained an 8-minute error.

His solution was to invent a parallax-free sight, wherein he observed a star through a pair of fine slits across opposite edges of a cylindrical brass pivot.[20] When he saw the star equally well through each slit on each pivot edge, he knew that he had a perfect alignment. This sight became standard on all of his instruments, and remained in use amongst astronomers for the next hundred years, until the development of the telescopic sight in the 1660s. Though Tycho had no way of knowing it, his improved sights and scale divisions exceeded in accuracy the capacities of the astronomers using them, for while his great quadrants carried divisions down to 10 arc-seconds, the unaided human eye cannot resolve angles less than one minute.[21] Tycho had [75] taken pre-telescopic astronomy as far as it could go, and any significant improvement had to wait until the resolving power of the human eye could be increased. It was a tribute indeed to Tycho's meticulous observing techniques that his constant revision of the same fundamental observations enabled him to get average values that vastly exceeded the angular resolution of the naked eye, for when astronomers in the nineteenth century reduced Tycho's determinations for the First Point of Aries, they obtained a value which was within 6 arc-seconds of the correct one.[22]

Though Tycho published detailed illustrated accounts of his instruments, it is unfortunate that he said nothing about the processes of manufacture on a workshop level. This should not be interpreted as secrecy on Tycho's part, however, for his stress was always on the openness of astronomical knowledge, but a lack of awareness that such things needed to be discussed at all.

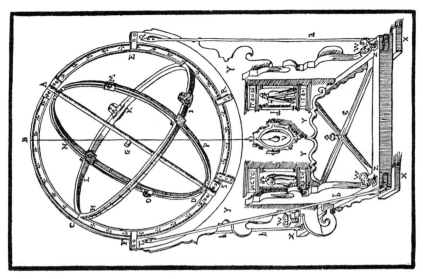

Figure 9(a). Tycho's Equatorial Armillary (*Mechanica*).

Figure 8. Assembled rings of Ferdinand Verbiest's half-completed Equatorial Armillary, after Tycho (*Astronomia Europaea, Museum of the History of Science, Oxford*).

Figure 9(c). Modern photograph of Verbiest's Equatorial Armillary, Peking.

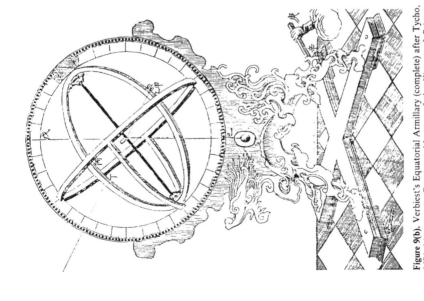

Figure 9(b). Verbiest's Equatorial Armillary (complete) after Tycho, 1674 (*Astronomia Europaea, Museum of the History of Science, Oxford*).

With the well-established metallurgical craft tradition which existed in northern Europe, he probably felt that when his designs and researches were published, existing craftsmen would have no trouble in duplicating them. This belief was no doubt well-founded, for Tycho's instruments had been made by craftsmen from various lands, for while his earliest pieces were made in Germany, and his main Uraniborg instruments in Denmark, he found no difficulty in getting the famous Czech craftsman Erasmus Habermal to make extra ones when he was exiled in Prague, shortly before his death in 1601.[23]

It is only now, after an interval of four hundred years, and the passing away of that tradition, that we ask exactly how did one make the rings for an armillary sphere or graduate a scale? Ironically, such information is only available from Oriental sources, for when European Jesuit astronomers re-built the Imperial Observatory in Peking in the 1670s, they made duplicates of Tycho's Uraniborg instruments copied from the *Mechanica*.[24] Because the Emperor, K'ang Hsi, was interested in western technology, the Jesuit scientist Ferdinand Verbiest obligingly provided a detailed set of 117 engravings and text, which included the main processes of workshop manufacture. These European missionary scientists would have been [76] sufficiently familiar with instrument-making processes to enable them to instruct Chinese craftsmen, and record them in a sumptuous work for the Emperor, whereas in Europe, the knowledge would have been commonplace enough to make 'workshop manuals' unnecessary (Fig. 8).

Political circumstances forced Tycho to leave Denmark in 1597, and it is sad to reflect that all his major instruments were lost soon after. It is in China, however, with its great sextants, armillaries and quadrants built by the Jesuits, along with their treatise *Astronomia Europaea* describing their manufacture, that one comes closest to a surviving physical memorial to Tycho Brahe today.[25] Fortunately, the old Imperial Observatory weathered war, invasion and the Cultural Revolution, to provide the only direct descendant of Tycho's Uraniborg to survive to the present day. These great bronze instruments have survived in astonishingly good condition, moreover, and make it possible to compare a modern photograph with a seventeenth-century engraving, and place both alongside Tycho's own illustrations of the *Mechanica* pieces (Fig. 9).

Yet Tycho's real memorial lies not in any set of artefacts, but in a new approach towards the study of the heavens based upon progressive instrumentation and systematic observation. Though he died nine years before the first astronomical use of the telescope by Galileo, it had been Tycho who provided so much of the context in which telescopic evidence could be intelligently interpreted. For when Tycho's eccentric and brilliant life ended prematurely following a bout of over-indulgence at a feast in Prague in autumn 1601, he had already created the foundation for a reformed science of astronomy.

References

1. A.C. Crombie, *From Augustine to Galileo* (Peregrine, 1969), vol. 1 pp. 89-110.
2. Tycho Brahe, *De Nova Stella* (1573). See also J.L.E. Dreyer, *Tycho Brahe, a picture of scientific life and work in the sixteenth century* (1890), pp. 38-69.
3. Tycho Brahe, *Astronomiae Instauratae Mechanica* (1598), translated by H. Raeder, E. and B. Strömgren as *Tycho Brahe's Description of his Instruments and Scientific Work* (Copenhagen, 1946), cited hereafter as *Mechanica*. Also Tycho Brahe, *Astronomiae Instauratae Progymnasmata* [Astronomical Exercises] (1610). Many of Tycho's principal works were published shortly after his sudden death.
4. *Mechanica*, pp. 84-7.
5. Victor E. Thoren, 'New light on Tycho's instruments', *Journal for the History of Astronomy* 4, 1 (1973), pp. 25-45. Also *Tychonis Brahe Dani, Opera Omnia*, edited by J.L.E. Dreyer (Copenhagen, 1923-6).
6. *Mechanica*, pp. 85-6.
7. *Mechanica*, pp. 80-3. [77]
8. *Mechanica*, pp. 72-5.
9. Dreyer, *Tycho Brahe* (n. 2), p. 351.
10. *Mechanica*, pp. 60-3.
11. *Mechanica*, pp. 64-7.
12. Allan Chapman, 'The astronomical art — the reconstruction and use of some Renaissance instruments', *J. Br. Astron. Assoc.* 96, 6 (1986), pp. 353-7.
13. *Mechanica*, pp. 88-91.
14. *Mechanica*, p. 141.
15. *Mechanica*, pp. 28-31.
16. *Mechanica*, pp. 36-9.
17. Euclid's *Elements*, Bk. IV, theorem 15.
18. Two of the straight edges of the square, enclosing the quadrant, carried a 5-figure sine table: *Mechanica*, p. 37.
19. This derived from an 8-minute-of-arc sight error which he found in one of Copernicus's instruments: *Mechanica*, p. 46.
20. *Mechanica*, pp. 141-3.
21. This was first demonstrated by Robert Hooke in *Animadversions on Hevelius, his 'Machina Coelestis'* (1674), p. 7. Hooke's value is borne out by modern physiology, as in Sir Stuart Duke Elder, *System of Ophthalmology* V (1958), p. 148.
22. Reductions of Tycho's observations were made by Peters, Argellander, Le Verrier and others: see Dreyer, *Tycho Brahe*, p. 351. Also G.L. Tupman, 'A comparison of Tycho Brahe's meridian observations', *Observatory* 23 (1900), pp. 132-5, 165-71.
23. The Habermal sextant still survives in Prague, and a photograph is included in Dr. Hubert Slouka (ed.), *Astronomy in Czechoslovakia from its Early Beginning to the Present Times* (Prague, 1952), p. 102.
24. Allan Chapman, 'Tycho Brahe in China: the Jesuit mission to Peking and the

iconography of instrument making processes', *Annals of Science* 41 (1984), pp. 417-43.

25. Ferdinand Verbiest, *Astronomia Europaea sub Imperatore Tartaro-Sinico ...* (1687), was the first European printing of this work, though it was based on an earlier text first produced in China. The sets of original Chinese plates are rare, though copies exist in the British Library (Oriental Division), the School of Oriental and African Studies, and the Museum of the History of Science, Oxford. Some copies also survive in European collections.

A note on Tycho's units of linear measurement

I have cited the dimensions of Tycho's instruments in English feet. These figures have been rounded up, however, as Tycho gave them in cubits, and there is some contradiction in the length of that value. Dreyer in *Tycho Brahe* (1890), p. 39, footnote, gives it as 16.1 English inches, while D'Arrest and Charlier in 1868 determined it to be 388 mm [15.25 inches]; see Raeder and Strömgren's translation of *Mechanica* (1946), p. 9. This would mean that a 'six-foot' instrument was in reality about 5ft. 4.4 inches.

Reprinted from Journal of the British Astronomical Association *99, 2 (1989), pp. 70-7. Numbers in square brackets indicate original pagination.*

IV

Tycho Brahe in China: the Jesuit Mission to Peking and the Iconography of European Instrument-Making Processes

Summary

In the late 1660s, Ferdinand Verbiest, a Flemish Jesuit missionary in Peking, was instructed to re-equip the Imperial Observatory. The new instruments which he caused to be built were modelled neither upon contemporary European prototypes, nor those of traditional Chinese astronomy, but on the pieces in Tycho Brahe's *Mechanica*, of eighty years before. The Chinese instruments were lavishly illustrated, moreover, in 105 woodcuts that contained detailed representations of their processes of construction. It is argued that these illustrations not only give us valuable insights into what the technical Jesuits did in Peking, but show how sixteenth- and seventeenth-century European craftsmen constructed their instruments, for while the location was Oriental, the technology was Western. They can also give important insights into how Tycho's prototypes had been built, and provide us with useful information regarding European instrument-making technology.

Contents

1. Introduction

Not least amongst Tycho Brahe's astronomical innovations was the provision of a detailed account of the instruments with which he made the observations for his catalogue. By describing each of his major instruments in detail and providing woodcut illustrations, his *Astronomiae Instauratae Mechanica* (1598) created something of a precedent in scientific exposition that was to influence many later generations of astronomers. Scheiner and Riccioli discussed their apparatus while Johannes Hevelius, in particular, followed Tycho's example by accompanying his observations with a treatise explicitly on his instruments. By the time of Picard, Cassini, and Flamsteed, in the late seventeenth century, the discussion of the instruments used had become a necessary preliminary to any major astronomical treatise.[1]

---

[1] Tycho Brahe, *Astronomiae Instauratae Mechanica* (Wandesburgi, 1598), translated by Hans Raeder, Elis and Bengt Strömgren (Copenhagen, 1946). Johannes Hevelius, *Machina Celestis*, I (Dantzig, 1673). John Flamsteed also provides exhaustive treatment not only of his own instruments, but also of those of his European and Arabic predecessors, thereby stressing the instrumental basis of astronomy. See *The 'Preface' to John Flamsteed's 'Historia Coelestis Britannica', 1725*, edited and introduced by Allan Chapman (National Maritime Museum Monograph No. 52, 1982).

Yet all these works present the instruments in a state of completion, and ready for use. There is virtually no information relating to the processes of their construction, or the graduation of their scales, nor are workshop methods described. Presumably, such information was not considered relevant to the astronomer's purpose, in addition to which, any serious reader might visit the workshops himself, for such processes ranked amongst the living arts and crafts of the time. To the historian of scientific instruments, however, these arts are lost, and we are often left with little more than speculation as to how Tycho and his successors actually fabricated their tools.

Only when it became incumbent upon a Western mathematical practitioner to transport his skills into an alien environment, where native craftsmen would have to do the work and technically-minded patrons would require explanations, was it necessary to produce detailed records of constructional processes. Such a situation occurred in China around 1670, when the Flemish Jesuit missionary, Ferdinand Verbiest (1623–1688), was given the Directorship of the Imperial Observatory.

While the Chinese possessed astronomical records extending back over several millennia, and were familiar with a variety of complicated instruments of indigenous design, their astronomy was in a state of stagnation when the first Jesuits arrived at the end of the sixteenth century. Indeed, the early missionaries quickly capitalized on the fact that the superior science and technology of Europe could be turned to advantage in their objective of converting the Chinese to Christianity. Astronomy, in particular, occupied a place of importance among the Jesuit plans, for it was through his ability as a calendar calculator that Verbiest was appointed Director of the Observatory, only to find it equipped with unwieldy instruments of native design:

> But Father Verbiest, when he undertook the survey and management of the mathematicks, having judged them very useless, perswaded the Emperor to pull 'em down, and put up new ones of his own contriving.[2]

It was the contriving of these pieces which obliged Verbiest not only to teach European workshop skills to the Chinese artisans, but in addition to produce an illustrated treatise on their manufacture for the delectation of his Imperial patrons. The Emperor K'ang Hsi, under whose authority Verbiest built the instruments, was a young and intellectually curious ruler in the 1670s. He was fascinated by European science and technology, and the Jesuits found him to be an eager pupil. In consequence, Verbiest was not only elevated to Mandarin rank, but often accompanied the Emperor on his progresses around the country. K'ang Hsi was proud of his European technical expertise, and delighted in showing it off before his courtiers. He had familiarized himself with Euclid, certain aspects of Western mathematics, and the theory and

[2] Louis Le Comte, *Nouveaux Mémoires sur l'état présent de la Chine*, 2 vols (Paris, 1696), anonymously translated into English as *Memoirs and Observations Topographical, Physical, Mathematical, Mechanical, Natural, Civil and Ecclesiastical, made in a late journey through the Empire of China* (London, 1697), p. 65. For a comprehensive account of Verbiest's life and career, see H. Bosmans, "Ferdinand Verbiest, Directeur de l'Observatoire de Péking", *Revue des questions scientifiques*, 21 (1912), 195–273. [Hereafter cited as Bosmans I.] Verbiest's rise to eminence as a calendar calculator under the Emperor K'ang Hsi during 1668–1669 is outlined by Bosmans I, pp. 235–50. Bosmans also produced a detailed critical bibliography of Verbiest's Chinese writings in the second part of his article, headed "Les ecrits Chinois de Verbiest", *Rev. ques. sci.*, 24 (1913), 272–98. [Hereafter referred to as Bosmans II.]

practice of a variety of scientific instruments. Verbiest appreciated the good fortune of the Emperor's scientific curiosity in the overall success of the Jesuit mission, and this must have played a major part in his production of *Astronomia Europaea*.[3]

2. The documentary sources

Verbiest's treatise included 105 woodcuts, the first a general prospect of the Peking Observatory not numbered, followed by the rest with illustrations numbered 1 to 117 in Chinese numerals, some occurring two or more to a plate. They were probably intended for his (*Hsin chih*) (*ling t'ai*) *I hsiang chih*, or 'Newly constructed observatory instruments', and subsequently copied for the Imperial Encyclopaedia *Thu Shu Chi Chhêng* (1726), as well as later treatises. Though this Chinese work has not been translated, Verbiest provided a brief Latin summary in *Astronomia Europaea*, which preceded the *Liber Organicus*, in which the plates themselves were published in Peking around 1674.[4] The Latin text of the *Astronomia Europaea* was printed on to ten folio sheets, recto and verso, to form eighteen pages of script. Its Western cursive handwriting had been engraved directly into the wooden printing blocks to produce the facsimile of a handwritten manuscript rather than a conventionally printed page. It refers to a variety of Western scientific instruments illustrated in the *Liber Organicus*, in addition to the Tychonic observatory pieces, although Verbiest provides no real discussion, as opposed to illustration, of the constructional processes. There are no references to any of the plates beyond number 46 in the sequence. The technical processes are, nonetheless, quite explicit in themselves, and to a European Latin reader

---

[3] Verbiest's interests and accomplishments in Western science are discussed in Jonathan Spence, *Emperor of China; a self-portrait of K'ang Hsi* (London, 1974). See also Joseph Needham, *Chinese astronomy and the Jesuit mission* (London, 1958).

[4] The bibliographical position of Verbiest's writings has been elucidated by a variety of scholars, such as Bosmans II, and L. Pfister, *Notices Biographiques et Bibliographiques sur les Jésuites de l'Ancienne Mission de Chine*, 2 vols (Shanghai, 1932). Verbiest produced at least two works with Latin titles, the *Astronomia Europaea sub imperatore Tartaro-Sinico Cam Hỹ appellato ex-umbra in lucem revocata a. P. Ferdinando Verbiest Flandro-Belgae Brugensi & Societate Jesu Academiae In Regia Pekinensi Praefacto Anno Salutis MDCLXVIII*, and *Liber Organicus Astronomiae Europaeae*. Although both of these works are dated Peking 1668, they are clearly backdated from c. 1674, as the instruments which they describe had not even been built in 1668. A copy of both works, bound up together in European covers of c. 1800 (until re-binding in the early 1980s), is in the library of the School of Oriental and African Studies, London, Accession No. 35409 (Ex 39). Whether this mode of binding or order of presentation was intended by Verbiest is not known, for the title page *Astronomia Europaea* precedes the Latin text, and the 117 numbered and one unnumbered xylographs of the *Liber Organicus* are printed on 105 separate sheets of folio-sized rice paper. In addition to the S.O.A.S. volume, there are two separate sets of the *Liber Organicus* plates in the British Library, Oriental Section, but they lack the Latin text. In 1977 another unbound set of the same plates, minus the text, were acquired by the Museum of the History of Science, Oxford. The illustrations that follow are taken from this set by kind permission of the Curator.

Verbiest's Latin text was later printed in Europe by Philippe Couplet in a small volume describing the scientific activities of the Jesuits in China. While Couplet's work has an almost identical title to Verbiest's, *Astronomia Europaea*...(Dilingen, 1687), it is somewhat confusing as the Verbiest text only occupies chapters XIII–XXVIII, Sig. 40 $^{E}4^{V}$, and Sig. 57^{H} of this 126-page quarto volume. The European edition, however, does not reproduce the plates from the *Liber Organicus*, with the exception of the re-engraved prospect of the Peking Observatory included as a frontispiece. Le Comte also includes the observatory prospect in *Nouveaux Mémoires*, I, p. 142 (footnote 2).

Joseph Needham also refers to Verbiest's Chinese publications in *Science and Civilisation in China*, III (Cambridge, 1959), p. 452. I am especially indebted, furthermore, to Mr John Combridge, of Ilford, Essex, for the latest bibliographical treatment of the text and plates. Mr Combridge's work still awaits publication, though he has been generous in imparting to me the benefits of his research, both by letter and in conversation. I wish to state, however, that I take full responsibility for any bibliography or technical errors committed in this paper.

familiar with the works of Tycho Brahe, would probably have been self-explanatory. They contain, between plates 23 and 57 in the Chinese numbering sequence, illustrations of the casting, finishing, graduating and assembling of the parts of an armillary sphere and a great globe, which are clearly modelled on the pieces in Tycho's *Mechanica*. It is my argument in this paper that Verbiest's work provides not only an insight into Chinese science, but an account of how a contemporary European would have built a major set of observatory instruments. These, in turn, might provide some insights into how Tycho's own instruments were built.

In his Latin text, Verbiest gives no explicit provenance for his instruments, although in a letter of 1670 he speaks of their graduations '[de] si grand usage chez Tycho', and the Tychonic association is substantiated by many of their technical features.[5] Not only are the dimensions of Verbiest's instruments the same as those of Tycho, but they are generally depicted from the same angles as their equivalent pieces in the *Mechanica* illustrations. Indeed, several of Verbiest's instruments are shown standing on the same type of black and white 'chessboard' pavement as the Tychonic pieces. Verbiest's representation of his large sextant, voluble quadrant, and globe is characteristically Tychonic, while it is the armillary spheres that best demonstrate the European prototypes. Not until the arrival of the Jesuits was the ecliptic armillary sphere known in China, as oriental astronomers worked exclusively from equatorial coordinates. Verbiest included a large ecliptic armillary in the newly-equipped observatory, its six-foot diameter rings disposed in the same way as in Tycho's illustration. The companion equatorial, or equinoctial armillary, which Verbiest constructed, not only duplicated Tycho's armillary in all its essential details, but differed significantly in its design from traditional Chinese equatorials. Instead of carrying the traditional double rings of the oriental pieces, the Verbiest equatorial had solid, single rings, like those of Tycho, implying that even when the Jesuit included instruments of a type already known in China, he preferred to model them upon European prototypes.

In spite of their obviously European technical features, the Verbiest instruments represent a curious cultural confluence, as the European circles and technical parts were mounted upon stands contrived in the form of lions, dragons, flaming pearls, and other oriental motifs. The technology is wholly European, while the decorative features are characteristically Chinese. Verbiest stated in a letter to Jacques Le Faure that the elaborate mounts of dragons and lions derived from the regal associations of these creatures, and the royal patronage of the observatory. Visually magnificent as they were, he pointed out that the stands constituted encumbrances from a technical aspect, often making important parts of the instrument inaccessible.[6]

In addition to providing detailed drawings of the individual new instruments, Verbiest also provided an overall prospect of the observatory with each piece set in its correct place.

3. The observatory instruments and their European prototypes

The observatory was built on the top of a short tower or bastion in the old wall of the city. Proceeding clockwise from the bottom right-hand corner of the plate (Figure 1), the instruments depicted are: equatorial armillary sphere, great stellar globe,

[5] Ferdinand Verbiest to Jacques Le Faure, letter 20/8/1670, in Bosmans I, p. 269. Bosmans himself, p. 254, also stresses the Tychonic provenance of the instruments.

[6] Verbiest to Le Faure, 20/8/1670, Bosmans I, p. 270. See also J. Needham, "The Peking Observatory in A.D. 1280", *Vistas in Astronomy*, 1 (1955), pp. 67–83.

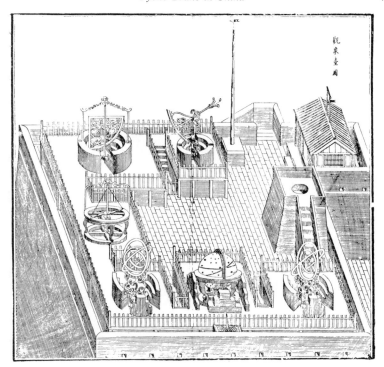

Figure 1.
Unnumbered in Chinese sequence.

ecliptic armillary sphere, horizontal azimuth, voluble quadrant, and sextant. Note the masonry viewing steps surrounding the armillary spheres, quadrant, and sextant, which were based on Tychonic prototypes. From the position of the sundial, outside the fence at the front of the globe, it is clear that the bottom part of the picture faced south. This dial, moreover, is a Western analemmatic type.

This was the only one of Verbiest's plates to be published in Europe during the seventeenth century, appearing re-engraved as the frontispiece to Philippe Couplet's *Astronomia Europaea* (Dilingen, 1687).[7]

The graduations on Verbiest's instruments are of great interest. All of them were engraved within 360° circles, as opposed to the 365¼ digit circles of Chinese usage, and both Bosmans and Needham agree that it was the Jesuits who imposed the 360°

[7] In Couplet's *Astronomia Europaea*, the plate is reduced in size from folio to quarto and bears the engraver's name, Melchior Haffner. This plate, however, was re-copied in several subsequent works on China, as in Le Comte's *Nouveaux Mémoires*, and its English translation, *Memoirs and Observations Topographical* (footnote 2), and in J. B. Du Halde, *Description géographique, historique, chronologique et physique de l'Empire de la Chine et de la Tartarie Chinoise*, 4 vols (Paris, 1735). In Richard Brookes' translation of Du Halde, *A general history of China*, 4 vols (London, 1736), portraits of Verbiest and his scientific predecessor, Adam Schall, are printed as frontispieces to volumes III and IV.

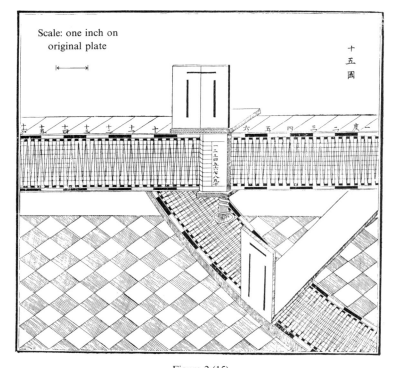

Figure 2 (15).

The Chinese number is given in parentheses.

sexagesimal system on China.[8] We are also informed by Louis Le Comte in his description of the Peking Observatory in 1696 that:

> The circles are divided both on their exterior and interior surface by cross lines [transversals] into 360 degrees and each degree into 60 minutes and the latter into portions of 10 seconds each by small pins [dots?].[9]

This was the same mode of subdivision, using dots and lines, as that used by Tycho and his followers, such as Johannes Hevelius.

---

[8] Needham (footnote 4), III, pp. 82, 203, 374, 381, states that China remained virtually untouched by Arabic and Indian influences, and failed to develop a geometry based on the 360° circle. Bosmans I, p. 250, also discusses the Chinese mode of dividing both the circle and the day on a decimal basis, while emphasizing the primacy of the Jesuits in the introduction of sexagesimal geometry. See also A. Damry, "Le Père Verbiest et l'Astronomie Sino-Européenne", *Ciel et Terre*, 34, no. 7 (1913).

[9] Le Comte, *Memoirs and Observations* (footnote 2), p. 67. Because I feel that the English translator was unable to visualize the technical features of what he was translating at this point, I cite the French original: "Les cercles sont divisez sur leur surface extérieure & intérieure en 360 degrez; chacque degrez, en soixante minutes par des lignes transversales & les minutes de dix en dix secondes par le moyen des pinnules qu'on y applique" (*Nouveaux Mémoires*, I, p. 144).

Figure 2 shows detail, with one-inch comparison scale, of Verbiest's graduated arcs. The scale running horizontally across the plate is one of the rings of an armillary sphere, divided both on its edge and on its flat plane. The lower scale is engraved on its flat plane only, and judging from the long alidade attached to the sight, is probably from the six-foot sextant.[10]

The transversal lines on both scales read to one minute of arc direct, but the reading edges on the alidades would have carried the delicately engraved 'dots' or 'pins' mentioned by Le Comte, that were capable of subdividing each minute into ten seconds. This would have been too fine to represent on a woodcut illustration, although they were known in the West.

The design of Verbiest's naked-eye pinnule sights are clearly shown in this plate (Figure 2), and were virtually identical to Tycho's. The sights incorporated a broad metal peg at the geometrical centre of the instrument, which was viewed through slits cut into the metal plate of the 'eyepiece' sighting arm, shown in Figure 2. When the alignment had been made, the sight would be secured with the locking screw, and the angle read. Such sights are to be seen on most of Verbiest's instruments, especially the armillary spheres, and represent the most advanced naked-eye sights before the invention of the telescopic sight.[11]

But Verbiest's instruments were not slavish copies of the *Mechanica* pieces, for in spite of an overall influence regarding dimensions and design, they contained original, albeit European, features of their own, indicating that Verbiest was cognizant of many innovations that had appeared in Europe between Tycho's death in 1601, and his own sailing to China in 1656. Verbiest's metal sextant was simpler, and probably lighter in design than Tycho's, while the absence of wooden parts would have made it less susceptible to warping when left in the open. Indeed, this sextant has several features that make it similar to the great sextant of Hevelius, and it is possible that Verbiest saw the Hevelian instruments (which were modelled in turn upon those of Tycho) before he left Europe. But Verbiest was well in advance of Tycho in providing geared setting adjustments and pulleys for his sextant, which showed a distinct improvement on the much cruder ball- and claw-mounts of his predecessors, and looked forward to the geared equatorial sextant of John Flamsteed.[12]

Though Verbiest's six-foot sextant, shown in Figure 3, is still mounted in the azimuth, the geared settings would have given greater stability and better control than the ball mounts of Tycho and Hevelius. The pulley and tackle would also have allowed more delicate manipulation in the tracking of objects than the iron stabilizing spikes shown on Tycho's sextant, and described by him in the *Mechanica*, p. 74. A less refined form of semicircular clamp mounting, but lacking any gear work, is used by Tycho on his bipartite arc, and may have provided Verbiest with the initial idea, Figure 4B.

In addition to his development of the sextant mount, Verbiest introduced other innovations in instrument design. These included the introduction of semi-circular

---

[10] Verbiest's plate 11 [not illustrated], also includes the same armillary scale and eyepiece pinnule. All of Verbiest's plates in the *Liber Organicus* are numbered in Chinese numerals, and even where he refers in his Latin text to a plate with a Roman or Arabic number, the plate itself would carry a Chinese numeral. Note also, that when [not illustrated] follows a number, it indicates that the plate in Verbiest's sequence has been referred to, but not reproduced in the present paper.

[11] Tycho Brahe (footnote 1); see "Addendum on the subdivisions and diopters of the instruments", Raeder and the Strömgrens' translation, pp. 141–4.

[12] For further comparative discussion of the sextants of Tycho, Hevelius, and Flamsteed, see Chapman (footnote 1), Figures 7, 8 and 14 respectively, and their related text.

424

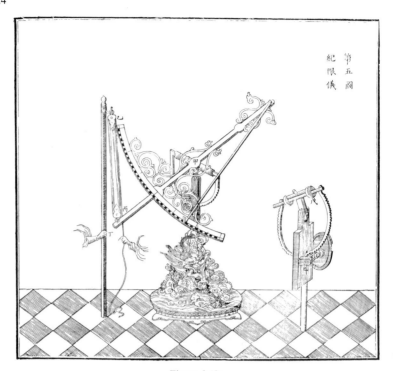

Figure 3 (5).

bracing rings on the colures of his armillary spheres, which were in advance of those of Tycho and gave a much less impeded view of the sky when observing near the equinoxes, along with his depiction of the horizontal azimuth [not illustrated]. This instrument has no equivalent in Tycho, and though Verbiest fails to give an account of its operations, its most probable use lay in taking horizontal or horizon 'fixes' of the Sun or Moon by shadow alignments on the six-foot circular scale. Tycho would have used one of his voluble, or rotating azimuth quadrants to obtain similar fixes, although Verbiest simplified the design. While Verbiest's Peking instrument collection included a voluble quadrant, it did not incorporate an azimuth circle, and would have been used to take alt-azimuth fixes out of the meridian. In his development of a specialist instrument to take the azimuths, Verbiest no doubt showed his awareness of the increasing European practice of avoiding multi-purpose instruments, which were likely to shift out of adjustment.

In Le Comte's view, the Peking instruments were the finest pieces of their kind to be found anywhere in the world. He applauded the fine designs and mode of execution, and only in one respect was it necessary for him to withhold his praise: he considered the quality of the scale graduations to be poor by European standards. Although he acknowledged that Verbiest had done his best to get the circles divided accurately, the

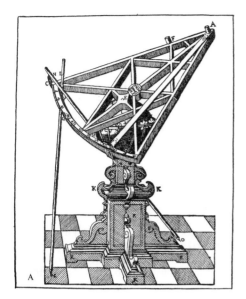

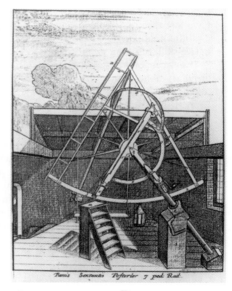

C   Figure 4.   D
The development of the astronomical sextant mount.

(A) Tycho's six-foot sextant, *c.* 1580, *Mechanica* [Figure 16], p. 72. Note the ball- and claw-mount and steadying spikes. (B) Tycho's lesser bipartite arc, *c.* 1585, *Mechanica*, Figure 15, p. 68. Note the ungeared semi-circular clamp to adjust and steady the arc in the lateral plane. (C) Johannes Hevelius's great brass sextant, *c.* 1655, *Machina Celestis*, I (1673), plate 'N'. Note the dis-assembled parts of the ball mount. (D) John Flamsteed's seven-foot sextant, 1676, *Historia Coelestis Britannica*, III (1725). Flamsteed developed the geared mount in the equatorial plane, so that by the turning of a handle, an assistant could make the instrument track objects across the sky. Note also Flamsteed's replacement of naked-eye sights with a telescope.

426

Chinese had failed to master the elaborate beam-compass geometry necessary to divide a circle in accordance with Western specifications.

> Though the Father was, no doubt, very careful of the division of his circles, the Chinese artificer was either very negligent, or very incapable of following his directions; so that I would rather trust a quadrant made by one of our good workmen at Paris, whose radius should have been one foot and a half, than that of six feet which is in this tower.[13]

Having outlined the iconographical position of the Verbiest plates and indicated their Western prototypes, it is now necessary to examine the constructional plates themselves. Before considering the details of these plates, however, it must be confirmed that the constructional processes illustrated within them are indeed European, and not Chinese.

When he visited the Peking Observatory, Le Comte saw a collection of discarded bronze instruments of great age that had formerly occupied the site before Verbiest took charge. The Chinese, moreover, were skilled metalworkers, capable of fabricating elaborate bronze castings, as the stands of the Verbiest instruments themselves testify. But this does not mean that they could produce precision devices, and one gets the impression from Verbiest and Le Comte that much of their native science and technology had been reduced to a matter of rote.[14] The Chinese craftsmen even seem to have lost their traditional ability to learn quickly, for according to Le Comte, the doubtful graduations on the armillary spheres would indicate not only a dearth of native skill, but even an incapacity to learn from Verbiest himself.

Fourteen of Verbiest's plates depict the processes of manufacture for a large armillary sphere and an astronomical globe. The plates do not follow a strict numbering sequence, however, and are scattered between 23 and 57 in the original Chinese numbers, though most of them fall between 35 and 45. Elsewhere, the seemingly logical construction process will often be interspersed with plates of drawing instruments, scales, tools, and other supplementary artefacts. Furthermore, it must be stated that plates representing the actual processes have themselves got out of sequence; i.e. the two plates 41 and 42, showing the preliminary handling of the rough-edged castings of the rings of an armillary sphere, are placed *after* the plates depicting their turning and polishing. But with a few exceptions, the overall sequence is quite explicit, and takes the reader from the initial casting to the final erection of the instrument under discussion. The plates shown here are placed in what would appear to be the logical order of construction, the figure number being followed by the original Chinese number in brackets.

Figure 5 shows the mould used to cast the bronze rings, which appear to be composed of about twelve trough-shaped segments, capable of being joined up to form a circle of the required diameter. Matching lids were also used to seal the mould, the whole probably being buried in sand or earth prior to the pouring in of the metal. Two rough rings of the same diameter as the mould are also represented. Only the upper section of this plate relates to the casting process, while the lower part seems to be

---

[13] Le Comte, *Memoirs and Observations* (footnote 2), pp. 66 and 69.

[14] Ibid., pp. 64, 70, 71, 72. The original Chinese instruments dated from the thirteenth century, and survived long enough to be photographed.

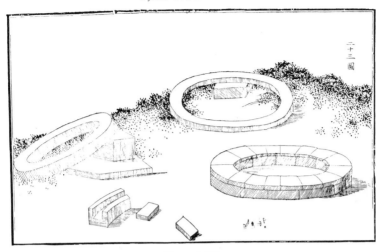

Figure 5 (23).

Figure 6 (36).

428

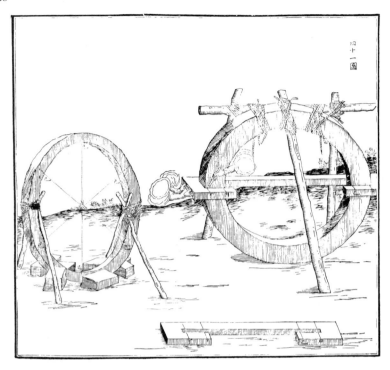

Figure 7 (41).

illustrating the volumetric proportions of the sphere and the cube. Several of the plates 'double up' on illustrations, and even represent different branches of science on the same piece of paper.

Figure 6 shows a donkey mill being used to rough-grind and polish the flat edges of the cast rings. The cutters are also displayed lying on the ground in the same disassembled manner as was characteristic in many European drawings at the time. On plate 37 [not illustrated] a rotary grindstone with a water reservoir is being used to sharpen the cutters of the donkey mill, while on plate 38 [not illustrated] the same donkey mill is being used to polish the freshly-ground rings with abrasive blocks.

Figure 7 shows the marking of the circumference lines upon the rings, in preparation for the turning and polishing of their edges. A gauge is shown lying on the ground, carrying the intended dimensions of the finished ring, while the right hand part of the picture shows the gauge applied to a ring, and a hand-held scriber transferring marks from it.

In the upper part of Figure 8 (39 right), wooden spokes are inserted into the ring to provide a centring, while in the lower part of 40 (left), beam-compasses are being used to strike off the intended circumference line. This line was probably intended to pass through the six preliminary points shown in 41 (Figure 7). Note the excellent detail of the beam-compass parts.

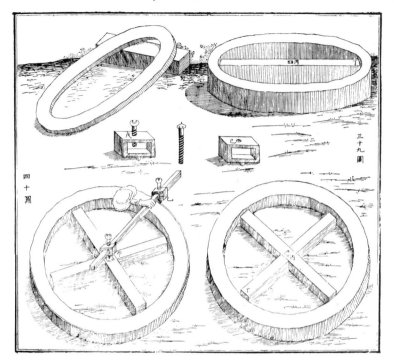

Figure 8 (39 right and 40 left).

Sawing around the scribed line, to 'true-up' the ring and give it a regular breadth, is shown in Figure 9. On the left hand part of the print, the freshly cut ring is being finished with an abrasive block and file. Several metal working tools, including files, hammers, chisels, and polishing blocks, lie scattered upon the ground.

In Figure 10 we see the turning-up of the edge of a newly sawn ring. For this operation, the wooden spokes are re-inserted and used to accommodate an axle, so that the ring can be lathe-turned and its edges polished. The placing of this plate, with a Chinese numbering mark 35, before the rough sawing only confirms the approximate character of the Chinese number sequence. Perhaps the plates were numbered by a scribe who did not understand the technical order of contents.

The testing of the ring for flatness of plane and balance is shown in Figure 11. In the right-hand part of the picture, the ring is being tested with three weighted strings with a fourth stretched across at right angles. When this fourth string is plucked, the others will only vibrate in sympathy if the ring is perfectly flat. The ring on the left-hand part of the plate is fitted with pivotal lugs, and is being tested for balance before being installed as a polar supporting ring, inside an ecliptic armillary sphere.

Figure 12 shows preparation for graduating the edge of a large ring, probably the ecliptic or meridian ring of a zodiacal (ecliptic) armillary. The artisan holds a right-angle marking gauge by which parallel lines may be scribed around the edge of the ring.

430

Figure 9 (42).

By this stage of construction, the ring would have been burnished to a high polish; hence the mat on the ground beneath to prevent scratching. Tycho graduated both the edge and plane surface of his ecliptic and meridian rings, and this plate indicates that Verbiest did the same. Verbiest's plate 15 (Figure 2) shows an arc graduated in this manner. Note once again the fine detail of the beam compasses.

Figure 13 shows the assembled armillary sphere, without its stand. Notice especially that it is an ecliptic armillary, possessing an inner ring, the pivots of which are offset at an angle of $23\frac{1}{2}°$ to the outer polar axis. This is an instrument of distinctly European form, and constitutes the only representation of an unfinished armillary with which the author is familiar.

Figure 14 is the complete ecliptic armillary sphere, which Verbiest claimed was used to measure planetary positions (whereas the equatorial armillary was used to observe the equinoxes).[15] Notice the characteristic Chinese decorative motifs, in contrast to the European functional parts. Considering the detailed attention lavished upon the 'scientific' parts of the instrument, one wonders why there is no mention of the elaborate foundry processes that must have been necessary to produce the stands. The

[15] Bosmans I, p. 273. In plate 71 [not illustrated], incomplete armillaries are depicted in the act of being hoisted up the observatory ramparts by means of pulleys.

Figure 10 (35).

二十五圖

Figure 11 (43).

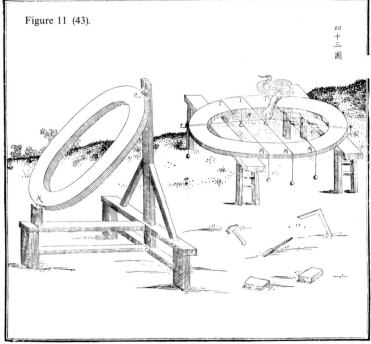

四十三圖

IV

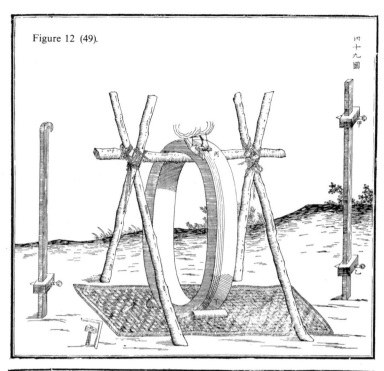

Figure 12 (49).

四十九圖

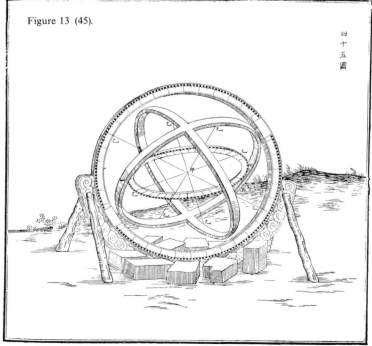

Figure 13 (45).

四十五圖

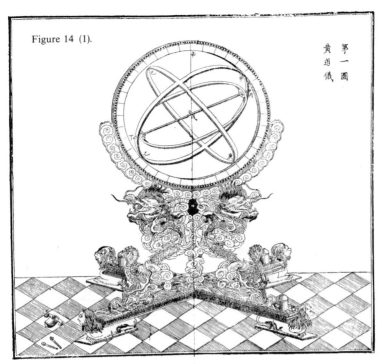

Figure 14 (1).

第
一
圖

黄
道
儀

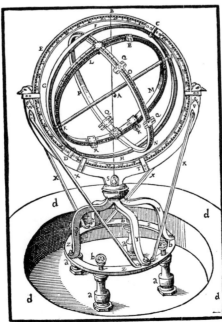

Figure 15.

IV

434

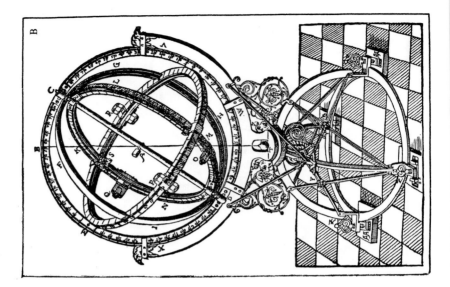

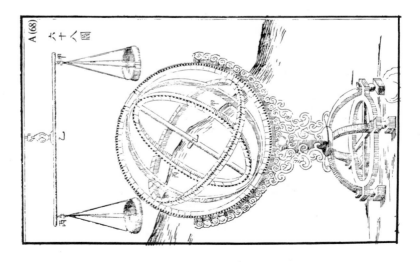

Figure 16.

reason for this omission is probably the same as that which caused the scientific constructional processes to be omitted from Western treatises: they were a familiar part of everyday technology to the Chinese. On the other hand, the construction of the alien Western parts demanded explicit discussion.

Figure 15 shows Tycho Brahe's ecliptic armillary sphere from the *Mechanica* [Figure 11]. While Verbiest's armillary occupies a much more elaborate stand than Tycho's prototype, the technical relationship is clearly visible. Both instruments follow the European iconographical convention of illustrating degree graduations by alternate black and white hatchings, while both are equipped with screw-set adjustments on their stands, to enable them to be brought into alignment. Unlike Verbiest, Tycho does not depict the ecliptic armillary standing on a black and white pavement, although he *does* depict the equatorial armillary doing so, in *Mechanica*, [Figure 13]. Instead, Tycho shows his equatorial armillary placed inside a circular stone stairway, the different levels of which permitted the observer to view high and low angles on the instruments. Both of Verbiest's armillaries have, in fact, been placed within similar stairways when in use, for they can be seen *in situ* in the Peking Observatory prospect (Figure 1).

Figure 16 (A) depicts an inset from part of Verbiest's plate 68, representing an equatorial armillary. The instrument in this drawing clearly possesses a different stand from the six-foot armillary which Verbiest built for the Peking Observatory, shown in his plate 94 [not illustrated], and which still survives. Its stand was formed from bronze dragons, and was similar to that of the ecliptic armillary shown in Figure 14.

Figure 16 (B), illustrates Tycho's equatorial armillary, from *Mechanica* [Figure 13]. It seems likely that the equatorial armillary included by Verbiest as part of plate 68 was never actually built, but simply copied by the Chinese artist—with a few decorative modifications—directly from the *Mechanica*. The angle from which the rings are drawn, the circular arched stand and levelling screws all relate back to the European original.

Figure 17 depicts the turning of the foundation ball for a six-foot diameter globe, to render it spherical. Motion is imparted by two labourers, one turning a crank handle, and a second who 'runs' the globe around while suspended in the roof of the work-shed. Two other workmen apply cutting and abrasive tools to the spinning globe. This procedure was probably influenced by Tycho's account of the construction of his own six-foot globe, as related in the *Mechanica*.

Graduating the globe with arched beam-compasses is depicted in Figure 18. The globe is now set within its meridian ring, and pivoted at the polar points, which pass through the two ungraduated polar points of the ring. That the object in question *is* a globe is substantiated by the narrow gap between the globe itself and the ring, through which details of the background landscape are visible and the concentric ring shading which is highlighted at one side. The collection of logs and blocks around the ring, moreover, implies their use to stabilize a large three-dimensional object. The meridian ring within which the globe is suspended would have been graduated in advance, so that its divisions could be copied on to the globe. The workman's hand appears to be transferring graduations from the meridian ring by means of a set square and scriber to trace parallel lines of latitude (or stellar declinations) around the globe when it was rotated. The arched beam compasses would probably have been used to locate the places of the longitude (right ascension) lines, set at right angles to them. The compasses are opened to an amplitude of 60° to the circle. In beam-compass division, the 60° angle was always struck off first, as it corresponded exactly with the radial length of the

IV

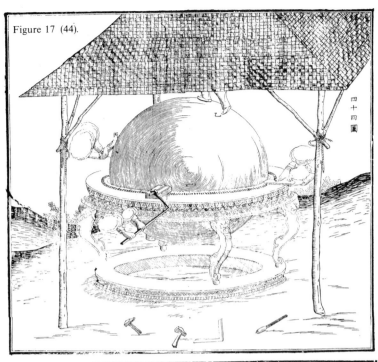

Figure 17 (44).

四十四圖

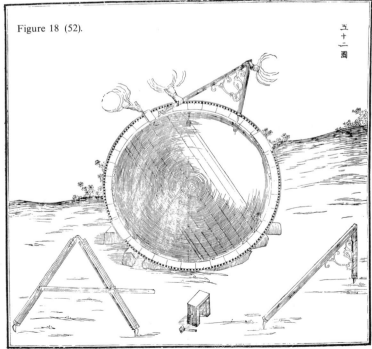

Figure 18 (52).

五十二圖

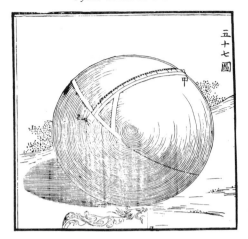

五十七圖

Figure 19 (57).

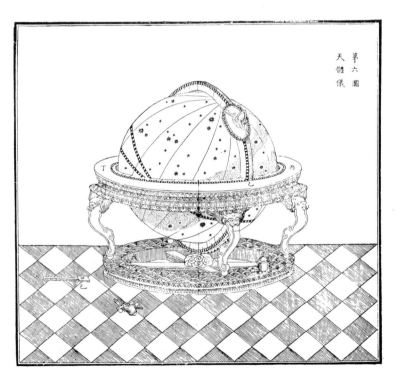

第六圖
天體儀

Figure 20 (6).

438

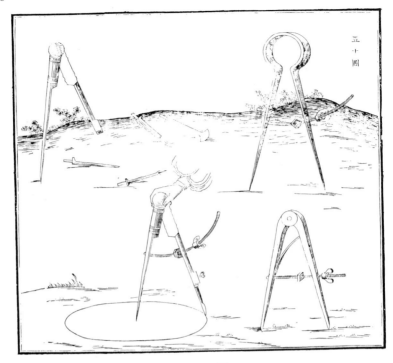

Figure 21 (50).

circle.[16] This is the earliest representation of the division of an astronomical circle known to the author.

Figure 19 shows a globe being subdivided in unit degrees by means of a 90° template. This spherical set-square is being applied to fill in the divisions of the large arcs drawn with the beam-compasses, and would also have served as a cross-check against the divisions drawn on different parts of the globe. This illustration is one of three separate plates numbered 56, 57 and 58, all printed on to the same folio sheet.

Verbiest's globe complete and assembled is seen in Figure 20. Verbiest claimed that the globe was an extremely useful instrument, and cost over '50,000 pièces d'argent chinois, ou *taes* prises toutes au trésor royal'.[17] Verbiest began his *Liber Organicus* with pictures of the instruments in their complete form, before going on to illustrate their construction; hence the low plate number.

Figure 21 shows compasses and dividers with screw adjustments and interchangeable parts. In addition to the beam-compasses, Verbiest included illustrations of a wide variety of draughtsmen's instruments amongst his plates, and numbers 44, 45 and 47 to

[16] Allan Chapman, *Dividing the Circle, the history of precise angular measurement in astronomy, 1500–1850*, forthcoming.

[17] Verbiest to Le Faure, 20/8/1670, Bosmans I, p. 270. Plates 82 and 75 [not illustrated] demonstrate the pulleys and hoists necessary to transport the incomplete globe up the observatory ramparts, and set it in place.

52 show different types of spring bows, map, proportional, and other compasses. These instruments are clearly of Western design, and possess close parallels in many Western museum collections. It might further be argued that the presence of these drawing instruments, especially in conjunction with scale and globe graduating scenes, substantiates the European provenance of the overall technology represented in Verbiest's plates.

4. Verbiest and the European astronomical instrument revolution

In the complete set of *Liber Organicus* plates, Verbiest depicted a wide variety of instruments and techniques drawn from the physical sciences as they were understood up to the 1650s, demonstrating such subjects as ballistics, mechanics, meteorology, navigation and optics, as well as astronomy. The nature of optical refraction is demonstrated, and there are several pictures of telescopes amongst the plates. Nowhere, however, does he discuss the telescope as a *measuring*, as opposed to a *viewing*, instrument, nor is there any mention of astronomical instruments incorporating the telescopic sight or micrometer. Indeed, Verbiest fails to progress beyond the instrument technology of Tycho in matters of astronomy, and his great bronze instruments would have been obsolete by European standards even before they were built. One is forced to ask, therefore, why Verbiest chose to equip the new Peking observatory with devices based upon archaic prototypes.

This question can be answered on two separate grounds. The first derives from the date when Verbiest left Europe for China; the second, from his role in Peking as both missionary and astronomer.

Verbiest sailed for China in 1656, on the eve of what would prove to be a series of major breakthrough in European instrument technology. At the time of his departure, astronomers were still faced with much the same problems that had beset Tycho. The measurement of the heavens was still a matter of erecting large arcs in the main celestial planes, and observing the stars against them through naked-eye sights. Tycho's 'peg and slit' sights were still the best aids to accurate alignments, whilst his method of subdividing scales by transversal dots and lines was the only effective way of delineating tiny fractions of linear or circular measure. The best available clocks were either refined versions of the same horological escapements that had been known since the fourteenth century, or the cross-beat mechanisms of Jost Bürgi from the time of Tycho himself. In consequence, European *positional* astronomy still operated in an essentially Tychonic context over fifty years after the Dane had died. When Verbiest left Europe, the latest instruments of Riccioli and Hevelius were still Tychonic in their essentials.

Yet in little more than a decade after Verbiest's departure, the whole scale of European astronomy had changed. In 1658, Huygens developed the first efficient pendulum escapement, which revolutionized the clock as a precision scientific instrument. This process of development was further accelerated just over a decade later, when Robert Hooke and William Clement explored the physics of the long pendulum, and invented the anchor escapement, thereby providing a simpler and more accurate method of taking right ascensions than the great sextants and armillaries of Tycho. These horological developments were contemporaneous with the invention of the micrometer in the 1660s, which allowed celestial angles to be read to a far higher level of exactitude than the Tychonic transversal scales. At the same time, the appearance of the telescopic sight transformed the working accuracy of the sextant and

IV

quadrant, as the removal of the physiological limitations of the naked eye extended research into a whole new field of accurate measurement.[18]

The result of these inventions was to open up many new areas of enquiry, and to transform the whole scope of astronomy. In particular, the new inventions drastically changed the design of astronomical instruments, along with the observatories in which they were housed. When one compares Verbiest's observatory in Peking with the contemporary observatories at Greenwich and Paris, all of which were equipped within less than a decade of each other, one appreciates the profound impact which these inventions had upon the hardware of seventeenth-century science. The new inventions made astronomical instruments smaller, more delicate, and vastly more accurate than the great circular instruments of Tycho. One might argue that the influence of the telescopic sight, clock, and micrometer on seventeenth-century science was similar to that of the micro-chip on electronics in the late twentieth century.

What is especially interesting, when reviewing the contents of all 117 of Verbiest's plates, is what he omits to discuss. The pictures reveal him to have been a well-informed general scientist up to the time of his departure from Europe, but he displays little knowledge of what went on afterwards. In particular, he shows no knowledge of what went on in the great scientific watershed of the 1660s. Not only do his astronomical writings fail to make reference to the telescopic sight and micrometer, but his treatment of horology, in plates 64 and 65 [not illustrated] shows no awareness of the pendulum clock. It is true that these plates depict horological devices on the one hand and free vibrating pendulums on the other, but he never discusses the pendulum as a device with which to govern a clock escapement.[19]

Likewise, his plates showing meteorological instruments make no mention of the emerging science of pneumatics, and, one wonders, had he left Europe six or seven years later, would Verbiest's illustrations have included air-pumps and experiments with the vacuum? The microscope is also conspicuous by its absence amongst the collections of optical demonstrations. One might surmise, therefore, that Verbiest's active knowledge of European science was cut off after 1656, for while he may have acquired more up-to-date scientific information from visitors and correspondents thereafter, he incorporated no such material into the *Astronomia Europaea* or *Liber Organicus*.

5. The scientific aspirations of the Jesuit observatory

The problem of Verbiest's failure to equip his observatory with more modern instruments is not only explained in terms of lack of European information, however, for it also related directly to his role in China, and what he was trying to achieve there.

Scientist as he might have been, one must not lose sight of Verbiest's primary task, which was to secure Christian converts in the Imperial Court, and not simply to act as a

[18] For an account of these inventions, see Allan Chapman, "The accuracy of angular measuring instruments used in astronomy between 1500 and 1850", *Journal for the History of Astronomy*, 14 (1983), 133–7.

[19] Free vibrating pendulums are demonstrated on plates 115, 116 and 117 [not illustrated] as devices to time the flight of projectiles and the crash of a lightning bolt, and to make an astronomical observation. Verbiest's failure to discuss the pendulum *clock*, however, is all the more curious when one considers the important 'ambassadorial' function played by the mechanical clock in the seventeenth century. It was the opinion of Matteo Ricci and other early missionaries that they had introduced the mechanical—as opposed to the sand or water clock—into China, and found a strong native interest in horology. See Joseph Needham, *Clerks and Craftsmen in China and the West* (Cambridge, 1970), p. 205. The Chinese fascination with the "self-ringing bells" is also discussed by Carlo M. Cipolla, *Clocks and Culture, 1300–1700* (London, 1967), pp. 76–103.

scientific civil servant. His astronomy was only a way of making himself useful to the Emperor's well being, and was pursued in relation to that end. After all, Verbiest was not beset with the same intellectual and technical problems that faced Flamsteed, Cassini, and Picard back in Europe. The Earth's motion in space, the longitude, the laws underlying planetary motion, and the stellar parallax were not part of his brief. All that the Emperor required was an accurate calendar, careful eclipse predictions, and stimulating after-dinner conversations on natural philosophy from his observatory director, and these could be provided in an almost routine fashion by an educated European with a large collection of Tychonic instruments.[20]

These conditions should not, however, be seen as minimizing Verbiest's achievement, but rather as placing it in a cultural context. K'ang Hsi was an intelligent absolute ruler who, as luck would have it for the Jesuits, was genuinely interested in European science. Yet K'ang Hsi was not part of the same overall culture that produced the Royal Society and the Académie. His world view was different from that of Charles II or Louis XIV, and his approach to European science not dissimilar to the interest displayed by European rulers in chinoiserie. It was a foreign curiosity to be used or enjoyed as circumstances required, not a source of fundamental social and intellectual revision. One must also bear in mind that while the Emperor may have found Verbiest and his science both useful and interesting on a certain level, it was still part of an alien and, to a Chinaman, inferior culture. In short, Verbiest's work in Peking was regulatory in character, and neither pre-supposed nor expected research in a contemporary Western sense. It was ambassadorial rather than exploratory, and in consequence, he had no need constantly to bring his methods up to date.

It is also probable that the Tychonic instruments were much easier to construct, given the resources placed at Verbiest's disposal, than more up-to-date European pieces. With a large labour force at his command, the construction of great armillae and sextants demanded no research and development comparable to the way in which Hooke, Tompion and Flamsteed conceived of instrumentation. With a good technical training, and a volume of the *Mechanica* to fall back upon, the fabrication of the Peking instruments required little more than a great deal of sawing, filing, and grinding for his labourers, work that would automatically come to fruition after the requisite energy had been expended. In a letter to Jacques Le Faure of 20 August, 1670, Verbiest claimed that the instruments were already half finished, and would be ready by the end of the year. In fact, they were not completed until 1673, having taken some four years to build.[21]

Cut off from European craft skills as he was, it would have been foolhardy for Verbiest to attempt the construction of the new precision instruments, even had he known precisely how they worked. Peking might well possess clever metal founders, but it did *not* possess optical lens makers, horological gear-wheel cutters, and men who could make accurate micrometer screws. There is no evidence that Verbiest attempted to construct instruments with telescopic sights, obtain the elements of the lunar orbit from micrometric measurements, or build zenith sectors with which to pursue the

---

[20] There are many references to Verbiest and his colleagues in the role of cultural and technical ambassadors to China in Spence (footnote 3). Spence also looks at the scientific Jesuits and the way in which they captured the imagination of the Emperor and his court in *The China Helpers; western advisors to China, 1620–1960* (London, 1969), pp. 3–33. For further references to the Jesuits entertaining the court with technical devices, see Bosmans I, p. 262. Also, Paqquale M. D'Elia, *Galileo in China* (New Haven, Conn., 1960).

[21] Verbiest to Le Faure, 20/8/1670, Bosmans I, p. 269.

stellar parallax. Without an indigenous and rapidly growing body of craftsmen who *understood* the problems of the scientists, such as were found in London, it would have been futile to expect Western-style scientific instruments to emerge from the Imperial workshops, no matter how well Verbiest chose to supervise the workmen. Casting a bronze ring or dragon was, after all, part of their culture, but grinding a twenty-foot-focus objective lens that could compare with a Yarwell or Campani was not. In this respect, Verbiest was wise to remain firmly within a conservative tradition, and keep K'ang Hsi happy without dwelling on the unobtainable.

The Jesuit observatory never produced any discoveries that were of significance to Western science. What its continuing existence did illustrate, however, was the gulf that had come to divide Eastern from Western science-based technology by the end of the seventeenth century. It is true that the observatory was also important in maintaining a Jesuit foothold in the Imperial Court, which continued long after Verbiest's death in 1688, although this particular advantage was political and religious in character, rather than scientific.

While the observatory continued as a Jesuit foothold into the mid-eighteenth century, and provided a scientific base for von Hallerstein, Kögler and others, the work it produced was of a sound but essentially antiquated character. It produced catalogues of star positions that were of very high quality by naked-eye Tychonic standards.[22] Yet such painstaking catalogues were in no way comparable in accuracy to those of Bradley or Tobias Mayer, made in Europe with telescopic quadrants built by John Bird, and capable of yielding values accurate to 2 seconds of arc. In spite of the care lavished upon it, the Peking Observatory was, and remained, a magnificent scientific dinosaur, and related to Western science in the same way as the slide-rule now relates to the computer.

PEKIN WALL WITH VERBIEST'S 4 FOOT QUADRANT.

Figure 22.

[22] The fruits of the later Jesuit observatory, under Kögler, Da Rocha, von Hallerstein and others were included in *I Hsiang Khao Chhêng* (1757).

IV

6. The present state of the Peking Observatory

It is fortunate, however, that the great bronze instruments in Verbiest's observatory are still preserved in Peking. In the early twentieth century they were taken to Potsdam, as trophies for the Kaiser, although they were returned to their original locations after World War I.

A photograph of the southern wall of the Peking Observatory, probably taken in the late nineteenth century, is shown in Figure 22. Though Verbiest's armillary spheres still occupy their original positions, as in the engraving reproduced in Figure 1, the great globe has been relocated, and replaced in the centre by a later quadrant, constructed by a Jesuit successor, Stumpf in 1714.

Part of the south wall of Verbiest's observatory was damaged by storms in the late 1970s, and although I was sent a photograph of the equatorial armillary perched precariously above the partially demolished wall, I understand that everything is now on the way to full restoration.[23] It is ironic, considering how quickly Tycho Brahe's Uraniborg and its contents were destroyed in early seventeenth-century Denmark, that it is through a set of instruments and accompanying illustrations, built by the Jesuits for the Emperor of China, that one finds the closest physical connection with the astronomical technology of Tycho Brahe.

Acknowledgments

My first acknowledgment must, without doubt, go to my friend Roger Mason of Oxford, who had the good fortune to purchase in an auction a set of Verbiest's plates. These were later purchased by the Museum of the History of Science, Oxford. It was from an initial examination of these plates, in the light of Tycho Brahe's *Mechanica*, on which I was then working, that my interest in the subject was first aroused. I am also indebted to Mr John Combridge, for advice on some of the Chinese sources, and for the breadth of his own experience in matters of Chinese horology. I also thank the Librarian of the Museum of the History of Science, Mr A. V. Simcock, for his unfailing assistance in my study of the Museum's Verbiest holdings. I also thank the Curator, Mr F. R. Maddison, of the same Museum, for granting me access to the collections, and for providing the illustrations for this paper.

[23] I am indebted to Mr Andrew George, formerly of Clifton College, Bristol, for kindly sending me these photographs, along with a description of the observatory in a letter dated 17 April 1980. Figure 22 is from a small collection of nineteenth- and early twentieth-century photographs of the Verbiest instruments in the Museum of the History of Science, Oxford. Verbiest's use of Western technology to equip the Peking observatory was also the subject of part of a BBC television Chronicle documentary, made by the author, in 1980. This programme, 'China—travellers in the Celestial Empire', became a topic of correspondence in *The Listener* magazine during November and December, 1980.

V

Jeremiah Horrocks, the transit of Venus, and the 'New Astronomy' in early seventeenth-century England*

The commemoration of a notable scientific event can be a potentially perilous exercise in so far as it assumes that an observation or sequence of ideas from a previous age possesses a legitimate ancestral relationship with what we do, and how we think today. In the confused scientific climate of the early seventeenth century, when Aristotelians, Copernicans, sceptics, experimental and mechanical philosophers vied with each other to devise plausible explanations of the natural world, traditional literary authorities and physical observations alike could be made to yield temporary advantages to the respective factions. Yet because an observation or chain of conclusions happened to prove 'right' when viewed with hindsight, we must be careful not to extract its initiator out of his contemporary context, and see him as a visionary or hero.

This was, however, a status long since conferred upon Jeremiah Horrocks, for even the first generation of Fellows of the Royal Society, when trying to construct an ancestry for the 'New Philosophy' in England, eagerly alighted upon him. The middle-aged John Wallis, following his migration to Oxford as Savilean Professor of Geometry, collected and edited the remaining papers of his old Cambridge undergraduate contemporary in 1673 (1). John Flamsteed saw Horrocks and his circle as the founding fathers of observational astronomy in England and an inspiration behind the creation of the Royal Observatory in 1675 (2). Even Newton was to pay him homage, as the man whose contributions to planetary and lunar theory formed the crucial connection between Kepler and his own work (3). Heroic references to 'our illustrious Horrocks' continued through the eighteenth century while in the nineteenth, John Herschel was to style him 'the pride and boast of British astronomy' (4). The earnest Victorians endowed him with all the virtues of a true Smilesian hero; a poor boy made good and a clergyman of the Church of England to boot! Horrocks's conveniently early death at the age of twenty-two further enabled Victorian scientific hagiographers to turn him into a character in the 'romance of science', whose genius, Godliness and hard work had brought him to an early grave.

These accumulated images still colour our perspectives of the man and his achievements down to the twentieth century and make the authentic Jeremiah Horrocks still more elusive. But one thing which is genuinely difficult for the historian to assess, hagiographic conventions notwithstanding, is the plain fact that his documented contributions to astronomy

---

* The text of a lecture delivered at the RAS Ordinary Meeting, 1989 November 10.

were formidable by any standard. Though it is not admissible for us to bestow an imaginary moral character upon him from the morsels of personal evidence that survive, he was one of the first men in England to grasp the significance of what was going on in contemporary European astronomy (5). Not only did he repeat many of the techniques of Kepler and Galileo, but he went on to develop the New Astronomy to produce conclusions which substantially advanced those of its continental founders. Herein, I suppose, lies the fascination which Horrocks has exerted upon the history of astronomy from the mid-seventeenth century onwards, and what I wish to do in this paper, is to assess his scientific contributions within the astronomical priorities of his time, and place him within an historical context which is as authentic as the sources will allow. Though Horrocks's scientific career was brief, covering the years 1635–1640, it was far-reaching in so far as it touched upon many aspects of theoretical and practical astronomy. But the event upon which his principal fame came to rest was the lucky prediction of the first observed transit of Venus, on 1639 November 24, and of which we here mark the 350th anniversary.

No baptismal record survives for Horrocks, and while one line of nineteenth-century writers argues that he was born in the Deane district of Bolton, Lancashire, his old Cambridge associate John Worthington, writing in 1659, stated that he came from Toxteth, in the same county, and now a suburb of Liverpool (6). Various dates have been put forward for his birth, but as John Wallis (who had known him) said that Horrocks died in his twenty-second year, it would appear that he was born in 1619 (7). His father was either a farmer named William or else a watchmaker named James, though sources agree that his mother was Mary Aspinwall of Toxteth (8); Horrockses and Aspinwalls appear to have been well-established families in the Toxteth district, the Aspinwalls being of some local note, while watchmaking and similar mathematically based handicrafts were becoming important local trades (9).

Just as a variety of speculations have been advanced about Horrocks's exact parentage and place of birth, so his early education has been the subject of conjecture. The eminent Puritan divine Richard Mather *might* have been one of his early teachers for he was certainly active as a schoolmaster around Liverpool during Horrocks's early years, though there is no documented connection (10). Similarly, we do not know how he got to Cambridge, though he was entered on the books of Emmanuel College as a Sizar on 1632 May 18 (11). In an age when the B.A. was often given at seventeen and the M.A. at twenty-one, Horrocks was not exhibiting too much precocity by commencing his undergraduate career at thirteen (12). It was at Cambridge, however, that he began to manifest a serious interest in astronomy, for in a surviving volume of Philip Lansberg's *Tabulae Motuum* which Horrocks autographed in 1635, he listed twenty-four European astronomical authors whose works he had read to date (13). It is significant to note that he mentioned no English authors, with the exception of the medieval Anglo-Irish writer, Sacrobosco.

Horrocks was fortunate in his entry into Emmanuel in 1632, for he was a contemporary of several men who would grow famous in their own rights, and recall his memory in later years. The mathematician John Wallis entered

Emmanuel in 1632 and, as mentioned above, would carry his memory into the Royal Society. The Cambridge Platonist Ralph Cudworth was also a contemporary, while John Worthington of Manchester came up in 1632, and as Master of Jesus College, would later correspond with Samuel Hartlib about his undergraduate contemporary (14). It was most likely Worthington who put Horrocks in touch with the Salford cloth merchant William Crabtree, who was to be his closest scientific confidant between 1636 and 1640, and was to be his co-observer of the transit of Venus. Worthington was also to be amongst the first to take steps to rescue the already voluminous correspondence which had passed between Horrocks and Crabtree, following the latter's death in 1644.

The work of Horrocks, and his mathematical friends William Crabtree and William Gascoigne, was seen as sufficiently significant within a decade of his death in 1641, to excite the Lancashire antiquary Christopher Towneley to try to rescue his papers, and take them to his residence at Carre Hall near Burnley (15). Here, they were first consulted by Jeremy Shakerley who used them in part to compute his *Tabulae Britannicae* (1653) and who acknowledged a massive intellectual debt to Horrocks, gained from the study of his papers at Carre Hall.

If one wants a starting point for the Horrocks hero worship, one might find it in Shakerley, for while he never knew the astronomer personally, he was extracting mathematical data from his papers by 1649 (16). Furthermore, he journeyed to India where, in 1651, he observed a transit of Mercury which, in all its details of technique and intention, was a clear emulation of the work which was to secure Horrocks's distinction (17). Shakerley's three published works and several surviving letters sing the praises of the 'Noble Genius of our Worthy Country-man Master Jeremy Horrox [sic]' prior to 1653 (18). Horrocks, in the mind of Shakerley, was already the English Galileo, whose transit observation had challenged the old order of astronomy no less than the Tuscan astronomer's Medicean Stars, and as a self-conscious fame seeker in his own right, Shakerley set out to emulate his hero and make equivalent discoveries. It is unfortunate that he died obscurely 'in the East Indies' sometime after 1653.

The transit of Venus took place within the last 14 months of Horrocks's life, and one might correctly assume that its writing up and interpretation occupied most of his remaining time. His *Venus in sole visa* was not only Horrocks's one complete piece of astronomical writing, but his final statement on the New Astronomy (19). It contained a mixture of observation, interpretation of data, and instrumental discussion, while the whole work of 116 printed pages is a sustained polemic which leaves the reader in no doubt about whom Horrocks's *own* heroes and villains were. In the tradition of Galileo's *Starry Messenger, Letter on Sunspots* and Tycho's *De nova stella*, Horrocks's *Venus* is as much a work on the philosophy of science as it is an observation of a particular event, as he used every new fact obtained from the transit to extract the maximum interpretative and polemical mileage. It shows Horrocks as the bold defender of the New Astronomy, and a sworn adversary of those conservative astronomers who preferred to correct existing tables rather than look at the sky direct. Intellectually far reaching as Horrocks's work up to 1639 November had been, the transit provided him with the platform he needed to focus his energies and assault the old boundaries with all the brash vehemence of a very clever undergraduate. If Shakerley and other astronomers were so willing to bestow heroic laurels on Horrocks, it could well have derived from the way in which he appeared to clutch them for himself. Like Galileo, Horrocks was ambitious and hungry for fame in the new science, and while both men, when viewed in hindsight, made an astonishingly large number of 'right' interpretations from the limited amount of observational evidence available, much of their success stemmed from their tendency to simplify the astronomical debate into stark contrasts of wise men and fools.

Perhaps one reason why Horrocks made so much of the transit of Venus was because it was the shining pearl which almost slipped through his fingers. A month before the event took place, he did not even realize that it was going to happen, and having predicted, observed and interpreted the transit, felt fully justified in castigating those astronomers who had missed it altogether. Not until 1639 October 26 did it dawn on Horrocks that at the forthcoming nodal crossing of the ecliptic, Venus *would probably* pass across the Sun's disk, and hastily dispatched letters to his brother Jonas in Toxteth and Crabtree in Salford. He also requested Crabtree to pass on the news to Samuel Foster of Gresham College, with whom we might presume he was already in correspondence, though nothing survives (20).

Though a transit of Venus had never previously been observed, one had been correctly predicted by Kepler in 1631, who had come part way towards explaining the elaborate orbital cycle of approximately 120 years which lay at the heart of the transits (21). While no transit was actually observed in 1631, this derived from the fact that it occurred after sunset for most European stations rather than from any fault of the theory (22). What Kepler missed, however, was the realization that transits fell in *pairs* 8 years apart, and that instead of having to wait until 1761, a second transit was due to take place at the inferior conjunction in autumn 1639. It was Horrocks who grasped this fact, though it came to him more as a lucky insight rather than at the culmination of a protracted investigation. Horrocks, who was already disappointed by the inaccuracy of the standard planetary tables of the day – especially those of Lansberg and Hortensius – was attempting to ascertain the exact time and place of Venus's nodal crossing for autumn 1639, and found that all the main tables gave conflicting values. Some of them could be made to suggest, in reduction, that Venus might pass across the solar disk *around* November 24 (or December 6 new style) though there was disagreement between them as to day and place. By a combination of averages from the tables of Lansberg, Longomontanus, Copernicus and Kepler, along with his own values based on five years of personal planetary observation, he decided that a transit was worth looking for (23). While he probably realized that, if such a transit did occur, it would be a mathematical feather in his cap in so far as no one else had announced the possibility, he had no reason beforehand to see anything of particular philosophical importance in the event. Only after it had happened, and various new phenomena had presented themselves, is it likely that he realized that the transit possessed a philosophical potential; as a bludgeon with which to beat more conservative astronomers, and a trumpet with which to draw attention

V

to his own initiative and methods.

The day on which he expected the transit, November 24, was a Sunday, and while claiming not to expect the transit to begin until after 3 p.m., he decided to keep intermittent watch from dawn. To be on the safe side, he had also observed through 'the greater part of the 23rd', and 'omitted no opportunity to observe her ingress'. The transit day seems to have been clear at Hoole, near Preston, from which he observed, for he was able to keep watch for the greater part of the morning with the exception of an interruption between 9 a.m. and 9.50 a.m. He had an unbroken stretch of observation from 10 a.m. to noon, and a few more interruptions between noon and 1 p.m. Horrocks was engaged between 1 p.m. and 3.15 p.m. and observed thereafter until sunset, during which time he actually succeeded in seeing the transit. This total 3-hour absence was occasioned by him being 'called away in the intervals by business of the highest importance which, for these ornamental pursuits, I could not with propriety neglect' (24). It has been popularly assumed that, following an unfounded tradition that Horrocks was a clergyman, these absences were necessitated by his clerical duties, but I will return to the point below.

When Horrocks did return to his telescope at 3.15 p.m., he found that the transit had just begun, with a small, circular and intensely black object ingressing the solar disk, 62·5° on the top, right-hand side of the image.

The method by which Horrocks observed the transit was the already well-established telescopic projection technique. It had been used by Gassendi to observe the Mercury transit of 1631, while Christopher Scheiner's extensive sunspot researches, published in 1630, had been conducted by projecting the solar image onto a graduated sheet of paper (25). Horrocks made his telescopic image fill a six-inch diameter circle, the circumference of which was divided into degrees 'more exactly' than if they were on the limb of a 50-feet quadrant, he boasted. The horizontal diameter was divided into thirty equal parts. Though he would have preferred to work with a larger projection disk, he tells us that the smallness of the observing chamber did not provide the space.

Unfortunately, Horrocks provides no information about the telescope with which he made the observation, nor about the actual projection distance employed. His telescope, however, was claimed to be 'of such power as to shew even the smallest spots upon the sun, and to enable me to make the most accurate division of his disk'. It may have been the 'half-crown telescope' mentioned elsewhere in Horrocks's writings (26). In the 1930s, S.B.Gaythorpe attempted to reconstruct the optical characteristics of Horrocks's transit observation, suggesting that the telescope employed magnified no more than two or three times. This low magnification, indeed, would have been in keeping with the known characteristics of other early-seventeenth-century telescopes (27).

Having explained how he saw the ingressing object on the solar disk, Horrocks went on to establish, in Chapter IV of *Venus in sole visa* that it was Venus. As a way of disarming potential critics, it was essential for him to do this, though in addition it provided opportunity for him to display his erudition about dark spots which had been reported upon the solar disk over the centuries. He mentions the pre-telescopic 'maculae' of AD 807, 1160 and

1607, some of which were thought to be Mercury in transit, though he was quick to discount this interpretation of the spots. As Gassendi had discovered when he observed the first telescopically authenticated Mercury transit in 1631, the size of the planet upon the solar disk was much smaller than expected, and far too small to have been seen with the naked eye (**28**).

Having satisfactorily disposed of one set of objections, Horrocks addressed himself to others. One of these dealt with the ancient belief that the planets (being made, as they were, of the fifth element, the 'Quintessence') did not possess massive reflective bodies, but shone by their own intrinsic light. Horrocks asserted that they must indeed be physical, for Venus, like Mercury, had been shown to be an opaque, black sphere when seen in transit, and must shine, therefore, by reflecting sunlight rather than by intrinsic luminescence.

While it is true that some of these ideas were not quite so radical in 1639 as Horrocks implied – Galileo, after all, had shown that Venus must shine by reflected light by 1610 – he was out to leave no polemical stone unturned in this, his great opportunity to make an original defence of the New Astronomy (**29**). What Horrocks did conclude from the transit, however, is that its perfect roundness, blackness and independent motion demonstrated that the object could not be a sunspot, and must be Venus. This important point was independently noted by William Crabtree in his own observation of the transit, reported by Horrocks in his *Venus* treatise, and by Crabtree himself in a letter to William Gascoigne. Furthermore, by measuring the linear size of the black dot, as a fraction of the projected solar diameter, he concluded that the planet subtended an angular diameter of no more than 1′ 12″ arc [or 1′ 16″ after correction], which was some ten times smaller than the figures commonly accepted for the Venusian diameter in inferior conjunction (**30**).

During the half-hour before sunset when Horrocks was able to observe the positions of the transiting Venus, at 3.15, 3.35, and 3.45 p.m. he was able to obtain important data about the planet's orbital elements *vis-à-vis* the much better known orbital elements of the Sun. They enabled him to calculate the times of ingress and egress (neither of which were observable), the exact position of the node, and the solar parallax. The latter, he fixed at 14 arcsec (**31**). The one thing which perhaps most impressed Horrocks in the entire transit was the unexpected smallness of Venus, for while he was prepared (knowing what he did about Gassendi's Mercury observation) to find a Venusian diameter less than the traditionally ascribed 3 arcmin, he was clearly astonished at what he found. Crabtree likewise was surprised by this revelation and estimated the Venusian diameter to be about 7/200ths of the solar diameter, or 1′ 03″ arc (**32**).

Venus in sole visa came as the culmination of several years of intense astronomical thinking and observing and, to put the transit in context, it is essential that the character and direction of his work during these years should be explained. When he first came to be acquainted with William Crabtree in 1636 he was already well read in the standard astronomical literature of the day, being a skilled calculator and familiar user of the principal tables, such as those of Lansberg and Kepler. How he had obtained this expertise is unclear for while he had been resident in Cambridge between

1632 and 1635, it is not known what official instruction in practical astronomy took place or would have been available (33).

But by 1636 he had become disillusioned with the tabular, computational approach to astronomy, as a result of the numerous errors which he had encountered. Whether these errors resulted from internal contradictions within the tables themselves, or from discrepancies between the tables and the observed heavens is uncertain, though by 1636 he was corresponding with Crabtree about their mutually discovered errors in the *Tabulae Motuum* of Philip Lansberg, with both men resolving to observe the heavens at first hand, rather than rely on tables (34). It was this resolution which gave rise to the substantial body of correspondence which passed between them over the next 5 years, and of which the transit of Venus was the hiatus.

Both men shared the same cosmological assumptions, being Copernicans, admirers of Tycho, Galileo and Kepler, yet in no way blind to their faults. Horrocks and Crabtree were skilled practical mathematicians, though it is not clear how much sustained observation either had done or what instruments they possessed in 1636 (35). At the outset of their friendship, it seems that Crabtree (1610–1644), as the older and perhaps more experienced man, was the main instigator of inquiry, though his initiative was quickly superseded by Horrocks once they started to operate on a regular basis. In addition to their expressed annoyance with the 'vain pretensions' of the tables, their correspondence revolved around three general topics: planetary theory, the lunar orbit, and instrumentation.

The first two of these topics can be subsumed within their wider interest in the importance of Kepler's law of elliptical motions and attempts to demonstrate such motions in the solar system. The motions and retrogrades of the outer planets were monitored and discussed in detail, but much of Horrocks's efforts were devoted to those investigations which led him in 1637 to the discovery that the moon moved in an elliptical orbit around the earth, and that the apside line precessed with relation to the position of the sun. Though less dramatic than the transit of Venus, it was in many ways a far more important discovery, and one which was to deeply influence John Flamsteed (36). What was truly remarkable, however, was the speed with which an 18-year-old, largely self-educated amateur, living in an intellectual backwater, came to such a conclusion. If the early Fellows of the Royal Society, in their quest for a father of the New Science in England, really wanted a 'hero', then one can hardly blame them for alighting on Jeremiah Horrocks.

Horrocks was not only a convenient inteliectual for the early Royal Society to admire, but he doubly fitted their bill by being a practical observer and experimenter as well. To a body of men who still excited popular ridicule by putting cats into air pumps and fleas under microscopes, Horrocks provided the Royal Society with a happy example of how observation and experiment could challenge and overthrow the Old Philosophy. Many of Horrocks's letters and observations which the Society published in 1673, and the *Venus* treatise already published by Hevelius (a Correspondent of the Society) in 1662, abounded with experimental data. One might say that the data were experimental rather than just observational because Horrocks and Crabtree often devised specific courses of observations as 'acid tests' to decide upon particular phenomena. While there is no evidence that Horrocks and Crabtree ever read the works of Francis Bacon, one can understand how the early Royal Society could admire their thorough-going scepticism towards scientific authority, and expressed preference for instrumental, measurable and experimental solutions to questions in nature.

In attempting to ascertain the angular diameters of the stars and planets, for instance, Horrocks states that (using a method mentioned by Galileo) he observed them through a pinhole of known size at a measured distance to create a set of trigonometrical proportions (**37**). Similarly, when he and Crabtree observed an occultation of the Pleiades on 1637 March 19, they watched how suddenly the stars were extinguished by the dark lunar limb, suggesting thereby that the stars were but points of light (**38**). In 1638, moreover, when Crabtree was requesting a physical model for the newly discovered motion of the lunar apside line, he sent details of an elegant demonstration, using a string and a weight, which would make it clear (**39**). At every turn, it appeared, Horrocks had an instinct for asking the right question, devising an investigative technique, or framing a physical model of the phenomenon under discussion.

What lay at the heart of Horrocks's achievement, and must have further added to his Royal Society appeal, was his skill as a deviser of instruments. Though the Victorians were fond of depicting Horrocks as an impoverished individual who was compelled to make his own instruments in the absence of better ones, one must remember that his 'self help' approach was not simply occasioned by poverty. Large, accurate astronomical instruments were scarce in the early seventeenth century, though it would be incorrect to ascribe this scarcity to the limited means of astronomers. Astronomical instruments, like all other scarce resources, are subject to demand as well as supply, and one reason why precision instruments were rare and expensive is because demand was not sufficient to stimulate and sustain regular commercial manufacture. It is true that wealthy patrons like Tycho Brahe could maintain craftsmen over many years to produce evolving designs of improved accuracy, but the demand for such pieces was insufficient to sustain a large body of men outside princely patronage (**40**).

There was, on the other hand, a thriving and well-established trade across Europe for the manufacture of more routine mathematical instruments such as astrolabes, dials and pieces for the surveyor and navigator. England had developed a thriving trade in the days of Elizabeth with Humfrey Cole and others, while Elias Allen of London was supplying quadrants and similar instruments to John Greaves at Oxford in 1637. But such skills were often conservative in character, cultivated to serve a market of draughtsmen, architects and teachers rather than pure scientific research (**41**). Not until the later seventeenth century did England's real ascendency in the research and development aspect of scientific instrument making come into its own, in the wake of the post-Restoration passion for the 'Experimental Philosophy' and the Boyles, Flamsteeds, and Sir Nicholas Gimcracks ready and willing to pay for it (**42**).

But Jeremiah Horrocks lived a half-century before this movement got off the ground. Because astronomy was still a relatively conservative, computational (as opposed to observational) discipline, few people required instruments of advanced design for there were few investigations which needed them. With the Vernier scale only just invented in 1631 and the pendulum clock, micrometer and telescopic sight not generally viable until the 1660s, no specialized trade existed to fabricate tools for men engaged in original research of the kind which interested Horrocks (43). In consequence, if one wished to measure the solar diameter, ascertain the apparent sizes of the planets, determine the node of Venus's orbit or obtain new values for the Sun's parallax, one had little alternative other than to make one's own tools (44).

One must further bear in mind that most of the instruments manufactured for the commercial market by the London and foreign trades tended to be multi-functional in intention. The astrolabe and Gunter quadrant, for instance, could be made to yield a diversity of astronomical, calendrical and horological values, and even relatively specialized instruments, such as those of the navigator or surveyor, were only expected to perform within traditionally prescribed tolerances. The modern concept of an *investigative* scientific instrument, designed to do one particular operation with a very high degree of accuracy in pursuit of hitherto unknown data, did not exist, because the generality of astronomers were not concerned with physical researches. The only two men of recent times who had addressed themselves to a fundamental revision of astronomy from original observations – Tycho Brahe and the young Johannes Hevelius – both found it necessary to invent and manufacture specialized instruments to suit their needs (45).

When Jeremiah Horrocks in his turn determined to pursue *investigative* astronomy, he found it necessary to do the same, though as he was a relatively poor man, he was unable to commission major artisans, and made his own tools. Yet Horrocks was a singularly gifted practical geometer who thought in terms of interrelated circles, straight lines and tangents, as well as being a man of great manual dexterity, and both of these attributes enabled him to extract significant data from what at first sight appeared to be a rudimentary arrangement of sticks and pins.

All of Horrocks's instruments hinged on the mathematical relationships which exist between the centre, radius and circumference points of a circle. Needless to say, Horrocks was not the first person to work thus, for it had formed the basis of celestial measurement since antiquity, although he found original ways of adapting it to his special needs while using the simplest equipment.

In particular, it formed the basis of his Astronomical Radius. This instrument consisted of a pair of stout wooden rods joined together to form a 'T' shape. The long axis of the 'T' was divided into 10 000 subdivisions, while the short crosspiece was graduated into degree and minute equivalents from Petiscus's *Table of Tangents*. Two metal sighting pointers could be made to slide along this crosspiece, so that when the whole instrument was held up, with the end of the long axis to the eye and the short arm about 3 feet away, they could be adjusted to enclose a pair of stars. Horrocks would then strike off the angular amplitude between the sights with a pair of dividers, and divide it into the already graduated length, from which he could compute the angle from a table (46).

The principle behind the instrument was not new, having been allegedly devised by Levi Ben Gerson some centuries before and simplified in the sixteenth century to form the mariners' Cross Staff, though Horrocks was one of its most careful users (47). The instrument was simple to make and if one possessed a natural dexterity and keenness of sight, as Horrocks clearly did, it could be made to yield very accurate measures.

Much of the early correspondence with Crabtree revolves around the performance of astronomical radii, and on 1636 July 25, at the very outset of their friendship, Horrocks spoke of using an instrument with a 3-feet long axis. This axis must have consisted of a flat board rather than a rod, for it had drawn upon it a diagonal scale divided into 10000 equal parts (48). If he was able to divide up an English yard into 10000 parts against which to measure tangental proportions, one can understand how he confidently claimed to make angular observations accurate to a single minute. While in practice, neither his scribed divisions nor angular alignments could have been so accurate as claimed in theory, a very high level of relative accuracy would nevertheless have been possible for a skilled user.

Horrocks was also aware of the possibility of instrumental errors creeping in, and he applied computed corrections to compensate for eye eccentricity and other quantities. Observing the angle between Venus and Saturn on 1637 November 23, for instance, he found it necessary to apply a correction factor of 7 arcmin to reduce the observed angle of $10° 29'$ to $10° 22'$ on his 10000 unit scale to eliminate the eccentricity error (49). It is in this context, indeed, that Horrocks makes reference to one of the few English astronomical authors encountered in his writings. This is Edward Wright, whose *Certain Errors in Navigation* (London, 1599) p. 213, discussed the need for eye eccentricity corrections to be applied by mariners using cross staves at sea (50).

If the astronomical radius formed his basic angle-measuring technique, he was quick to adapt the principle which it embodied to obtain values of increasing specialization and delicacy. In 1636 December, Horrocks and Crabtree were corresponding about a form of radius with an 11-feet axis to measure the solar diameter by enclosing the disk between two upright pointers 11 feet from the eye and dividing the resulting tangental amplitude into the long axis (51).

Over the course of 5 years, Horrocks accumulated a large number of solar, planetary and lunar measures, and one can understand how, in increasing disillusionment with the existing tables, he found it necessary to make his own independent observations if original conclusions were to be drawn. One also understands why the transit of Venus formed such a climacteric event for Horrocks, providing as it did such a wealth of new data, the interpretation of which hinged upon a set of consistent measures for the apparent solar diameter. There was no way in which Horrocks could extract precise vital statistics about Venus's angular diameter and node if he did not know the exact apparent diameter of the sun, while he further realized that if Kepler's elliptical orbits existed in nature, and the earth was closer to the sun at some seasons than at others, then its angular diameter should vary in accordance with exact geometrical postulates.

Since 1636, Horrocks, in cooperation with his friend Crabtree, had been regularly measuring the solar diameter, and noting its seasonal changes.

These values provided substantiation to the theory of elliptical orbits and, by the time of the transit of 1639, he had plenty of measurements available to enable him to establish the solar diameter for November 24, from which he could extract the new planetary elements (52).

Horrocks obtained solar diameter measures using two techniques: the Upright Wires and the Foramen. The Upright Wires constituted nothing more than a specialized astronomical radius, in which the sun was viewed between two metal pointers at the end of a graduated rod, in the manner of the 11-feet radius mentioned above. The Foramen was a pinhole camera method, whereby he produced a prime focus image in a darkened room, carefully struck off the linear size of the disk with a pair of compasses, and divided it into the projection distance. This method was, in fact, identical in principle to the radius, with the difference that in one instrument one looked directly at the sun – probably through a smoked glass – while in the other one produced a projected solar image.

On 1636 December 2, using a pinhole five units in diameter, for instance, and at a projection distance of 4100 units, a solar image of 42·5 units was produced, which reduced to a solar diameter of 31′ 26″ arc (53). On 1636 November 15, he had obtained a value of 31′ 34″ arc by this same method, while on 1637 March 10, his Upright Wires gave a solar diameter of 32′ 10″ arc (54). Horrocks rarely provided dimensions in inches for any of these measures, or tells us how big in linear terms his 'units' were, though this was not necessary (55). It was from the proportions of the projection distance and image that he extracted his data, so that precise linear dimensions were not important in themselves, provided that the chosen units were used consistently. In his *Venus* treatise, he gave the diameter for the transit day, November 24, as 31′ 30″ arc, 'an estimate which I adopt, not from regard to the idle adage *medio tutissimus ibis*, but because I have found it, from my own repeated observations, to be very close to the truth'. Though he had only divided the solar image drawn on his projection screen into thirty equal parts, he corrected all the Venusian elements for 31′ 30″ in his eventual results, which increased the planet's diameter from 1′ 12″ to 1′ 16″ arc (56).

I have on several occasions replicated Horrocks's Foramen technique to measure the solar diameter, and found that his value of 31′ 30″ arc is a very good one for midwinter. For January 2, for instance, I obtained an average diameter of 33′ 43″ arc for the *horizontal* diameter from twelve observations which spread between 32′ 17″ and 34′ 38″ arc. When measuring the *vertical* diameter, however, I obtained a more consistent set of readings, giving an average value of 31′ 55″. This diminished vertical diameter was occasioned by the apparent elevation of the lower limb of the sun due to refraction. The *Astronomical Ephemeris* diameter for January 2 is 32′ 35″ arc (57). Unfortunately, Horrocks does not tell us which solar axis he measured to obtain his 31′ 30″ (though he does in some other cases), but I would suspect it may have been the vertical. This is, after all, much easier to measure, for the vertical diameter remains fairly constant on the projection screen, while the horizontal diameter shifts with the setting sun.

In the light of their concerns with accurately measuring the solar diameter one can understand the importance of the work of the Yorkshire astronomer and inventor William Gascoigne to Horrocks and Crabtree in 1640.

Gascoigne was the inventor of the telescopic eyepiece, or filar micrometer, who had the insight to take the Foramen method one stage further. Gascoigne came to realize that with a Keplerian optical system, the prime focus image fell within the telescope tube. Instead of measuring off a pinhole solar image with dividers from a screen, Gascoigne measured the prime focus image *inside* the telescope. By introducing two fine-pitched screws into the eyepiece, so that each one moved a knife-edge pointer, one could 'enclose' the lunar or solar image produced telescopically, and knowing the exact focal length of the object glass and the linear pitch of the screw, calculate the angle subtended to a hitherto unobtainable degree of precision (**58**).

But Gascoigne was not involved in the 1639 transit of Venus observations. He is not mentioned as an alerted potential observer in October that year and, from surviving correspondence, does not seem to have been known to Horrocks and Crabtree until 1640. Gascoigne, moreover, does not seem to have invented his micrometer arrangement until late 1639 or 1640, though it must have been soon afterwards that he began to correspond with Crabtree, and Crabtree in turn alerted Horrocks to the potential of the micrometer (**59**).

In autumn 1640, Crabtree visited Gascoigne at his home near Leeds and, in a letter sent following his return, expressed Horrocks's interest in obtaining a micrometer for himself (**60**). Horrocks was at this time engaged in the composition of his *Venus* treatise, and the procurement of an instrument which might enable him to obtain yet more precise solar and planetary diameter measures could influence the proportions which he would ascribe to the Venusian orbital and solar parallax values.

One must never forget that Horrocks's astronomy dealt almost entirely in *proportions* rather than absolute dimensions, and he was constantly searching for reliable measurements whereby he could translate the proportions into physical quantities. The much greater accuracy and ease of operation made possible by the micrometer above the pins and graduated rods he had used for 5 years would have provided a fundamental breakthrough in this respect, and one can understand how the news of it 'ravished' Horrocks's mind, making Crabtree ask Gascoigne if he 'could purchase it with travel, or procure it with gold' (**61**).

Crabtree's correspondence with Gascoigne about the micrometer seems to have been conducted over the last few months of 1640, about a year after the transit, and when Horrocks was busy composing *Venus in sole visa*. There is no evidence, however, that Horrocks ever acquired a copy of Gascoigne's invention prior to his death in January 1641, for all the planetary diameters which Horrocks measured in 1640 were obtained by his usual methods (**62**). Crabtree and Gascoigne were to correspond extensively on astronomical matters, however, from their acquaintance sometime before October 1640 to the disruption of their investigations in 1643, following the outbreak of Civil War (**63**).

Horrocks's *Venus* treatise is an ingenious and original book, and much more than a straightforward description of a celestial event. But I believe that it is in the last two chapters, where Horrocks enters upon a discussion of planetary diameters and cosmological dimensions, that it becomes genuinely far-reaching. In addition to continuing to display a remarkable erudition, and making it clear that 'John Kepler, the prince of astronomers' headed his own personal list of scientific heroes, he used the transit to provide evidence showing that the solar system could be ten times larger than traditionally believed (**64**). He was one of the first astronomers, moreover, to advance arguments in favour of a physically vast universe that hinged not on points deduced from philosophical postulates, but from measured results.

Horrocks's starting point in this respect was the smallness of Venus when seen in transit. If Venus and the other planets were much smaller than traditionally believed from naked eye observation, then it was inevitable that the parallax values attributed to them in the Rudolphine and other tables, must be reduced by several times once the telescopic images revealed their smallness. He applauds Gassendi's discovery that Mercury, when seen in transit in 1631, was only a fraction of what was expected, and as a man well trained in the arts of dialectic promptly took issue with those astronomers who had tried to explain away Mercury's smallness in terms of optical distortion. Schickard, for example, had argued that the planet's edge had refracted the sunlight passing over it, thereby making it appear smaller, as the 'umbra' of a shadow (**65**). In many ways, this explanation harked back to the Quintessence theory of the planets which made them out to be semi-transparent, self-radiant bodies enveloped in a lucid medium.

But Horrocks would have none of it. He first demonstrated experimentally that hard, solid objects never refracted light around their edges so that their shadows or silhouettes were smaller than themselves. Secondly, Horrocks went on to demonstrate that both Mercury and Venus – like all the other planets – possess observable physical bodies just like the earth, and are not partially composed of irridescent substances. Just as a circular stick held up edge-on against a candle flame will appear hazy at the edges and diminished when looked at with the naked eye, it immediately assumes a hard, sharp, correct size when viewed through a telescope. Similarly, Horrocks pointed out that the moon, which always seems diminished in size against the sun when seen in eclipse, was immediately restored to its correct dimensions when viewed telescopically, as he had already noted during the eclipse of 1639 May 22 (**66**).

The final blow against the 'transparent medium' idea of the planets had been demonstrated to Horrocks, however, by the occultation of the Pleiades by the moon's dark edge on 1637 March 19, when each star in turn had been instantaneously snuffed out (**67**). For this to happen one had to acknowledge (*a*) that the stars *must* be only points of light lacking linear diameters, and (*b*) that the lunar limb was solid and opaque.

From this ingenious line of experiments and demonstrations, augmented by others discussed elsewhere in the *Venus* treatise, Horrocks concluded that the naked eye could be extremely deceptive when used to ascertain the sizes of bodies seen in contrast against the surrounding background. This was occasioned by the 'moisture of the beholder's eye' which scattered incoming light so that bright objects appeared magnified when seen against a black background while dark objects against a bright one were diminished (**68**). The problem could be easily solved, however, by viewing both classes of objects through the telescope, which, because it sharply focused the light and magnified the image entering the eye, prevented the scattering from taking place. This discovery should also be seen as part and parcel of Horrocks's deeply instrumental view of science, in which potentially deceptive natural faculties could be corrected by the use of specialized instruments and nature shown in her true condition. This instrumental concept of scientific enquiry also lay at the heart of the Royal Society's approach to physical investigation as mentioned above, and one can understand how Horrocks's work so impressed the early Fellows.

In the light of his realization about Venus and Mercury, Horrocks next set out to investigate the apparent diameters of the other planets to see if the same rule held. But because the outer planets could never be seen in transit and no other way of measuring the angular sizes of telescopic images was yet known to Horrocks, he used a variation of his Foramen technique. Starting with Venus 'whose diameter I have observed most accurately', he decided to compare the new morning star of 1640 January 7 with her transit image of 6 weeks earlier. Horrocks took 'an iron needle whose diameter was 8 parts at a [viewing] distance of 4300 [and which] covered the planet; therefore, the diameter was 0·38″ [arc]' (69). Using the Foramen, or pinhole in a card technique, he went on to measure the Venusian diameter on 1640 January 29 and found it to have diminished to 27″ arc, as the planet continued to move to eastern elongation. In addition to the pinhole technique providing a set of geometrical proportions from which he could calculate an angular diameter, it also removed those ocular distortions which occurred when one looked at a bright object in an open sky, for by this:

> method alone, even on a dark night, the diameters of the planets appear to be wonderfully reduced: so that, unless you are very strong-sighted, you can scarcely discover either the planets or the fixed stars which deceive the naked eye from their rays being entirely cut off by the narrow opening (70).

Knowing the proportions in accordance with which Venus subtended particular angles, he then tried to ascertain the angular diameters of the other planets by comparison. In that part of its orbit where Kepler said Jupiter subtended 50″ arc, Horrocks found the planet yielded 37″ arc, and on 1640 February 24 and March 2, was smaller still, and even less than Venus, which had by this time diminished to 24″ arc. Saturn was confessedly difficult to measure, but his diameter was somewhere between 30″ arc and one minute on 1639 September 6, while Mars varied greatly in luminosity and size due to a constantly changing distance. Early in 1640 March, for instance, Mars seemed smaller than Venus and Jupiter (measured at around 24″ arc), though at other times the planet approached two arcmin (71). Horrocks liked to measure other planetary diameters at such times when Venus also happened to be visible, so that he could use one value, for which he felt very confident, as a yardstick for the others. During 1640, indeed, he seems to have been occupied in determining planetary diameters either by direct pinhole measurements or else by comparisons with other bodies, in an attempt to relate visible diameters to orbital locations, *vis-à-vis* the Sun and Earth. It is easy, moreover, to understand how his mind must have been 'ravished' when he received news of Gascoigne's micrometer from Crabtree towards the end of 1640, promising as it did, a much more straightforward telescopic method of measuring small angular diameters.

While his value for the diameter of Venus in transit had been a good one, it must not be forgotten that his pinhole technique of measuring, by computing the proportions obtained by holding a card pierced with a hole at the appropriate distance from the eye, was a very awkward one. Quite apart from the problem of estimating precisely when a bright planetary image 'fills' a pinhole 6 feet from the eye (as I have discovered at first hand in replicating his technique), Horrocks seemed ignorant of the fact that the physiological resolving power of the naked eye is only one arcmin (72). He seemed to assume that once the extraneous 'rays' had been removed by observing through a pinhole, then one could proceed to measure angles down to seconds. Nor did he take account of other correction factors, such as refraction and atmospheric conditions in any of these measures, the omission of which must have resulted in inevitable errors of which he was not aware.

Yet in spite of the reservations which we must bear in mind when considering Horrocks's *exact* numbers, much more important was his realization that, when compared with the silhouetted Venus, all other planetary diameters were very small. This realization stimulated in his mind a much wider cosmological concern which *may* have been present before the Venus transit of 1639, but which certainly received a major impetus after it. This was the problem mentioned above concerning the proportions existing between planetary sizes, distances and orbital velocities.

One fact of great significance which Horrocks extracted from the 1639 transit of Venus was a greatly reduced value for the solar parallax. Hitherto considered to be between one and two arcmin, Horrocks found it to be no more than 14″ arc, which is respectably close to the 8·88″ arc accepted today (73). He was quick to perceive the consequences of this discovery for Keplerian astronomy and the overall dimensions of the solar system.

Having already dispensed with the classical idea that the planets were carried around on crystalline spheres, he argued that their motions were occasioned by 'natural and magnetical causes' in which a force from the Sun acted upon the planetary masses and caused them to move (74). While he could not say what these 'magnetical causes' were, he knew that solar attractive and planetary axial rotative forces were somehow involved, and that they worked in accordance with Kepler's laws. But where Horrocks disagreed with Kepler was in the German astronomer's use of proportions based upon planetary *diameters* conjoined with orbital *semidiameters*. To Horrocks's mind, the aesthetic of the argument required semidiameters of planetary bodies to be computed in accordance with semidiameters of orbits, or else full planetary diameters with whole orbital ones.

Seen in transit, Venus revealed a diameter of 1′ 16″ arc. From the Keplerian proportions of the Earth–Venus–Sun ratio, Horrocks then computed that, to an observer on the surface of the Sun, Venus would only reveal a diameter of 28″ arc. Using Gassendi's surprisingly small value of 20″ arc for the diameter of the transiting Mercury, as seen in 1631, Horrocks proceeded to calculate that, seen in the same alignment *from the Sun*, Mercury would also subtend a diameter of 28″ arc, if it occupied the place assigned to it by Kepler. As I have already discussed above, Horrocks was

V

348

busy in 1640 attempting to obtain angular measurements for Mars, Jupiter and Saturn and, on the limited data which he could obtain, believed that these in turn would reduce down to 28″ arc each as seen from the Sun.

As one proceeded out from the Sun, therefore, the planets became larger, with physical volumes and diameters that were in exact proportion with their Keplerian orbital areas. While unable to explain what it was about the power of the Sun which made them move, he was convinced that it was natural (as opposed to magical) and physical. It was clearly related to their volumes, velocities and distances, growing proportionately weaker with distance, and needing a bigger mass to act upon. If Venus, along with the rest of the planets, revealed a 28″ arc diameter in the 'mean distance' of its orbit, then its semidiameter, in proportion to the radius vector which connected with the Sun, must be 14″ arc. This was used by Horrocks to compute an Astronomical Unit of 15 000 terrestrial semidiameters, which indicated in turn that the solar system and observable universe must be vastly larger than formerly believed (75).

Though we today no longer accept the broader geometrical aesthetics upon which Horrocks based his conclusions, knowing what we do about the sizes and distances of the outer planets, we are compelled to acknowledge the brilliance of his interpretation from the data available (76). Knowing two planetary values (Mercury and Venus) that were conveniently congruent with the overall theory of planetary geometry, he did his best, with the techniques available, to fill in the rest.

But where he was quite dramatically correct was in his realization that the solar parallax was up to ten times smaller than commonly believed, the implications which this had for the dimensions of the solar system, and the consequent necessity to observe and measure the heavens afresh. In the last pages of *Venus in sole visa* Horrocks concluded by addressing himself to the wider implications of the solar parallax, and planned a treatise to be entitled *De syderium dimensione*. Unfortunately, he never lived to write it (77).

In retrospect, one can fully understand why that body of men who would form the Royal Society came to view Horrocks in the way that they did, and see him as England's first astronomer of continental stature. Various efforts had been made to rescue his and Crabtree's papers, following Crabtree's death in August 1644, and Christopher Towneley, Jonas Moor, John Worthington and others all played their parts (78). On 1659 April 28, however, Samuel Hartlib wrote to Worthington (Horrocks's old undergraduate contemporary and now Master of Jesus College, Cambridge), enquiring after a copy of *Venus in sole visa*. Worthington replied that he had two incomplete manuscript copies and, while some years ago a friend had promised to edit and publish a complete version of the text, nothing had come of it (79). Perhaps such a manuscript passed into the hands of Johannes Hevelius in Danzig, who published it in 1662 (80). The Royal Society obtained a substantial cache of Horrocks astronomical letters in 1663, and John Wallis and Christopher Wren were invited to examine them. The result, after many delays and some losses in the Great Fire of 1666, was the *Opera Posthuma* (81).

At the outset of this paper, I mentioned that by the nineteenth century, Horrocks was being regularly referred to as 'Reverend', and spoken of as the

Curate of Hoole, Lancashire (**82**). The original source of this derivation is obscure and probably stemmed from the vague reference to the 'higher things' which he claimed occupied part of his time on Sunday, 1639 November 24, when he was keeping watch for the transit. But when one looks at how he spent that day, he appears to have been away from his telescope for no more than three daylight hours, including mealtimes. This 3-hour period, moreover, would have been no more than any serious Christian – especially a schoolmaster – would have devoted to his normal Sabbath devotions in the early seventeenth century. Indeed, one might safely say that any man who could spend between 4 and 5 hours looking through a telescope on a short autumn Sunday was *not* a clergyman, let alone a busy curate! (**83**).

I have found no reference to Horrocks possessing clerical status in the writings of those who knew him. William Crabtree gave no such indication, even in the obituary which he wrote to his friend's memory in 1641 (**84**). Nor did John Wallis and John Worthington, in spite of the fact that both of these men were in orders themselves, and had known Horrocks as an undergraduate (**85**). Jeremy Shakerley, it is true, had never known Horrocks personally, but was publishing liberal acknowledgments to his work as early as 1649 without any reference to clerical status (**86**). Edward Shereburne and John Flamsteed, who also came to know of Horrocks's work via the Towneley connection, also fail to mention holy orders, though Flamsteed in particular was profoundly influenced by his surviving astronomical manuscripts and was himself a cleric (**87**).

Perhaps most telling of all, however, is the simple fact that, according to law, Horrocks was, at twenty, much too young to have been a deacon, let alone a full priest (**88**). The only way around this minority would have been in the granting of a special Faculty by the Archbishop, though this would have been a complicated and expensive business. It would have taken place only if a rich patron had a benefice with which to present a young man. But Horrocks was never Rector of Croston – the Mother Church to Much Hoole in 1639 – nor did the local patrons of the district, the Stones family, ever present Horrocks with another benefice. It would have been both irregular and absurd to attempt to have a man 4 years under canonical age without an M.A. ordained by Archepiscopal Faculty merely to make him a curate – which was the only clerical title popularly ascribed to Horrocks. We know that by the summer of 1640, he was living back with his family in Toxteth with no known clerical connections (**89**). Much more likely is the fact that Robert Fogg, who was given the newly created living of Hoole by the Stones family in 1641, had served as curate previously, for we know that he had lived in the district during the 1630s (**90**). As pointed out above, I suspect that Horrocks was a schoolmaster, perhaps combining the office with that of lay Bible clerk. On the other hand, as a Godly youth with a Cambridge education behind him, he would probably have intended to get ordained, maybe through his college, when he reached the canonical age of twenty-four, and thereby make himself eligible for ecclesiastical patronage, as Worthington and Wallis did (**91**). It was perhaps this expressed intention, passed on at third and fourth hand, which led to the title 'Reverend' being attached to his name after his death.

Bearing all the hagiographical caveats in mind, however, it is difficult to deny the importance of Jeremiah Horrocks to seventeenth-century English astronomy. This importance, when recognized a few years after his death, caused many to speculate what might have been achieved had he lived even into early middle age, and his older contemporary James Gregory put his finger on the point when, in 1672, he bemoaned to John Collins: 'It was a great loss that he died so young, [when] many naughty fellows live till eighty' (**92**).

Jeremiah Horrocks was, perhaps, the first man in England to comprehend fully the astronomical revolution going on in continental Europe. He was the first Englishman (along with Crabtree and Gascoigne) to recognize the investigative power of the telescope in original research and draw conclusions that went beyond those of Galileo and Kepler. He was the first of his countrymen to examine independently the Keplerian astronomy, improve upon it, and produce an analysis of the Lunar orbit which would provide a working model for the rest of the solar system. But he was possibly the first man in Europe to recognize the methodological faults of post-Tychonian astronomy, discard tabular computation, and stress fundamental observation with specialized mathematical instruments as the only way to answer the dominating questions of the age. And these conclusions were reached, moreover, not in a university city or intellectual centre, but in the relative isolation of a few square miles of Lancashire countryside.

But the high point of Horrocks's short though remarkably fruitful scientific career, to both the first and subsequent generations of his admirers, lay in those 30 minutes before sunset on 1639 November 24, when he observed the transit of Venus.

ACKNOWLEDGMENTS

I wish to thank the staff and librarians of many institutions who have helped me in researching this paper: The Royal Society Library, Chetham's College, and Municipal Local History Library, Manchester, and the Lancashire County Record Office, Preston. In Oxford, I thank the staffs of the Bodleian and Museum of the History of Science libraries. I am also indebted to Miss Rachel Woodrow, of the Classics Office, University of Oxford, for assisting in the decipherment of some of Jeremy Shakerley's intractable Greek.

NOTES AND REFERENCES

(1) John Wallis entered Emmanuel in 1632 and edited *Jeremiae Horrocci Liverpoliensis Angli ex Palinatu Lancastriae Opera Posthuma* (London, 1673). Editions of Horrocks's works were issued in 1672, 1673 and 1678. In this paper, I have used the copy in the Bodleian Library, Art 4° K 15, of 1673, hereafter referred to as *Op. Post.* A full bibliographical exposition of the *Opera Posthuma* is found in J.E.Bailey's 'The writings of Jeremiah Horrox [*sic.*] and William Crabtree', *The Palatine Note-Book*, III (Manchester, 1883), pp. 17–22.

(2) Flamsteed was deeply influenced by the works of Horrocks, especially his lunar theory, in which he had a 'paternal pride'; see R.S.Westfall, *Never At Rest* (Cambridge, 1980), p. 547. Flamsteed also contributed an exposition on the Horrocksian lunar theory to *Op. Post., op. cit.* (1). Sir Jonas Moor, the motivating force behind the founding of the Royal

Observatory, may have known Horrocks as a young man in Lancashire, and was certainly a friend of Christopher and Richard Towneley, who preserved the Horrocks papers after Crabtree's death in 1644; see Eric G.Forbes, *The Gresham College Lectures of John Flamsteed* (London, 1975), pp. 8–9. Flamsteed, whose appointment as Astronomer Royal in 1675 owed much to Moor, had seen and transcribed sections of the Horrocks correspondence at Towneley Hall in the early 1670s. They are now preserved in the Flamsteed papers at the Royal Observatory, vols 40, 68, the latter of which contains a Latin manuscript of Horrocks's *Venus in sole visa*. Flamsteed's interests in the Horrocks circle are evident in Francis Baily's *An Account of the Revd. John Flamsteed* (London, 1835), which contains Flamsteed's journals.

(3) *Principia* (1687), edited by Florian Cajori as *Sir Isaac Newton's Mathematical Principles*, (Berkeley, 1934). Newton especially commends 'Our Countryman *Horrox* [who] was the first who advanced the theory of the moon's moving in an ellipse about the earth placed at its lower focus', Book III, Scholium; p. 475. Numerous references occur to Horrocks in *The Correspondence of Sir Isaac Newton*, eds Turnbull, Scott *et al.* (Cambridge, 1961 ff.), see I, III, IV, V. See also, H.C.Plummer, 'Jeremiah Horrocks and his "Opera Posthuma"', *Notes & Records of the Royal Society* (1940), pp. 39–52.

(4) I do not know the source of Herschel's quote, but take it from A.B.Whatton, 'A Memoir of his [Horrocks] Life and Labours', which prefaces the English translation of his *The Transit of Venus across the Sun* (London, 1859), x. Hereafter as Whatton, 'Memoir' and *Venus*.

(5) One might argue that Thomas Harriot (1560–1621) was the first Englishman to take serious interest in the New Philosophy in some ways, though he never possessed the impact of Horrocks upon either his own, or succeeding generations, nor did he publish his work; see Jean Jacquot, 'Harriot, Hill, Warner and the New Philosophy', and John North, 'Thomas Harriot and the first telescopic observations of sunspots'. In *Thomas Harriot, Renaissance Scientist*, pp. 107–165, ed. John Shirley, Oxford, 1974.

(6) J.E.Bailey, 'Jeremiah Horrocks and William Crabtree, Observers of the Transit of Venus, 24 November 1639', *The Palatine Note-Book*, II (Manchester, 1882) pp. 253–266; see pp. 254–255 for detailed treatment of Horrocks's lineage.

(7) *D.N.B.* gives his birth date as '1617?'. John Wallis, however, in his 'Epistola Nuncupatoria' in *Op. Post.*, sig. b₃, v., states that at his death in January 1641, Horrocks was 'sub vicesimum secundum', or under twenty-two. In fact, Horrocks would have been at the beginning of his twenty-second year.

(8) W.F.Bushell, 'Jeremiah Horrocks; the Keats of English astronomy', *Mathematical Gazette*, **43** 343, (February, 1959), 1–16.

(9) For the importance of the Aspinwalls, see F.A.Bailey and T.C.Barker, 'The seventeenth century origins of watchmaking in west Lancashire'. In *Liverpool and Merseyside*, ed. J.R.Harris (London, 1969) pp. 1–15.

(10) According to *D.N.B.*, Mather spent fifteen years in Toxteth, which would have covered the whole of Horrocks's childhood and youth.

(11) See the 'Admissions at Emmanuel College, Cambridge, 1610–1723'. In *The Palatine Note-Book*, IV (Manchester, 1884) pp. 78–81, especially p. 79.

(12) His friend Worthington came up to Cambridge at 14; *The Diary and Correspondence of Dr John Worthington*, I, ed. James Crossley, Chethams Society XIII, (Manchester, 1847), p. 3.

(13) J.E.Bailey, *Palatine Note-Book*, II (1882), *op. cit.* (6), pp. 256–257. The Lansberg volume is preserved in the library of Trinity College, Cambridge.

(14) Worthington's *Diary* I, *op. cit.* (12), pp. 130–131. Also, Whatton's 'Memoir', *op. cit.* (4), p. 60.

(15) Flamsteed to Molyneux, 1690 May 10, in E.G.Forbes, *Gresham College Lectures*, *op. cit.* (2), pp. 8–9. Flamsteed states that after Crabtree's death, Towneley and Jonas Moor went to visit his widow and obtained his surviving mathematical papers.

(16) Shakerley's astronomical and astrological papers are in the Bodleian Library, Ashmole 242, 243, 333, etc. He praised Horrocks in *Tabulae Britannicae* (London, 1653), see 'To the reader'. Also, *The Anatomy of 'Urania Practica'* (London, 1649), p. 18, and other works. See also Allan Chapman, 'Jeremy Shakerley (1626–1655?), astronomy, astrology and patronage in Civil War Lancashire', *Transactions of the Lancashire and Cheshire Historical Society* (1986), pp. 1–14.

(17) Shakerley to Henry Osborne, Surat, 1652–3 15 January 15, Bodleian MSS, Ashmole 242, fos. 95–96. Also, Vincent Wing, *Astronomia Britannica* (London, 1669), p. 312.

(18) Shakerley, *Anatomy of Urania Practica, op. cit.* (16), p. 18.

(19) The *Venus* treatise was not Horrocks's only sustained piece of writing, though it was his most original and complete within itself; see the edited 'Astronomia Kepleriana Defensa & Promota' which takes up a large section of *Op. Post., op. cit.* (1).

(20) Horrocks, *Venus*, 43. See also, Horrocks to Crabtree 1639 October 26, *Op. Post.*, 331. Horrocks also asks Crabtree to inform Dr Foster, and while he and Jonas Horrocks may have been alerted, there is nothing to suggest that they saw anything.

(21) This had been Kepler's tract *Admonitio ad Astronomos rerumque celesti studiosos, de mirisque rarisque anni 1631 phaenomensis, Veneris puta et Mercurii in solem incursu* (Frankfurt, 1630). Also, Robert Grant, *History of Physical Astronomy* (London, 1852), p. 415.

(22) Horrocks, *Venus*, pp. 133–134. Horrocks points out that in 1631, the beginning of the transit would just have been visible in Western Europe, though best seen in America.

(23) *Venus*, pp. 161–187. Though the tables of Lansberg and Rheinhold could be used to extract a conjunction, they were up to a day in error, thereby making it impossible for an accurate transit to be predicted; see Duncan MacNaughton, 'Horrocks's Observations & Contemporary Ephemerides', *Journal of the British Astronomical Association* 47, 4 (1937 February), pp. 156–157. Also, Betty Davis, 'The astronomical work of Jeremiah Horrocks', University of London unpublished M.Sc. Thesis, 1966, pp. 94–98.

(24) *Venus*, p. 123.

(25) Pierre Gassendi, *Mercurius in sole visus* (Paris, 1631). Horrocks was familiar with Gassendi's observation, as he noted in *Venus*, pp. 116–118, though he does not seem to have read the book in November 1639, for on 1640 April 20, he wrote to Crabtree expressing his desire to see it before completing the Venus treatise; *Op. Post.* p. 333. Horrocks had tried unsuccessfully to observe the Mercury transit a year before; *Op. Post.*, 'Observations', 1638 October 21, p. 382. Christopher Scheiner's *Rosa Ursina* (Bracciani, 1630) deals in detail with sunspot observation and illustrates the sunspot projection technique as used by Horrocks; see Plate 77. *Rosa Ursina* was not one of the books with which Horrocks noted his familiarity in his list of 1635, so one presumes that he encountered it after leaving Cambridge, J.E.Bailey, *The Palatine Note-Book*, II (1882), *op. cit.* (13), pp. 256–257.

(26) *Venus*, p. 119. Horrocks speaks of a half-crown telescope 'but superior to others I have handled' to Crabtree, 1638 June 6, *Op. Post.* p. 309. Half-crown telescopes seem to have been popular, for Jeremy Shakerley also had one; Shakerley to Osborne 1652–3 January 15, Bodleian MSS Ashmole 242, pp. 95–96.

(27) S.B.Gaythorpe, 'Horrocks's observation of the Transit of Venus 1639, November 24, O.S.' *Journal of the British Astronomical Association*, 47, 2 (1936 December), 60–68. In this article, Gaythorpe points out that Horrocks makes no refraction correction. Gaythorpe made several attempts to reconstruct the characteristics of the telescopes of Horrocks and his circle; see 'On the Galilean telescope made about 1640 by William Gascoigne, inventor of the micrometer', *Journal of the British Astronomical Association* 39 (1929 June), 238–241.

(28) In his observation of Mercury in transit, 1631 October 28, Gassendi found that the planet only revealed a 20″ arc disk; *Mercurius, op. cit.* (21), p. 7, *Venus*, p. 208.

(29) *Venus*, pp. 136–145. In this respect, Horrocks is firmly in the footsteps of Galileo, who had dealt with these issues in his *Starry Messenger* (1610), *Letter on Sunspots* (1613) and *The Assayer* (1623); see texts included in *Discoveries and Opinions of Galileo*, ed. Stillman Drake (New York, 1957) pp. 47, 94 & 252.

(30) *Venus*, p. 187. Horrocks obtained a Venusian angular diameter of 1′ 12″ arc on his thirty-digit solar image projection screen. When corrected for a solar diameter of 31′ 30″ arc, he derived the Venusian diameter of 1′ 16″ arc. Once again, he was in the wake of Galileo, who in his *Letter on Sunspots* had asserted Venus's diameter to be less than the 3′ arc ascribed by Apelles, 'not even the sixth part of one minute'; *Discoveries and Opinions, op. cit.* (29), p. 94.

(31) *Venus*, p. 212.

(32) Crabtree's observation of the transit is recorded in Horrocks's *Venus*, pp. 128–130. On the 1640 August 7, Crabtree wrote to William Gascoigne independently recording the blackness, smallness etc., of Venus in transit. Letter published by William Derham, *Philosophical Transactions*, 27 (1711), 287–288.

(33) It has been much debated by historians as to how much scientific material penetrated the Jacobean universities; see F.R.Johnson, *Astronomical Thought in Renaissance England* (New York, 1968), pp. 10–13. The eighty-year-old John Wallis was quite adamant in a letter to Dr T.Smith, 1697 January 29, that there was little mathematical interest at Emmanuel in his undergraduate days; Mark H.Curtis, *Oxford and Cambridge in Transition* (Oxford, 1959), pp. 244–245. John Worthington may have introduced Horrocks (whom he had known at Emmanuel) to Crabtree between 1636 June 1 and July 21, when he was visiting his family in Manchester; Worthington's *Diary, op. cit.* (12), p. 4.

(34) Many of the early letters contain references to the unreliability of Lansberg, especially those written in 1636; *Op. Post.* p. 247 mis-paginated 311 in Bodleian copy. See also pp. 121–155.

(35) Horrocks to Crabtree, 1636 July 25, expressed a desire to possess instruments like those of Crabtree, *Op. Post.* p. 247.

(36) Horrocks to Crabtree, 1637 November 23, *Op. Post.* pp. 294–300; 295. Also, 1638 July 25, *ibid.*, pp. 309–313; Flamsteed also wrote a detailed critique of the Horrocksian lunar theory in *Op. Post.* pp. 467–472; see also pp. 475–476.

(37) On 1640 January 7, he observed Venus through a pinhole of 8 'units' diameter, and at a distance of 4300 units computed a planetary diameter of 38″ arc or less; *Op. Post.* p. 395. Similarly, 1640 January 29, p. 395.

(38) *Venus*, pp. 198–199. He briefly describes this observation in *Op. Post.* p. 287, and recorded in detail the stages of the occultation in 'Observationes', pp. 352–353. He also claimed that Galileo believed the stars subtended an angular diameter of 5″ arc; *Venus*, p. 199, and to Crabtree 1639 September 28, *Op. Post.* p. 330, though in the *Starry Messenger*, Galileo states 'The fixed stars [unlike the planets] are never seen to be bounded by a circular periphery', thereby implying that they only subtend points of light; *Discoveries and Opinions, op. cit.* (29), p. 47. Horrocks discussed 'De Magnitidinem Fixarum' in a section of his defence of Keplerian astronomy, *Op. Post.* p. 61, taking up a point from Kepler, 'Fixarum diametrum omnino insensibilis & inobservabilis evadet'.

(39) Horrocks to Crabtree, 1638 July 25, *Op. Post.* pp. 309–313; for Apside pendulum demonstration, see p. 312. Also p. 295.

(40) Tycho Brahe, *Astronomia, Instaurate Mechanica* (1598), translated and edited by H.Raeder, E.Strömgren and B.Strömgren (Copenhagen, 1946).

(41) E.G.R.Taylor, *The Mathematical Practitioners of Tudor and Stuart England* (Cambridge, 1968). John Greaves's instruments, including a $6\frac{1}{2}$ feet radius brass-limbed quadrant and sextant and a 2-feet quadrant of 1637, are preserved in the Museum of the History of Science, Oxford. See, R.T.Gunther, *Early Science in Oxford*, II (Oxford, 1923), p. 79. Also, Gunther, 'The first observatory instruments of the Savilean Professors at Oxford', *The Observatory*, **60** (1937 July), 189–197. In 1977, I made detailed micrometric measurements of the scale graduations of these instruments, and found that these apparently exquisitely engraved divisions were only of a middling accuracy; A.Chapman, 'The design and accuracy of some observatory instruments of the seventeenth-century', *Annals of Science*, **40** (1983), 457–471. Horrocks was doing better with his Astronomical Radius.

(42) Taylor, *Mathematical Practitioners, op. cit.* (41), pp. 99–132. Robert Hooke's work best captures the central importance of instruments in the late-seventeenth century; *Micrographia* (1665), *Attempt to prove the motion of the Earth* (1674) etc.

(43) The connections between the nature of astronomical problems and the research thresholds imposed by existing instruments form a major theme in my *Dividing the Circle*, forthcoming, Ellis Horwood Ltd, Chichester, 1990.

(44) William Gascoigne, Crabtree's correspondent 1640–42, was in the same position, leading to his invention of both the telescopic sight and micrometer around 1640; see, W.Derham, 'Extracts from Mr Gascoigne's and Mr Crabtree's letters', *Philosophical Transactions* **30** (1717), 603–610. Gascoigne to William Oughtred, 1640 December 2, *Correspondence of Scientific Men of the Seventeenth-Century*, 1, ed. S.J.Rigaud (Oxford, 1841) pp. 33–34. Also, Gascoigne to Oughtred, Feb. (?) 1641, *ibid.*, pp. 35–59.

(45) Johannes Hevelius, *Machina Celestis* (Danzig, 1673). This work, like Tycho's *Mechanica*, *op. cit.* (40), describes an original set of major astronomical instruments developed and built for one man's specific researches, the likes of which could not be purchased commercially.

(46) Horrocks to Crabtree, 1636 August 30, *Op. Post.* pp. 250–251; 1636 December 30 for 11-

V

354

feet Radius, p. 252; 1637 January 4, p. 255. See also Crabtree's observations for 1637 February 1, p. 424.

(47) Tycho described a brass Radius which he had built in *Mechanica* (1946 edn), pp. 96–97. There was a substantial literature on the Radius (alias, Cross Staff, Baculus etc.) from the sixteenth century, with which Horrocks could have been familiar; Gemma Frisius *De Radio Astronomico* (1545) Petrus Apian, *Cosmographicus* (1524) and others; see John J.Roche, 'The Radius Astronomicus in England', *Annals of Science*, **38** (1981), 1–32; 27. B.Goldstein, 'Levi Ben Gerson', *Journal for the History of Astronomy*, **8** (1977), 102–112.

(48) Horrocks to Crabtree, 1636 July 25, *Op. Post.* pp. 247–248. Also, 1636 August 30, p. 250.

(49) Horrocks regularly applied eye corrections to his Radius observations; 1635 December 2, *Op. Post.* p. 418, as in an observation made before he knew Crabtree. On 1636 September 5, he applied a refraction correction, though it is unclear whether it was for the atmosphere or eye displacement, *ibid.*, p. 408.

(50) *Op. Post.* 1637 November 23. See eye eccentricity table, pp. 362–363.

(51) Horrocks to Crabtree, 1636 December 9, for 11-feet Radius, *Op. Post.* pp. 252, 424.

(52) Horrocks to Crabtree, 1636 December 9, for observation of 1636 November 15, solar SD = minimum 15′ 41″ arc; maximum 15′ 53″; media 15′ 47″. Observation, 1636 November 18, solar SD = 15′ 53″ arc. *Op. Post.* p. 252. On 1637 April 29, he reported to Crabtree results obtained on March 10: Apogee = 30′ 30″ arc, Perigee = 31′ 30″, average = 31′ 00″. *Op. Post.* p. 268. In *Venus*, p. 146, he cited various values for the solar D in current use, such as Kepler = 31′ 01″; Tycho = 31′ 54″; and Lansberg 35′ 00″. Horrocks's own accepted value was 31′ 30″, which he also cited in his discussion of SD variations. *Op. Post.* p. 141.

(53) *Op. Post.*, pp. 140–141, Observations made with Crabtree. Scheiner's *Rosa Ursina, op. cit.* (25), title page, contains a delightful comic depiction of a smiling bear observing the solar disk projected into his cave, with dividers ready to take the diameter.

(54) Horrocks to Crabtree, 1636 December 9, Obs., November 10, *Op. Post.* p. 252.

(55) If a 'common needle' is about 0·03 inches in diameter, and it represents 5 'units', then one unit must represent about 0·015 inches. Thus, 4100 units projection distance = 27·33 inches and the resulting SD at 42·5 units = 0·28 inches.

(56) *Venus*, p. 146. The value for the Venusian diameter at 1′ 12″ [1′ 16″] was larger than the 1′ 10″ arc recorded for the transit in *Op. Post.*, 'Observationes', p. 393; 'Diametrum Veneris fuit scr[upuli] 1′ 10″ qualium Sol habet 30′, certe non major'. It is possible that he derived his published value after taking further factors into consideration which he had not accounted for in his original notes.

(57) The *Astronomical Ephemeris* value for the solar diameter on November 24 (December 6 N.S.) is 32′ 31″ arc. My own projections were made at distances around 130 to 140 inches, which were considerably larger than those suggested by Horrocks's 'units'. I have tried his short distances, but obtained erratic results. In the hands of an expert who used it daily, however, as Horrocks probably did, I am sure that the Foramen could provide consistent results. See also, David C.Lindberg, 'The theory of pinhole images from Antiquity to the Thirteenth Century', *Archive for the History of the Exact Sciences*, **5**, 1 (1968), 154–176.

(58) The invention of the micrometer excited great interest in the 1660s, when Richard Towneley published 'An Extract of a Letter…touching on the division of a Foot into many thousand parts…', *Philosophical Transactions*, **2** (1667), 457–458, where he gave priority to Gascoigne. Similarly, when French claims were being advanced for priority of the invention, William Derham published his 'Extracts of letters…', *Philosophical Transactions* (1717), *op. cit.* (44). See also, Robert McKeon, 'Le débuts de l'astronomie précision: Histoire de la réalisation du micrometro', *Physis*, **13**, 3 (1971), 255–288. The treatment of the micrometer in Grant's *History of Physical Astronomy, op. cit.* (21), still stands up well.

(59) Derham, 'Extracts of letters…', *Philosophical Transactions*, (1717) *op. cit.* (44), pp. 603–610. Unfortunately, Derham did not print all the letters in his possession, but only selected parts, as with the unpublished letter of Christmas Eve, 1641, wherein Gascoigne 'describes the Wheel Work of his Micrometer and Shews how he could apply it to the taking of three Points; and Specifies his Observations of the Sun and Moon', *ibid.*, p. 604. Gascoigne also described his work on telescopic measuring devices to Oughtred, in an undated letter, probably written in 1641 February in S.J.Rigaud, *Correspondence of Scientific Men…*, I, *op. cit.* (44), letter XX, pp. 35–59.

(60) Crabtree to Gascoigne, 1640 October 30 and December 28, *Philosophical Transactions*, (1717), 606–608.

(61) *Ibid.*, pp. 607, 608. On 1641 December 6, Crabtree was discussing the 'screws' of Gascoigne's micrometer by which 'the Divisions of a Circle should be measured to Seconds without the Limb of an Instrument', *ibid.*, p. 609.

(62) Crabtree to Gascoigne, 1640 December 28, *Philosophical Transactions* (1717), *op. cit.* (44), p. 608. None of Horrocks's planetary diameter values indicates the use of a micrometer.

(63) Derham describes Crabtree's letter to Gascoigne, 1640 October 30 as his second letter; *Philosophical Transactions* (1717), *op. cit.* (44), p. 606. Flamsteed began his *Historia Coelestis Britannica*, I (1725) by printing five folio pages of Crabtree and Gascoigne letters and observations made between 1638 and 1643. Some of these observations were made independently, before the two men became acquainted, but Flamsteed was at pains to stress that in their use of the telescopic sight and micrometer (both of which were Gascoigne's inventions), they represented – along with Horrocks – the beginning of serious observational astronomy in England. Flamsteed printed in these pages eighty micrometric measures of the lunar diameter found in Gascoigne's papers, *Historia*, I, p. 5. Some years ago, Dr George Wilkins of the RGO Time Department very kindly undertook the reduction of these timed and dated observations for me, the results of which indicated the remarkable accuracy of Gascoigne's measures. They produced a mean error of 20″ arc and a root mean square of 35″ arc. I have also built a copy of Gascoigne's micrometer and observed with it through a Keplerian optical system of 6-feet focal length, and was able to come close to both Gascoigne's reduced diameters and modern Astronomical Ephemeris values; A.Chapman, 'Gauging angles in the seventeenth-century', *Sky and Telescope*, (April, 1987) pp. 362–364.

(64) *Venus*, p. 202.

(65) Gassendi, *Mercurius, op. cit.* (25), p. 7. Schickard took issue with Gassendi on the optical distortion of opaque objects seen against a bright light, in his *Pars Responsi ad Epistolas P. Gassendi...De Mercurio Sub Sole Viso* (Tübingen, 1632), p. 13.

(66) *Venus*, pp. 189–194. Horrocks had devoted great attention to this eclipse and its consequences, which he observed 'Per Tubam Opticam, in Camera Obscura', or by projection. He divided the solar image on the projection screen into thirty parts, and measured the lunar shadow. In many ways, it provided an unexpected trial run for the transit of Venus, which fell 6 months later. It certainly gave Horrocks an opportunity to perfect his solar observing technique; *Op. Post.*, 'Observationes', pp. 385–387.

(67) *Venus*, p. 198. *Op. Post.*, 'Observationes', 1637 March 19, pp. 352–353.

(68) *Venus*, p. 193.

(69) *Venus*, p. 208, Horrocks made this observation at 7.05 a.m., as Venus rose before the Sun; *Op. Post.*, 'Observationes', p. 395.

(70) *Venus*, p. 201.

(71) *Venus*, pp. 209–211. Horrocks cites Kepler's *Epitome Astronomiae Copernicanae* (Frankfurt, 1635), p. 484, for the original discussion on planetary proportions.

(72) Robert Hooke had been the first to investigate naked-eye angular resolution about 1670, obtaining the one arc minute figure still largely accepted today; see his *Some Animadversions...on Hevelius...* (London, 1674), p. 7.

(73) *Venus*, p. 188. Horrocks cites Tycho's Venusian diameter of 3′ 15″ arc, and on p. 207, says that Kepler attributed 1 arcminute to the solar parallax. See also Harry Woolf, *The Transits of Venus* (Princeton, N.J. 1959), pp. 10–13.

(74) *Venus*, p. 181. In his letter to Gascoigne, 1640 August 7, Crabtree spoke of the Sun's 'Magnetical or Sympathetical Rayes', after Kepler; *Philosophical Transactions*, (1711), p. 286.

(75) *Venus*, pp. 208–214. Gassendi, *Mercurius, op. cit.* (25), p. 7, describes Mercury as 'hoc est minuto tertiam partem, sui manis secunda 20″ [arc]'.

(76) *Venus*, p. 212. Horrocks does not cite his value for the terrestrial SD, although several authorities had placed it at just under 4000 miles. In 1635–37, for instance, Richard Norwood had computed 69·54 English miles to a meridional degree. This would have produced an SD of 3985·15 miles; Norwood, *The Seaman's Practice...touching to the compasse of the Earth and Sea and the quantity of our English degree* (London, 1637), pp. 38–46. From this figure, Horrocks *could* have calculated an Astronomical Unit of 59 777 250 English miles.

(77) *Venus*, pp. 214–215. It is difficult to know what progress may have been made on this by Horrocks up to his death in 1641 January.

(78) Crabtree's Will, dated 1644 July 19, in which he was described as 'sick in body but of sound

mind', was proved at Canterbury, under 'Rivers'; PRO, Prob. II, 194–240. He was buried on 1644 August 1 at Manchester Collegiate Church; see *The Registers of the Cathedral Church of Manchester*; *Burials, 1616 to 1653. Lancashire Parish Registers Society*, **56** (1919), p. 548.

(79) Mentioned in Whatton's 'Memoir', *op. cit.* (1), pp. 60–61. On 1659 May 5, Hartlib acknowledged the receipt of the *Venus* MSS; 'It is a very accurate piece & pity that it is not quite finished...', Worthington's *Diary*, *op. cit.* (12), pp. 130–131.

(80) *Venus in Sole Pariter Visa, Anno 1639, d. 24 Nov. St. V. Liverpoliae, a Jeremia Horroxio*, was published as a supplement to Johannes Hevelius' *Mercurius in Sole Visus Gedani* (Danzig, 1662), pp. 111–145.

(81) Wallis, 'Epistola Nuncupatoria', *Op. Post.*, sig. b₃. J.E.Bailey 'The writings of Jeremiah Horrox and William Crabtree...', *Palatine Note-Book*, III (1883), *op. cit.* (1), p. 18.

(82) When a Mr Holden erected a memorial tablet to Horrocks at Toxteth in 1826, his inscription made no mention of clerical status, but when some years later, the Reverend Robert Brickel placed one in Hoole church, Horrocks was described as curate of the Parish. Texts of both cited in Whatton's 'Memoir' to the 'Rev. Jeremiah Horrox [*sic*] Curate of Hoole near Preston', *op. cit.* (1), pp. 78, 80. My suspicion is that the Reverend Mr Brickel – who suffered from what his friends called 'Horrox fever' – played a major part in his hero's posthumous 'ordination', and the memorials which he erected successfully established Horrocks as 'Curate of Hoole'. See also W.F.Spalding, *The story of Hoole Parish Church* (Hoole, 1985), p. 10. Horrocks appears as Reverend in *D.N.B.* But there are other incongruities about his supposed Curacy. J.E.Bailey, in 'Jeremiah Horrocks and William Crabtree, Observers of the Transit of Venus, 24 November 1639', *The Palatine Note-Book*, II, (Manchester, 1882), pp. 253–266, maintains that Horrocks was an impecunious Curate struggling on £40 per annum, with no source cited. Yet in 1640, this sum would have been more than the living of most beneficed country vicars, while even in 1674, the Ordained Provost of Eton only received £50 p.a. See John Burnett, *A History of the Cost of Living* (Harmondworth, 1969), pp. 114, 116.

(83) *Venus*, p. 123.

(84) *Op. Post.*, sig. a₄ v, Whatton's 'Memoir', *op. cit.* (1), p. 58. *Op. Post.* p. 338.

(85) In his 'Epistola Nuncupatoria' Wallis speaks of Horrocks being resident in Hoole between 1639 June and 1640 July, but says nothing about his reason for being there. In particular, the words 'Sacerdos' and 'Scriba' (Priest and Clerk) are not used with any reference to him. See sig. a₄. v.

(86) Shakerley, *Anatomy of 'Urania Practica'*, *op. cit.* (16), p. 18 (see also Ref. 17 above). Miss Betty Davis in her excellent 'The astronomical work of Jeremiah Horrocks', University of London M.Sc. Thesis, 1966, pp. 20–21, states that Horrocks was first given the title 'Reverend' by Shakerley in his *Tabulae Britannicae* (1653). I have, on careful examination of the Bodleian Library copy, been unable to find such a reference, or in any other of Shakerley's writings, which only speak of Horrocks as 'Mr'. The nearest reference to Reverend which I have been able to find in any of Shakerley's works is in his *Anatomy, op. cit.* (16), p. 18, where the obscure and probably misspelled word in Greek characters *tō theocharitō*(?) – 'graced by God' – is added as a suffix to Horrocks's name. I have also seen it stated that John Gadbury in his *Ephemerides* (London, 1672) calls Horrocks 'Reverend'. I have looked carefully through the Bodleian Library copy, Ashmole 307, and found no reference to Horrocks whatsoever, though 'the Ingenious Shakerley' is named, sig. c₃ v.

(87) Edward Shereburne, *The Sphere of Marcus Manilus* (London, 1674). See his 'Catalogue of Famous Astronomers', p. 92. In his edition of *The Gresham College Lectures*, the late Eric Forbes speaks of 'the Rev. Jeremiah Horrox', p. 51, though as far as I can tell, this ascription is not backed up by Flamsteed in his text.

(88) *Constitutions and Canons Ecclesiastical, 1604*, ed. J.V.Bullard (London, 1934), Canon XXXIV, p. 38. I am indebted to the Reverend Dr Geoffrey Rowell, Fellow Chaplain, Keble College, Oxford, for advice and information pertaining to the ordination of seventeenth century Anglican clergy. Also, see the *Oxford Dictionary of the Christian Church*, ed. F.L.Cross (Oxford, 1961), under 'Age Canonical' p. 25.

(89) The Hoole letters cover the period 1639 June 8 to 1640 April 20. His Hoole 'Observationes', *Op. Post.*, pp. 390–398, cover the dates 1639 June 25 to 1640 June 29.

(90) W.F.Bushell, in 'The Keats of English astronomy', *Mathematical Gazette* (1959) *op. cit.* (8), p. 13, states that Robert Fogg was mentioned as Curate of Hoole in 1632 and 1639, and Rector in 1641, but fails to cite a source. I have been unable to find any trace of Fogg

or Horrocks in Diocesan and other records, though these are far from complete for the pre-Civil War period. See also correspondence between myself and the Lancashire County Council Archivist, County Record Office, Preston, January–February, 1990. Fogg is mentioned as the first Incumbent of Hoole, when it was created as a new parish from Croston, in 1641; See J.E.Bailey, 'Jeremiah Horrox ...', *Palatine Note-Book*, III (1883), *op. cit.* (1), p. 258.

(91) Horrocks's friend, John Wallis, was ordained in his twenty-fourth year in 1641, though Worthington did not take Deacon's Orders until he was twenty-eight years old; *D. N. B.*

(92) James Gregory to John Collins, 1673 March 7, in Rigaud, *Correspondence of Scientific Men*, II, *op. cit.* (44), p. 248.

VI

JEREMY SHAKERLEY (1626-1655?) ASTRONOMY, ASTROLOGY AND PATRONAGE IN CIVIL WAR LANCASHIRE

The civil war period witnessed a remarkable activity in the pursuit of astronomy and allied subjects in the northern counties of England, and well over half a dozen mathematical practitioners were active between 1635 and 1650. Perhaps the best known of these men was Jeremiah Horrocks and his circle, including William Gascoigne and William Crabtree who were active around 1640, and made contributions of international importance in celestial mechanics and instrument design.[1] Though working some years later, Jeremy Shakerley was deeply influence by the work of Horrocks, and in many ways, saw himself as continuing in the same tradition. While Shakerley worked in greater isolation in many respects, he did maintain an active London correspondence, and often made reference to fellow astronomers in the Pendle district, where he originally resided. Shakerely's historical importance lies in the nine substantial letters which he exchanged with the London astrologer, William Lilly between 1648 and 1650, along with others to Henry Osborne and John Matteson. Most of these letters, now preserved in the Ashmole manuscripts in the Bodleian Library, are rich in information about the aspirations and problems of a provincial mathematical practitioner. He was an admirer of the theories of Copernicus and Kepler and argued for strictly natural causes in celestial phenomena. Yet he also perceived a hierarchy of correspondences and astrological demonstrations behind the physical laws, whereby man could interpret God's design, as a guide to conduct.

Even a couple of decades after his death around 1655, little was remembered of the details of Shakerley's life, and when the Dictionary of National Biography article was written, it was based on material contained in his three published works, and could provide no dates for either his birth or death. In a letter to Lilly, Shakerley stated that he was born in North Owram, Halifax, Yorkshire in November, 1626,

© 'Jeremy Shakerley (1626–1655?) Astronomy, Astrology and Patronage in Civil War Lancashire', *Transactions of the Historic Society of Lancashire and Cheshire*, vol. CXXXV, 1985 (Liverpool, 1986), pp. 1–14.

and proudly pointed out that this was the 'same parish wherein John de Sacrobosco was'. At the age of twelve he was taken to Ireland, where he remained until the outbreak of the Rebellion, presumably in 1642, after which he returned to England. Around the age of twenty he says, 'I gave my mind to the mathematicks',[2] but as he was then living in Pendle Forest, bewailed that he was in a 'country where good words are as rare as good wits and ignorance seales the lips of compliment'.[3]

Shakerley's letters to Lilly are mainly concerned with astrology and patronage, as well as containing a good deal of autobiographical material. In the pre-1650 phase of his career, he was not so much interested in the physical phenomenon of the heavens as what they portended. This theme is very much mixed in with demands for help and encouragement, for as he laments to Lilly, 'Astronomie is not in league with fortune' and so far, his mathematical activities had only brought him frustration. He emphasised to Lilly that with some encouragement, he might grow to something, although at present, 'necessity is my jailer'.[4] Shakerley was an ambitious man, impatient for advancement, and his letters are full of flatteries and fawning obeisances as he tried every device he could think of to gain crucial metropolitan patronage from Lilly. By 1649, he had entered the protection of Christopher Towneley, for by February of that year, he was using Carre Hall as his address, and claiming to live with Mr Towneley.[5] But this arrangement was still unsatisfactory, for in the same letter he ungratefully complains 'meate and drink is all I can expect' and 'you may justly imagine that I have to desire a better fortune'. After relating some derogatory remarks about Towneley, he continued to fawn to Lilly 'for in you the deadness of my hope do live'.[6] Although Lilly never extended to Shakerley the formal patronage which he craved, he does seem to have been his main supplier of books, for on one occasion, he was urged to forward on works by Tycho Brahe and Gassendi which do not seem to have been available at Carre Hall.[7] In the long run, however, it is clear that Shakerley was no more satisfied with Lilly than he had been with Towneley, for he makes reference to other potential patrons in his letters, including the Parliamentarian Major General Lambert. Shakerley claimed that one of his mathematical friends in Pendle, Mr Nathan Pigholls, had an attorney brother in London who was trying to obtain Lambert's patronage through Parliamentary connections.[8] Shakerley left no stone unturned in his search for influential support, and was willing to cast over all shades of religious and political opinion to get it, for Lilly and Lambert obviously occupied diametrically opposite positions from the Catholic, Royalist Towneleys.

At the outset of their correspondence, William Lilly must have seemed highly promising as a potential patron to Shakerley, for the

London astrologer was reaching his zenith of influence as the oracle of the Parliamentary side in the civil war.[9] More significant intellectually, he was involved in the attempted restoration of astrology from corrupt charlatans and other base 'nymphes who . . . prostitute themselves to vulgar suitors' and bring disrepute upon the art. At this period too, Shakerley was much concerned with elevating the status of astrology as a scholarly discipline, along with clarifying the points at which it was at variance with astronomy proper.[10]

One important aspect of Shakerley's career, as revealed in his correspondence, is the increasing importance which he came to attribute to new knowledge. From one of his earliest letters to Lilly, one finds that he was surprisingly timid in his approach to certain forms of new knowledge, which in itself, was a world removed from the fierce iconoclasm of Jeremiah Horrocks. Shakerley described how, on one occasion when he was still living in Pendle Forest, he beheld the rare meteorological sight of two mock suns, one on each side of the true sun, about 10° apart. The mock suns were joined by a brilliant rainbow of red and green, a drawing of which he included in the letter to Lilly. Instead of revelling in the novelty of the event, as Horrocks would most probably have done, Shakerley confessed that he preferred to learn of such things from the writings of the ancient authors, for he clearly felt reservations about what they could imply. Of particular interest was Shakerley's explanation for the phenomenon, which he nonetheless regarded as quite natural. He told Lilly that the local people had been full of an 'abundance of terror and amazement' at the occurrence, but when they asked him what it was, his explanation was physical and straightforward. The mock suns, he told them, were occasioned by 'the brightness of the sun's beames from a cloud reflected on the earth', and similar events had been recorded in both ancient and modern times alike. Natural as the mock suns may have been, though, he still implored Lilly to pronounce 'astrological judgement' upon them, to reveal a significance that may have gone beyond common cause and effect.[11]

By 1649, however, Shakerley was coming to entertain serious doubts about the validity of astrological predictions, though a close reading of his remarks indicates not so much a scepticism about astrology as a science, but rather a dismay about the debased state into which it was thought to have fallen. He related to Lilly that he felt uncertain about some aspects of it, for whilst 'many things may be deduced to confirm the verity of astrology . . . I sometimes lament the manifold imperfections of the art'.[12] His growing doubts were also discussed with his friend John Matteson, as when in March 1649, he wrote 'yet if there be any help from astrology (which I will not deny maybe) I believe that not only the art, but even the key to the art is locked up. Astrology consists of too much uncertainty to inform us of anything therein'. Later in the same letter, Shakerley expressed his intention to deduce a more perfect astrology from philosophical principles.[13] Indeed, his doubts about astrology seem to have been crystallising so rapidly that he even feared that much of what would appear in his 1650 *Almanack* would be conjectural. He promised to discuss his doubts with Lilly, though unfortunately, this letter does not seem to have survived.[14]

During the first two or three years of his career, Shakerley does not seem to have been interested in practical observation, for nowhere in his correspondence does he make reference to detailed personal observation of the heavens. To some extent, one can also infer this reluctance to observe nature direct in his comments about the mock suns. Most of his data, instead of being derived from the heavens direct, was obtained from published astronomical tables such as those of Johannes Kepler. Quite probably, in his early days, he was not attracted to the meticulous and delicate cumulative work of the observational astronomer. Like most astrologers and almanack makers, he had a detailed knowledge of the ephemeris tables and used them as a basis for the three published works he was to issue between 1649 and 1653. He mentions by name the tables of Lansberg, Longomontanus, Argoll, Kepler and the *Prutenic* tables of Erasmus Reinhold.[15]

Shakerley's astrological doubts date from 1649, the same year that he came to reside at Carre Hall with Towneley. From the same year, one also sees the genesis of several other new currents in Shakerley's thought, and a broadening of his natural curiosity. His Copernicanism became more strongly defined, he started to make original observations of the heavens, and grew ever more abusive towards conservative astronomers and astrologers who derived all of their conclusions from printed tables. Each of these causes had also been important elements in the thought of Jeremiah Horrocks, ten years before and, following Horrocks' death in 1641, his papers had been in the hands of Christopher Towneley.[16] Considering the enormous reverence in which Shakerley came to hold Horrocks, it seems plausible to argue that his major shift in intellectual stance had something to do with him reading the Horrocks papers at Carre Hall.

In each of his three published works, Shakerley was to express his indebtedness to Horrocks in no uncertain terms. In his *Anatomy of Urania Practice* (1649), where he attacked Vincent Wing's treatment of the lunar theory, he stated that with the exceptions of Kepler and Bullialdus (Ismael Boulliau) few others had understood the lunar theory so well 'as that Noble Genius, our worth countryman, Master Jeremy Horrox (sic) from whose remains I have gathered most of what I shall write in this chapter'.[17] Similarly, he acknowledged his debt to Horrocks in both his *Almanack* for 1651 and also in the introduction to his *Tabulae Britannicae* (1653).

Shakerley's first published work, the *Anatomy*, or dissection of Vincent Wing's *Urania Practica* (1649), was probably written at Carre Hall soon after the appearance of the book itself, with the impact of the Horrocks papers still fresh in his mind. Wing's *Urania*[18] was a good and comprehensive astronomical textbook in English, dealing with the necessary mechanics of eclipse prediction, the use of tables and astrological computation. It did, however, contain certain assumptions which provided Shakerley with a golden opportunity to fulfil his desire, confessed to Lilly, to 'perform something against Wing'.[19] The *Urania* sometimes implied a fixed earth cosmology, while Wing was accused of being imprecise in some of his definitions. One cannot help feeling, though, that the book was, on the whole, a sound guide to the English-reading amateur astronomers for whom it was intended, and that Shakerley's *Anatomy* of it was over zealous and excessively polemical. What Shakerley found in Wing's *Urania Practica* was a convenient platform from which to trumpet his new ideas and ridicule conservative astronomers:

> This last age . . . doth enjoy the benefit of more admirable and useful Inventions than any or almost all before it . . . indeed, what shall we mortals now despair of? within what bounds shall our wits be contained? We have seen the spots of the Sun . . . the horns of Venus . . . the mountains and seas of the Moon . . . the generation of comets. O heaven and stars! how much hath our age triumphed over you! Neither doth our victory end here, still new miracles adde to the numbers of the old, and no day passeth without a triumph.[20]

Wing took Shakerley's broadside with surprising good humour, later remarking that 'there is in it more malice than matter', although it did not prevent him from composing a suitable retort entitled *Ens fictum Shakerleii, or the annihilation of Mr. Jeremie Shakerley* (1649).[21]

It was most probably the anti-Wing tract which finally severed the relationship between Shakerley and Lilly, for a letter was obviously sent from the London astrologer to inform him that further correspondence was no longer welcome. In his best self-abasing style, Shakerley wrote back, 'Now that you have rid yourself of mee, you no longer need be separated from the friendship of Mr Wing . . .', but there were clearly other factors involved in terminating the relationship. Ostensibly, Lilly had accused Shakerley of plagiarising the astronomical tables of another Pendle Forest mathematician, Jonas Moore, although one can see how the two men were growing apart in basic interests.[22] Shakerley was abrasively challenging the very precepts upon which Lilly's reputation stood when he argued that astrology was uncertain, while his solicitations to a much higher ranking Parliamentarian, General Lambert, must have annoyed Lilly's sense of pride as a patron.[23] Wing, on the other hand, was a sound astrologer and much more loyal to his protectors.

6

Around the same time as he was producing the *Anatomy of Urania Practica* (1649), Shakerley was also working on his *Almanack*. Several references to almanack compilation occur in his letters, though only one such publication can be unequivocally attributed to him. *Synopsis Compendiana; or, a brief description of the yeer of humane redemption MDCLI*, was an almanack of some significance, for in the concluding section of the work appeared 'A short astronomical discourse — De Mercurio in sole vivendo', in which he announced his calculation of the forth-coming transit of mercury across the Sun's disk for October 24th, 1651. Only one Mercury transit hd been calculated and observed to date, when Pierre Gassendi described that of 1631. While he was aware that Mercury transits were of less scientific significance than those of Venus, he realised that they could still be used to rectify knowledge of the planet's orbit. By observing Mercury on the Sun's disk in transit, astronomers could accurately define the point of intersection between the planet's orbit and that of the Sun — the node of the orbit — better than by any other method. Also, when cast against the Sun's light in transit and relieved of that glare which made it appear excessively large in the night sky, it was possible to measure Mercury's correct angular diameter, as a proportion of the known diameter of the Sun.[24] Throughout this section, Shakerley borrows considerably, and with due acknowledgement, from the published account of Gassendi's observation of 1631. More important, in terms of his own intellectual development, was his familiarity with the observation of the transit of Venus made by Jeremiah Horrocks in 1639. Horrocks, unlike Gassendi, never lived to publish the observations and conclusions obtained from the Venus transit, and they had to wait some twenty years after Horrocks's death, before being printed by Johannes Hevelius in Danzig in 1661. Shakerley most probably used a manuscript copy of what would later become *Venus in sub sole visa* that was in Towneley custody at Carre Hall.[25]

As in his attack on Vincent Wing, Shakerley lost no opportunity to use the *Synopsis* to wax lyrical on the achievements of the new science and pour scorn on more traditional practitioners. Attacking those mathematicians who did not accept the doctrine of elliptical orbits, he proclaimed himself a follower of Kepler. Then he took several Continental astronomers to task, particularly Longomontanus and Keckermann for not believing Mercury to be a solid body but rather 'subtile and rare' and hence unable to stand out in relief as a dark body when in transit across the sun.

What one finds, indeed, in both the *Anatomy of 'Urania Practica'* and the *Synopsis* for 1651 is the heady enthusiasm of a man whose whole thought had recently undergone drastic revision and had come to be re-established upon new and exhilarating foundations. The sheer pugnacity with which he was not attacking more conservative astronomers and defending the new science of observation and enquiry was very reminiscent of the language of Jeremiah Horrocks. Shakerley's recommended method of observing the transit was similar to that used by Horrocks, and consisted in projecting an image of the Sun with a telescope into a darkened room, so that the solar image and the transitting Mercury would exactly fill a circle divided into 360° from which measurements could be made. But all of this description was something of an anticlimax, for he goes on to state, 'we in England shall in vain expect anything of this appearance', for it will only be possible to observe the transit to any effect in China, the Antipodes or other Eastern lands. *Synopsis* confirms the impact which the Horrocks papers, especially the unpublished Venus transit observations, must have had on Shakerley. It also explains why Shakerley made such a fuss about the impending Mercury transit, for he beheld in it an opportunity to emulate his new hero and observe the passage of a planet across the solar disk. One can also understand why he was falling increasingly out of joint with William Lilly. For whatever reason, one must assume that Shakerley's fortunes failed to im'prove at home, especially after his estrangement from Lilly. By 1651 he was bound for India. No record of his passage seems to have survived, nor the capacity in which he sailed.[26]

The brief Indian phase of his career is most provocative, for in his one long letter to Henry Osborne written in 1653, one sees the emergence of a new intellectual confidence and curiosity, rooted in a firm scientific grasp, which had been absent or at best unformed in his Lancashire writings. He informed Osborne that he was already in India by the Autumn of 1651, for on October 24th that year he observed the much awaited transit of mercury in Surat.[27] Shakerley's observation of this event was probably the first planned telescopic astronomical observation to be made in India, and after Gassendi, he was probably the second man anywhere to see Mercury in transit across the sun. The observation was made under difficult conditions 'in which my chiefest infelicity was want of intruments, for I had only a small and little better than a half-crown perspective'. The technique employed to observe the transit was the one recommended in his *Synopsis* and which had also been used by Horrocks to view Venus in 1639 — by projection onto a graduated sheet of paper. He goes on to say that the event began at 6.40 a.m., but the room in which he was obliged to observe was incommodious, being 'encumbered with the neighbourhood of other buildings', and afforded him only one single sighting of the disk. Inset into the letter is a drawing of the solar projection, upon which the planet's position is noted. Mercury appeared as a tiny round dot, 'the colour of a brownish black and the diameter very small, as I am confident that it did not exceed half a minute'. Whether Shakerley employed the transit observation to derive the orbital elements of Mercury, as suggested in the *Synopsis*, was not stated.

The Osborne letter was clearly intended as an interim report on his astronomical work in India, and not only implied that more information would be forthcoming, but stated that two previous letters had already been sent. One of these had been entrusted to one Henry Wentworth sailing on the *Eagle*, while another had been sent via Aleppo. Both of these letters had contained astronomical information and Shakerley was interested in knowing whether they ever reached their destination. It is unlikely that they did so, for when Vincent Wing provided an account of the Mercury transit in his posthumously published *Astronomia Britannica* (1669), he gave details very similar to those contained in the Osborne letter, for we know that copies of this document were made and circulated by John Booker.[28]

In spite of the *Urania Practica* conflict in 1649, Wing soon forgave his rival and not only gave the transit observation a detailed treatment in his book, but had already in 1656 spoken generously of Shakerley's *Tabulae Britannicae*.[29] Although Wing had claimed Shakerley's Indian journey to have been scientific in its motivation, one suspects that at best, science was only one of a complex of motives which encouraged him to leave a home where his prospects seem to have been bleak. From the account of his transit observation, it is clear that Shakerley was not in a position of leisure in Surat, for he was only able to snatch one relatively fast sighting of Mercury. Nor does his lack of adequate instruments or cramped place of observation suggest a deliberately planned expedition. He writes of spare hours only devoted to mathematics. Most likely, he had sailed out to India in the hope of making his fortune, and took what time he could to observe the transit. It is interesting to note, however, that half-crown telescopes seem to have been popular amongst impoverished astronomers, for Jeremiah Horrocks observed the 1639 Venus transit with an instrument costing the same amount. Nevertheless, we can form an opinion of Shakerley's accuracy as an observer, and of his own objective assessment of his work from the same letter to Osborne, January 15th, 1653. Shakerley related that on September 7th, 1652, he observed a lunar eclipse, which started when the Moon stood at 50° 45' altitude, the duration of which he timed and measured. To make this observation, he also determined the latitude of Surat 'which I have observed as 21 degr. and betwixt 10' and 15' north, but my instrument is so rude, I dare not trust it to a nearer determination'. The modern determination for the latitude of Surat is 21° 12'.

The sophisticated astronomical knowledge of the Brahmins greatly fascinated Shakerley, for 'this I will say of them, that they predicted the last lunar eclipse . . . with . . . little or no errour'. It was his intention to examine 'the records of astronomy as I shall find amongst the Indian libraries which are very compleat in their own way and contain a great variety of learning' and he became acquainted with the Brahmin astronomers, though he needed 'a better interpreter than myself' to understand them fully. The learning of these sages so impressed him that by way of innuendo against the European practitioners he stated that 'I believe I should procure from them something more satisfactory than from many others who wear better clothes than they'. Then he related his intention to 'pick up something out of their Star-Book or *Juttack*', which they revered as much as their works of Divinity and were written on plaintain leaves. The Indian calendar aroused Shakerley's curiosity, for he told Osborne that they divided their day into thirty equal parts, or *garees*, and their night hours likewise, similar — he claimed — to the planetary hours in astrology. Indian years were a mixture of Lunar and Solar reckoning, being 360 days long, to which they added thirty days every three years, although he stated that it would have been more accurate had they added thirty days every six years after the manner attributed to the ancient Egyptians.

India was stimulating to Shakerley, and conducive to him making his own independent observations, for on December 8th, 1652, he recorded the appearance of a comet. It was 'first observed to blaze amongst the stars of Procyon in the lesser dog, with a quick and lively flame, the tayle of which extended itself northwards to the quantity of three or four degrees', though this soon diminished as it grew fainter, eventually disappearing in the constellation of Cygnus. Although it was his 'misfortune not to have instruments to observe to any purpose' he was able to determine that the comet's head never exceeded 5' of arc in diameter. Shakerley also enquired of Osborne whether the comet had been visible in England, and what predictions European astrologers may have drawn from it. Yet by 1653, his comments about astrology were much more guarded and sceptical, and his only other astrological query in the Osborne letter was to discuss its use in weather prediction. Shakerley pointed out that maximum rainfall in India came in summer, contrary to the astrological writer, Sir Christopher Heydon, who said that it should have come in winter.[30] In India, Shakerley was eager to observe the natural world, and his fellow men direct, and there is no regret that such things could not be found in ancient histories. All that he regretted now was a lack of adequate instruments and research facilities to discover new knowledge and win fame by communicating it to the world. His journey to India had become an intellectual as well as a physical adventure, and he had exchanged the persona of the cautious astrologer for that of the self-conscious discoverer.

With the exception of a short note to Towneley on January 14th, 1653,[31] the letter to Osborne was the only Shakerley piece from India either to reach its destination or to be taken notice of. There must have been many others written, however, for in addition to the letters to Osborne which got lost, he had a book brought out in

London in 1653 which must have occasioned correspondence. This was his *Tabulae Britannicae* (1653), probably seen through the press by 'R and W. Leybourne, for Robert Boydell neere the Tower', and calculated from the tables of Kepler Bullialdus *and* Jeremiah Horrocks.[32] Although the tables of Kepler and Bullialdus were available in published form, from which Shakerley could compute the new values for his *Tabulae*, nothing by Horrocks was yet available. Shakerley must, therefore, have made transcriptions from the Horrocks manuscripts in his days at Carre Hall, and taken them with him to India. *Tabulae Britannicae* was certainly known to the London astrologer John Booker by 1655, for in that year he wrote to Shakerley in India, a copy of the letter surviving in the Ashmole papers.[33] Booker, who was an associate of William Lilly, had accidentally come into possession of the letter which Shakerley had sent to Osborne. Incidentally, Shakerley's 'reply' and notes written on a separate sheet, casts an interesting insight into the problems of international communication in the 1650's.[34] Booker was so impressed by the original observations of Shakerley, recounted in the waylaid letter to Osborne, that he distributed copies of it to his friends, all of whom were interested in the discoveries. One might indeed speculate how many other letters Shakerley sent from India, only to be thrown into a London fireplace by a frustrated messenger!

It is not known whether Shakerley ever received the original of Booker's 1655 letter, or indeed, if he was still alive when it was written, for nothing more is known of Shakerley after he wrote to Osborne in January 1653. It is unlikely that Vincent Wing would have used a sixteen year old letter as his latest source on Shakerley in *Astronomia Britannia* (1669), had the astronomer returned to England. Edward Shereburne, when he wrote the short biography of Shakerley in his Catalogue of British Astronomers in 1674 had nothing further to add beyond that he had 'dyed in the East Indies'.[35]

Shakerley's intellectually productive career of five years, 1648 to 1653, was of precisely the same duration as Jeremiah Horrocks's a decade before, although in many ways, Shakerley was a more dynamic and interesting figure. While Horrocks's letters to Crabtree from 1636 contain a vigour and clarity of thought to which even Newton paid homage, they are less interesting insofar as they reveal a mind which was already formed and to which development was a matter of progressive growth. In Horrocks's letters there is no quest for patronage, no self-doubt, no astrology and after leaving Cambridge in 1635, they represent a man whose whole life and work took place within a few square miles of Lancashire countryside. Shakerley, however, was a very different figure from the astronomer he admired. He was restless, socially ambitious and willing to court Parliamentarian astrologers, Catholic squires and Puritan generals if they could help him. Either his intellectual curiosity, urge for distinction, or both, eventually drove him to India, from which he never returned. But most important, we see in Shakerley the evolution of a scientific mind of remarkable potential. After first learning how to calculate horoscopes, he came to doubt astrology. In its place, he became a staunch supporter of the new science, began to make observations and defend his position with all the zeal of a convert, and once in India, became fascinated by the mathematics and culture of the Brahmins.

Shakerley's career also tells us something about the astrological community of the time, its lines of patronage and rules of controversy. Moreover, one must recognise that, in spite of the seemingly vitriolic *Anatomy of 'Urania Practica'*, Vincent Wing was in Shakerley's debt for weaning him away from Ptolemaic to Copernican astronomy at the outset of his career before this 'archaism' could have attracted heavier fire than that first provided by Shakerley. Wing not only forgave Shakerley his insults, but came to develop an admiration for him which he indicated in several subsequent books. Perhaps most significant, Shakerley was the first man to 'discover' Jeremiah Horrocks, and announce his work to the world over a decade before it was formally recognised in London and Danzig. As one of the most vociferous and restless scientific Englishmen of his day, Shakerley casts a great deal of light onto the circle of men who pursued the new science in Lancashire, with regards to both the early Horrocks circle, and his own associates such as John Stephenson, Pignolls, Matteson and Towneley. But it was not until another member of this circle, Jonas Moore, who survived into the Restoration to make an independent London debut, that their work came before the Royal Society, and that its significance came to be recognised. Horrocks and his circle enjoyed the posthumous honour of having their papers published by the Royal Society, became the subject of correspondence in *Philosophical Transactions*, and had their observations printed at the opening of the Astronomer Royal's Catalogue of British Astronomy. At the Restoration, Moore rose to high office under King Charles II, won a knighthood, and not only became initial patron of John Flamsteed, the first Astronomer Royal in 1675, but was the effective founder of the Royal Observatory, Greenwich.[36] By this time, however, Jeremy Shakerley had long since disappeared into the East, in pursuit of that fame and fortune which he was never to find.

NOTES

1   A. B. Whatton, *Memoir of the life and labours of the Reverand Jeremiah Horrocks*, published with the English translation of Horrocks's *Venus in transit across the Sun* (London, 1859). Also, Allan Chapman, *Three North Country Astronomers* (Manchester, 1982).

12

2   Bodleian Library, Ashmole MS 423, 117r, Shakerley to William Lilly, 10th Feb 1649. The date given by Shakerley for his birth is confirmed by the Yorkshire Diocesan Parish Register transcripts, Borthwick Institute, York, thus, — 'Jeremie filius Willm Shakerley North', (i.e. North Owram, Halifax) November 1626. The association of the eminent 13th century mathematician Sacrobosco, or John of Holywood, with Halifax, is wholly apochryphal.

3   S. to Lilly, 26th Jan. 1648. Ashm. MS. 423, 114r.

4   S. to L. 26 Jan. 1648; S. to L. 4/6 Mar. 1648, Ashm. MS. 423, 114r, 116r.

5   S. to L. 10th Feb. 1649, Ashm. MS. 423, f117r.

6   Shakerley makes derogatory remarks about Towneley in several letters to Lilly; 10th Feb. 1649 Ibid., about his poor maintenance. In 3rd Apr. 1649, Ashm. MS. 423, 121r, he ridicules his mathematical enthusiams. On 26th May, 1649, Ashm. MS. 423, 124r, he askes Lilly not to inform Towneley of his (Shakerley's) tables, 'for he is Moorish'. Nowhere does Shakerley specify the Mr. Towneley about whom he was speaking. Almost certainly, it was the antiquarian and astrologer, Christopher Towneley (1604-1674).

7   Shakerley asked for books from Lilly in several letters, those of 10th Feb. 1649 and 26th May, 1649, cited in note 6.

8   S. to L. 10th Feb. 1649 for discussion of the Pigholls connection. By 8th Mar. 1649, Ashm. MS. 423, 119r, he is speaking of 'My worthy patron Major Generall Lambert' though he does not specify the nature or extent of his support. Shakerley also mentioned another mathematical friend in Pendle, Mr. John Stephenson, who also desired Lilly's support, 26th Jan. 1648. Ashm. MS. A23, 114r.

9   William Lilly, *Mr. William Lilly's history of his life and times* (London, 1715), see also, Derek Parker, *Familiar to all* (London, 1975), for a modern study of Lilly's career and political influence.

10 Lilly, *History of his life*, makes reference to the dishonest characters and practices of contemporary astrology. Also, S. to L. 28th Apr. 1649, Ashm. MS. 423, 123r.

11 S. to L. 4/6th Mar. 1648, Ashm. MS. 413, f116r.

12 S. to L. 28th Apr. 1649; 25th Jun. 1649, Ashm. MS. 423, 123r, 126r.

13 S. to John Matteson, 5th Mar. 1649, Historical Manuscripts Commission, *Various Collections*, VIII (1913) p. 61.

14 It is uncertain which Almanack Shakerley may have been referring to here. He had expressed intention to Lilly to write an Almanack on 4/6th Mar. 1648, Ashm. MS. 423, 116r, and produced an elegant manuscript dated 'Sept. 9 1648' entitled 'Anni a Nato Christo 1649, Synopsis Compendiania, Or, an Almanack for the yeare of Humane Redemption 1649'. This work was probably sent to Lilly, though it does not appear to have been published, and is now in Ashm. MS. 333, 166-192. On 28th Apr. 1649, Ashm. MS. 423, 123r, he again wrote to Lilly 'I send you also herewith my Almanack for 1650', though this likewise probably failed to secure a publisher, while the manuscript seems to be lost. Not until 1651 did Shakerley produce an Almanack which got published.

15 S. to L. 26th Jan. 1648, Ashm. MS. 423, 114r.

16 John Flamsteed to William Molyneux, 10th May, 1690, Southampton Civic Record Office, d.D/M. I/I. ff. 142 r.v. In this letter, Flamsteed stated that after the deaths of Horrocks and Crabtree, their papers were obtained by Christopher Towneley. Flamsteed also examined and transcribed Horrocks, Crabtree and Gascoigne material, as well as seeing the remains of Gascoigne's sextant, when staying with Richard Towneley (Christopher's nephew) in 1672; F. Baily, *Account of the Revd John Flamsteed* (London, 1835) pp. 30-32.

17 Jeremy Shakerley, *The Anatomy of 'Urania Practica'* (London, 1649) p. 18.

18 Vincent Wing, *Urania Practica, or practical astronomy in VI parts* (London, 1649).

19 S. to L. 10th Feb. 1649, Ash. MS. 423, 117r.

20 Shakerley, *Anatomy*, 'To the Mathematicall Reader'. It is quite likely that at this stage in his career, Wing held to the fixed earth doctrines as put forth by Ptolemy and refined by Tycho Brahe, and certainly implied such a cosmology in his unpublished manuscript almanack for 1641 in Ashmole MS. 190, 90ff. In the 'Mathematicall Reader' of his *Anatomy*, Shakerley openly attacked Wing's apparent earth-centred cosmology when he stated, 'Ptolemy, that founded his hypothesis upon observation, would not be angry if our observations perswade us to another hypothesis than he hath constituted'.

21 Wing to Lilly, 28th Jul. 1650, Ashm. MS. 243, 174r.

22 S. to L. 25th Mar. 1650, Ashm. MS. 423, 128r.

23 Amongst these slights, Shakerley dedicated the *Anatomy* to General Lambert, rather than Lilly.

24 The *D.N.B.* article on Shakerley states that no copies of the Mercury 'Discourse' are known to exist. In fact, one is bound into the Bodleian Library copy of Shakerley's *Synopsis Compendiana; or a brief description of the yeer of humane redemption MDCLI* (1651).

25 Horrocks's work on the transit of Venus was published posthumously by Johannes Hevelius in a Latin translation, *Venus in sub sole visa*, included as a supplement to Hevelius's own *Mercurius in sole visus anno 1661* (Danzig, 1662). It was translated into English by A. B. Whatton in 1859 (see note 1), although Horrocks's original Ms. is lost. Shakerley first made reference to Horrocks observing the Venus transit in his unpublished 'Almanack' for 1649, in Ashmole, 333, which he dated Sept. 9, 1648. In the section entitled 'System of the world' he gave full credit to Horrocks, although he does not describe the technique or conclusions of the observation. This dates, however, from before he took up residence at Carre Hall with Towneley, and probably knew of Horrocks only by repute, and had not yet studied his papers.

26 E. Sainsbury and W. Foster, *The English factories in India 1651-1654* (Oxford, 1913). A search for Shakerley material in the India Office Library and Commonwealth Office has been made without success. Shakerley's name does not appear on any official reports, servant or passenger lists for the India sailings for 1650-51.

27 S. to Henry Osborne, 15th Jan. 1651, Ashm. MS. 242, 95 r.v. This letter, from Surat, is around 1500 words long.

28 V. Wing, *Astronomia Britannica* (London, 1669) p. 312. The account of the transit *via* Wing was also cited by George F. Chambers, *A handbook of descriptive and practical astronomy* (Oxford, 1889) p. 341.

29 V. Wing, *Astronomia Instaurate*, (London, 1656), 'To the Reader'. See also Wing to Lilly, 28th Jul. 1650, where Wing says of Shakerley's reaction to *Ens fictum Shakerleii*, 'but I hope his ingenuity is such hee took it not unkindly . . . I wish him well'.

30 Christopher Heydon *An astrological discourse* (1608) (London, 1650), pp. 45-47. He cites Josephus Acosta as his source.

31 S. to Christopher Towneley, 14th Jan. 1653, Ashm. MS. 242, 94b, from Surat.

32 J. Shakerley, *Tabulae Britannicae*, (London, 1653). See p. 92 for the Horrocs acknowledgement. The Leybourne brothers Robert and William produced their own works on surveying and practical mathematics, and collaborated with Vincent Wing over his books. I have been unable to find any correspondence between Shakerley in Surat, and these three, or Boydell, in London.

33 John Booker to S. Dated in astrological cypher 'Sol intrante Aries' (21st Mar.) 1655, Ashm. MS. 242 93r.v. In Ashmole 348, 53b, there are several rough notes in John Booker's (?) hand, referring to Shakerley's calculation of the Sun's entry into the Cardinal signs in his *Synopsis*, or Almanack for 1651.

14

On the same sheet, Booker (?) compares the same calculation for the astronomers Eichstade, Kepler, Bullialdus *and* Horrocks (Horrox). The notes are not dated, but are adjacent to horoscope drafts for 1659.

34 Shakerley entrusted the delivery of the Osborne letter to one William Dynes, steward of the *Smirna Merchant*, but on arriving in London he was unable to trace the intended recipient. In a fit of anger in a London alehouse, Dynes swore to destroy the undeliverable letter, but happily, a friend of Booker's happened to be present and probably upon hearing either of the names 'Shakerley' or 'Osborne' saved the letter from destruction, and conveyed it to the astrologer. The whole story is unfolded by Booker in his long letter of reply to Shakerley, cited in note 33.

35 Edward Shereburne, *The sphere of Marcus Manilus* . . . (London, 1674), see the 'Catalogue of famous astronomers'.

36 John Flamsteed *Historia Coelestis Britannica*, I (London, 1725), pp. 1-5. Also, *The 'Preface' to J. Flamsteed's 'Historia Coelestis Britannica' 1725*, ed. A. Chapman, (National Marit. Mus. Monogr. No. 52, 1982) pp. 1, 111, 113. Jeremiah Horrox (sic) *Opera Posthuma* (London, 1673). See also, *Philosophical Transactions* for 1667, 1711, 1718, etc.

VII

The Design and Accuracy of Some Observatory Instruments of the Seventeenth Century

Summary

The graduated arcs of some seventeenth and early eighteenth-century observatory instruments have been examined in order to estimate the accuracy of the angular divisions. In addition, the design of the frameworks supporting the graduated arcs has been studied from existing instruments and from contemporary engravings. The analysis attempts to assess the skills of the craftsman rather than the perspicacity of the astronomer.

Contents

1. Dividing the arc

Relatively little information is available to the historian who attempts to establish the accuracy which astronomers could work to in the past. There is no shortage of claims by the scientists themselves as to the accuracies which they believed they could achieve, but these are often doubtful, and are difficult to verify. Instrument scales were generally read at face value, so that if an astronomer possessed a quadrant graduated down to 10 seconds of arc, he generally assumed that he could make reliable observations to that quantity, irrespective of other features in the quadrant's construction that may have introduced larger errors. It is not uncommon to find astronomers of the period claiming accuracies that are now known to have been totally outside the technical resources of the period. Because the rate of astronomical advancement was often governed by the graduation accuracy of observatory instruments, it is important for historians to know as much as possible about the ceiling, or uppermost limit of precision, to which astronomers could measure at any one time.

Several astronomical scales of large radii survive in museum collections. Though the complete instruments of which they were once part have in many cases been severely mutilated, some of the scales themselves remain geometrically intact. An attempt was made, therefore, to examine these scales, using modern techniques of measurement and analysis, in the hope of learning something about their method of construction and working accuracy.

VII

458

Six instruments, one of which carries two scales, constructed between *c.* 1637 and *c.* 1710, and with well-divided graduations in a good state of preservation, were chosen as suitable for study. They were as follows:

No.	Instrument	Maker	Date	Location	Accession No.	Radius (inches)
1	Equatorial	Sharp	*c.*1700	Greenwich	A73-10L/00/R22	9·31
2	Quadrant	Anon.	*c.*1637	Oxford	36·3	24·12
3	Quadrant	Butterfield	1700	London	1876/1530	25·00
4	Quadrant	Sharp	*c.*1710	Greenwich	NAA5008/OM/MI	59·00
5	Sextant	Anon.	*c.*1640	Oxford	36·2	73·26
6	Quadrant	Allen	1637	Oxford	36·1	78·05

The locations are: Greenwich, National Maritime Museum; London, Science Museum; Oxford, Museum of the History of Science.

Table 1.
Instruments chosen for study.

Two instruments constructed by Abraham Sharp warranted close attention, for Sharp had been the constructor of John Flamsteed's mural arc, and it was hoped that something might be learned, by analogy, about the Royal Greenwich Observatory's first meridian instrument, which has been lost since 1722.

None of the selected instruments was, however, suitable for examination on the plate of a dividing engine, as had been the technique of examination employed elsewhere on medieval astrolabes.[1] The long radii of the observatory instruments made them too large for conventional engines, while the frail condition of three of them made it unwise to attempt removal from their present mounts. An alternative method was, therefore, adopted. If accurate measurements could be made from the linear size of each degree space upon the instrument limb, it would be possible both to compare the degrees one against the other, and to compare the same degrees with the size of an 'ideal' degree, calculated as a function of the instrument's radius.

As none of the instruments possessed clearly defined centre points, and some were so mutilated that their original centre-work was missing altogether, careful measurements were taken of the chord 0° to 60°, which was always the first chord struck off when dividing, possessing the same length as the radius itself. To be certain of this length, the chord was measured two or three times with a high-grade engineer's steel rule, after which the measures were reversed, i.e. 60° to 0°. When any disagreement occurred between consecutive measurements, the radius was cross-checked by measuring the radius 30° to 90° and its reversals. By multiplying the radius by $2\pi/360$, the exact linear size of each degree could be found, from which it was possible to express minute and second fractions in terms of thousandths of an inch.

The use of linear equivalents in the study of instrument accuracies contains several drawbacks for, unlike dividing engine measurements, there are no external cross-checks that can be applied. There is no dividing engine screw or reversible turntable to

[1] Allan Chapman, 'A Study of the Accuracy of Scale Graduations on a Group of European Astrolabes', *Annals of Science*, 40 (1983), 473–88. See also Allan Chapman, 'Dividing the Circle' (unpublished D.Phil. thesis, University of Oxford, 1976), chapter 1; and Allan Chapman, *Dividing the Circle* (Amersham, forthcoming), chapter 11.

set as an independent standard, and the necessary reliance upon the 60° chord to fix the proportions of the radius means that the comparison, or 'ideal' degree calculated from it, is itself subject to the original errors of the 60° chord. The range of information that can be obtained from the resulting analysis is, therefore, more limited than with the astrolabes. Because a comparative zero point is lacking, against which to relate all subsequent measures, it is only possible to compare one degree against another. This comparative method allows calculation of the accuracy of each scale, but makes it impossible to discover error patterns that might have revealed the geometrical construction technique. As all but one of these instruments were quadrants or sextants, it was not possible to discover symmetry patterns in different parts of the scale, as may be done when assessing full circles.

To measure the degrees on each scale, an engineer's G micrometer was used, set in a special horizontal stand, so that when it was laid down upon the limb of an instrument the anvil and stem of the micrometer both stood at an equal distance from it. In this position, they were viewed through a magnifier of two inches focus, and adjusted so that the anvil and stem enclosed exactly one degree space upon the limb, thus making it possible to read the linear size of a degree division to one-thousandth of an inch. At a radius of six feet, one-thousandth of an inch corresponds to an angle of 2·88 seconds, an angular amplitude well beyond the skill of any seventeenth-century graduator.

Particular care was taken to establish the errors of the micrometer apparatus, especially the parallax error that can occur when the fiducial edges of the anvil and stem stand about one-fifth of an inch above the scale being measured.[2]

To make the linear measurements from the scales themselves, the micrometer and its stand were placed upon the limb of the instrument under examination, and the degree measurements read off in thousandths of an inch. But on two of the instruments, No. 4, the Sharp mural quadrant, and No. 6, the Elias Allen quadrant, it was impracticable to move the instruments from their wall mounts to examine them. Instead, fine paper rubbings were made from their scales, which it was then possible to examine in a horizontal position. To ensure maximum stability, rubbings were never made from measured arcs of more than 5° at a time, so that instead of taking measurements from one large rubbing, they were taken from eighteen rubbings which together comprised 90°.

To determine whether or not the rubbings differed in accuracy from the scale itself, a set of comparison measures were made from the half-inch divisions of a modern steel rule, the surface appearance of which resembled the graduated scales. The comparisons indicated that carefully made rubbings are capable of bearing accurate measurable impressions from a graduated scale.[3] The rubbing must, however, be kept dry in a stable temperature, and must be measured soon after being made.

In an attempt to apply some external standard, measurements taken from a quadrant of the author's manufacture were included on the grounds that, with knowledge of how it had been graduated, interpretation of the error patterns displayed in the seventeenth-century instruments would be made easier. The method of analysis was the same for all the instruments—the seven early scales and that of the author—

---

[2] Details of the technique can be found in Chapman (footnote 1), pp. 144–6.
[3] Ibid.

460

and proceeded in the following manner. The quadrant No. 6, by Elias Allan is used in the present example:

Radius   = 78·05 inches
therefore   1° = 1·362 inches and 1° contains 1,362 thousandths of an inch.
As there are 3,600 seconds of arc in a degree, one thou corresponds to 2·64 seconds.

The degree sizes of this instrument were next plotted in the form of a histogram. It will be seen that the smallest degree space on the quadrant contains 1,330 thousandths of an inch, and the largest 1,382. Most of the degree spaces on this instrument are smaller than the calculated 1,362 thousandths that they should have contained as a function of the radius. The mean point falls around 1,356, indicating the arc to be 6 thousandths, or 16 seconds of arc, less than a true quadrant in its degree distribution. In practical terms, however, this error is of little consequence; it is well within the instrument's plus or minus scale accuracy, and may even be a product of damage, shrinkage of the bimetallic limb, or experimental error in making the micrometer readings.

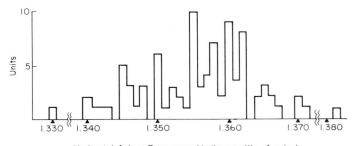

Horizontal Axis = Error spread in thousandths of an inch
Vertical Axis = Units of occurrence

Figure 1.

Distribution of the width of the one-degree divisions on the Elias Allen quadrant, No. 6 in this study. Horizontal axis: error spread in one thousandths of an inch. Vertical axis: units of occurrence.

As can be seen, the degrees cover a distribution range of 52 thousandths, whereas 71 out of the 91 fall within a range of 18 thousandths. By multiplying the angle-to-thousandths equivalent by each of these numbers in turn, one can ascertain the angular range over which the instrument can be considered accurate, thus:

$52 \times 2·64 = 137·28$ seconds (total scale error)
$18 \times 2·64 = 47·52$ seconds (71° best scale error)

Over its entire scale the Elias Allen quadrant, No. 6, is accurate to about 2' 16", which is not particularly high for an instrument of such a large radius, where direct division on the scale purports to measure down to 2 minutes and to 12 seconds by transversals. But over a range of observations restricted to the 71 more accurate, or 'best scale' degrees, reliable values down to 47 seconds would have been possible, although the early Savilian Professors who used the instrument would not have been aware of it. On the contrary, they would have confidently read the scale to the nearest 12 seconds that the

No.	Instrument	Maker	Date	Radius in inches	1° in inches	one thou. in seconds	Error over 90° Thous	Error over 90° Seconds	Average accuracy — Best Accuracy Degrees	Best Accuracy Thous	Best Accuracy Seconds
1	Equatorial	Sharp	c.1700	9·31	0·162	22·22	13	288	350	9	199·98
2a	Quadrant— inner scale	Anon.	c.1637	24·12	0·421	8·55	28	239	64	12	102·6
2b	Quadrant— construction dots	Anon.	c.1637	25·00	0·436	8·25	36	297	70	15	123
3	Quadrant	Butterfield	c.1700	25·00	0·434	8·29	11	91	73	4	33·17
4	Quadrant	Sharp	c.1710	59·00	1·03	3·49	16	55	77	8	27·92
5	Sextant	Anon.	c.1637	73·26	1·27	2·81	26	73	45	13	36·5
	Quadrant	Allen	1637	78·05	1·36	2·64	52	137	71	18	47·5
6	Quadrant	The author	1976	6·14	0·107	34·20	43	1470	63	13	444

Table 2.

A table showing the characteristics of all the instruments examined in this study. The 'Averaged Accuracy' contains (*a*) the total error over the entire instrument, expressed in thousandths of an inch, with a seconds equivalent, and (*b*) the 'Best Accuracy', or most consistent degree range in which the errors fall.

transversals purported to display, unaware that it was necessary to multiply this quantity between four and twelve times to obtain the true measure of the quadrant. Also, it is important to remember that Elias Allen was considered to be one of the best craftsmen of his day, when assessing the accuracy claims of astronomers who worked with instruments built by lesser men.

The details of all the instruments measured in this way, with their total and best scale degree ranges are given in the following table.

As one might expect from such a collection, the largest instruments were capable of measuring the smallest angles, although the progression is by no means uniform. What does become apparent is the way in which a large instrument, such as No. 6, shows a total error spread—expressed in thousandths of an inch—that is much wider than No. 3, less than one-third its size. Although the 'best scale' degrees for both instruments fall much closer together, the scale accuracy of the larger instrument is distinctly inferior. It will be noted that the author's quadrant, of 6·14 inches, displays a scale error of 24·5 minutes. This accords well with a set of reduced Pole Star observations made with the same scale, which produced an error value of 19 minutes for the instrument.

An Inter-comparison

In an attempt to compare the different instruments more evenly, all the errors were reduced to a uniform scale, in order to assess the skills of different craftsmen as if they were all graduating instruments of the same radius. Taking the author's quadrant, which has the smallest radius of all the quadrants measured, it was easy to scale the other instruments in terms of this base radius. By multiplying the 'best scale' accuracy of each instrument by the number of times that its radius exceeded the radius of the base quadrant, some estimate was obtained of the error that the craftsman would have produced had he been graduating at the base radius. For example, the Sharp quadrant, No. 4, of 59 inches radius, was found to be 9·61 times larger than the radius of the base. Multiplying by the scale error, 27·92 seconds, the result is 268 seconds. It would appear, therefore, that if Abraham Sharp had used the same geometrical techniques used on No. 4, but to graduate a scale of 6·14 inches radius, he would probably have produced a best scale error in the region of 268 seconds. This probable error of 4′ 28″, when working on such a small radius, would provide independent confirmation of Sharp's skill as a graduator.

Using this method, proportions were calculated for each of the other instruments. The results were as follows, arranged in order of increasing radius:

No.	Instrument	Radius (inches)	Multiple of 6·14 inches	Best accuracy error (seconds)	Product in seconds
	The author's	6·14	1	444·6	444
1a	Sharp equatorial	9·31	1·51	199·98	301
2a	Quadrant inner scale	24·125	3·92	102·6	399
3	Butterfield quadrant	24·90	4·05	33·17	134
2b	Quadrant outer scale	25·00	4·07	123	500
4	Sharp quadrant	59·03	9·61	27·92	268
5	Sextant	73·26	11·93	36·53	434
6	Allen quadrant	78·05	12·71	47·52	603

Table 3.
Comparative scaling errors between different instruments.

It will be noted that the graduator of the Butterfield quadrant, No. 3, was probably the most skilled craftsman in the present selection, capable of obtaining a very high scale accuracy and narrow error spread on the quadrant itself, 33 seconds at two-foot radius, plus the highest comparative accuracy. Abraham Sharp emerges as a very consistent workman, obtaining the highest scale accuracy of the selection—27 seconds for a 59-inch radius—and an extraordinary 199 seconds for a nine-inch radius. Indeed, Sharp's dividing of the small equatorial circle was exemplary in its homogeneity, providing an example of an instrument where the best accuracy of the main degree divisions was more accurate in its dividing than could be utilized by the transversals; the degree marks were accurate to 3′ 19″ (199 seconds), whereas the transversals only allowed the scale to be read to 5 minutes. When both of the Sharp scales were compared on the standard radius, they gave values of 268 and 301 seconds or 4′ 28″ and 5 minutes respectively.

These two examples of Sharp's work give a good indication of the possible accuracy of Flamsteed's mural arc. Assuming that this instrument had its errors spread across a similar range to the 59-inch quadrant (8 thousandths), and that at 84-inch radius one thousandth of an inch equals 2·10 seconds, the expected error would be in the range of 16·8 seconds. This comes remarkably close to the 15 second error that James Pound claimed to have detected in the divisions of the mural arc in 1722.[4]

The beautifully engraved quadrant No. 6, bearing the inscription '*Elias Allen,* F[?e]ecit *Londini, 1637*', scarcely lives up to its maker's reputation when examined, with a best scale accuracy that comes fourth on the list at 47 seconds, and a comparative accuracy that comes bottom. It ranks as inferior to instruments of only one-third its radius, and gives a poor comparative performance alongside the author's quadrant with its error of 444 seconds against Allen's 603 seconds.

The two-foot iron quadrant, No. 2, generally called the 'Greaves quadrant' because of its alleged association with John Greaves (1602–52), who held the Savilian Chair of Astronomy at Oxford University from 1643–8, was a particularly interesting instrument. Of all the instruments examined in this study, it was the only one to display distinct evidence of construction marks used in the process of graduation. The quadrant was divided directly to 10 minute intervals, and to single minutes by transversals across the 1·75 inch-wide limb. In the middle of this limb, running along the fifth transversal line, is a row of punched dots, 10 minutes apart. As the transversal lines pass *between* the dots, and only the full degree lines pass *through* every sixth dot, it is likely that they were put onto the limb as a guide to the drawing of the transversals and degree lines, thus: (1) shows the single row of dots drawn along the centre of the limb; (2) upper and lower terminal lines for the transversals are drawn in, and the 10 minute dots transferred to (3) transversal lines drawn in, linking upper and lower dots. Careful measurements were made, both of the row of construction dots at 25-inch radius and the inner graduated arc of 24·125 inch radius apparently drawn from it. It was not possible to measure the extreme outer edge of the scale because of its closeness to the edge of the metal, which made it difficult reliably to adjust the micrometer.

---

[4] This information is contained in a letter from Joseph Crosthwait to Abraham Sharp, January 27, 1721–2. It is bound in date order in a manuscript volume entitled 'Sharp Letters', in the Royal Society Library, ref. *FlSh xxiv d.* Part of the letter is also printed in Francis Baily's *Account of the Revd. John Flamsteed* (London, 1835), p. 346. The Mural Arc, which encompassed 140°, came into use in 1688, and remained the principal instrument until 1719. It was sold following Flamsteed's death, and its whereabouts unknown after about 1722.

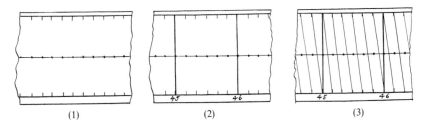

Figure 2.
The construction of the scale divisions on quadrant No. 2.

It is particularly interesting to note that the construction dots are less accurate in their positions than the degree lines drawn from them. While it is true that the dots, which measured around 0·15 to 0·20 inches across, were often difficult to bisect with the fiducial edges of the micrometer, it soon became clear that they were very irregular in their layout. As one may see from the table, even the 'best accuracy' dots displayed an error of 123 seconds in their spread, while the whole scale was almost 5 minutes out of true. The inner line of degrees, which runs parallel to the construction dots, has a smaller 'best accuracy' error of 102 seconds, yet with a maximum error of almost 4 minutes. The comparative weightings for the construction dot and degree scales, of 123 and 102 respectively, come close enough to show that tolerable homogeneity of degrees could obtained if a craftsman was willing to adjust his dots and lines before engraving the final marks. Indeed, the inner line of degree marks indicates that they had been adjusted or altered in their places from the 'primitive' line of original dots. The craftsman quite possibly used the dots as a guide to the drawing of the other two scales, and not as definitive marks. Adjusting was probably performed with a pair of screw-set dividers after laying down the preliminary degree marks, but before the final deep incision into the metal. On the outer limb of the transversals, where it was not possible to measure, there is a 10° run of trial dots, drawn at what would have been the 90° to 100° position, which were probably intended as a check on the scale before final engraving.

A careful search was made for any construction marks on the other instruments included in this study, but none were found. As it would have been impossible to lay off all the degrees on a quadrant or sextant without some adjustment, it might be suggested that one of two things took place: firstly, that the small scratches made by the beam compasses to lay off the 30°, 60° and other openings have been worn away by polishing, which is quite feasible, considering wear and tear over the centuries; secondly, the trial proportions for the radii were laid out on another piece of brass—since discarded—and only transferred on to the instrument when settled and perfected, thus leaving no marks. Most probably, this is the correct explanation, and one which makes the Greaves quadrant so interesting, in so far as it contains all its construction marks intact.

2. The structural design of observatory instruments

It is also interesting to observe the connection existing between the quality of the graduations, the rigidity of the limb on which they are engraved, and the overall strenght of the instrument frame. Considering the size of their radii, it is astonishing how weak and flimsy are the frames of the sextant, No. 5, the quadrant, No. 6, and the quadrant, No. 4. The brass limbs of Nos. 4 and 5 consist quite simply of a strip of one-sixteenth gauge sheet, riveted onto an iron sub-limb, which itself was no more than half an inch in thickness. These limbs are supported on weak frames: the main wrought-iron bars of No. 5 measuring $1 \times \frac{3}{4}$ inches, those of No. 6, $2\frac{3}{4} \times \frac{1}{2}$, whilst the frame of No. 4 is made of thin iron plates and small iron beams measuring $\frac{5}{8} \times \frac{1}{4}$ inches. Though the Elias Allen quadrant, No. 6, shows the presence of jointing lugs into which other bars may once have been fixed they are only four in number, and ostensibly designed to accommodate bars that were flimsier than the ones which have survived.

The anonymous sextant of 28-inch radius, *c.* 1640, currently on loan from the Museum of the History of Science to the Oxford City Museum,[5] also conforms to the above-mentioned design characteristics. Its frame, consisting of two flimsy bars of iron, indicates that only one bar has been lost, from the vacant joint positions, and even when complete the instrument must have been subject to innumerable flexures. Because of the damaged state of the limb, it was not possible to measure the accuracy of the sextant for, along with severe abrasion of the degrees, the thin brass limb had 'crept' and buckled upon the iron sub-limb to which it is riveted, the result, no doubt, of the different thermal coefficients of brass and iron.

It is likely that these weak frames could have been supported by wooden beams now lost. The Sharp quadrant, No. 4, has such a wooden outer frame, though on the Oxford instruments, Nos 5 and 6, it is difficult to see where the timber would have joined to the iron, for the bars show no evidence of screw holes, lugs, or anchor points. Indeed, to have used timber in the frame of any major instrument would already have been an anachronism by the mid-seventeenth century, for Tycho Brahe himself had pointed out the errors that inevitably occur when timber frames shrink and warp. Yet timber frames were still being used well into the eighteenth century, a good example being the so-called Napier quadrant in the Royal Scottish Museum, Edinburgh.[6] This anonymous 3-foot quadrant of *c.* 1730 has a boxwood lattice frame, which appears somewhat incongruous alongside the refinements of a limb-adjusting screw and telescopic sights. Almost certainly, it was a teaching instrument, and formerly belonged to the University Department of Natural Philosophy.

Although the mural arc built by Sharp for Flamsteed no longer exists, the engraving made by Bowen[7] indicates that the frame (figure 4 (*a*)) conformed in essential details to the frames of the Oxford instruments shown in Figures 3 (*d*) and 3(*e*). The 140° limb of the mural arc is supported upon nothing more than four bars, approximately $2\frac{1}{4} \times \frac{3}{4}$ inches in section. Large sections of the arc were totally without support, and the great brass and iron limb hung from seven unequally distributed suspension points. The whole instrument, and the limb especially, must have been subject to considerable lateral shake as Flamsteed racked the alidade along its screwed track, which probably contributed to that distortion which Flamsteed himself attributed to the sinking of the supporting wall. There is no evidence of wood being used to give extra support to the

[5] Museum of the History of Science, accession no. 36.4.

[6] The quadrant's current accession no. is 1975.54. I am indebted to Dr A. D. C. Simpson, of the Royal Scottish Museum, for further information about this instrument. It was included in 1876, in the exhibition of scientific apparatus at South Kensington as Item 1773. See *Catalogue*, third edition, p. 398. See also David Bryden's "Britain's First Observatory", *Journal for the History of Astronomy*, 3 (1972), 205.

[7] Originally published in John Flamsteed, *Historia Coelestis Britannica*, III (London, 1725), and reproduced in *The 'Preface' to John Flamsteed's 'Historia Coelestis Britannica'*, edited with Introduction by Allan Chapman, National Maritime Museum Monographs and Reports No. 52 (London, 1982), figure 19.

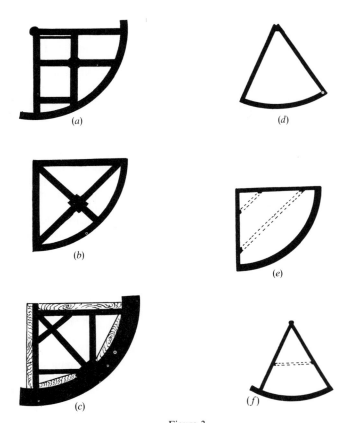

Figure 3.

Instruments surviving in museum collections. (a) Oxford quadrant, No. 2, c. 1637; radius 24 inches. (b) Butterfield quadrant, No. 3, c. 1700; radius 25 inches. (c) Sharp quadrant, No. 4, c. 1710; radius 59 inches. (d) Oxford sextant, No. 5, c. 1640; radius 73 inches. (e) Elian Allen quadrant, No. 6, 1637; radius 78 inches. Missing bars indicated from lug positions. (f) Oxford sextant, c. 1640; radius 28 inches. Missing bar indicated from lug positions. With the exception of (a) it will be noted that all of the instruments consist of radial frames, in which the limbs hang from the radial spokes. Only in instruments (a) and (c) is any rigid bracing attempted by the use of bars meeting at right angles to form a square.

mural arc, unlike the quadrant No. 4, which Sharp made some years later (Figure 3 (c)). Wooden blocks do seem to have been used to adjust the large quadrants, however, for in Place's engraving of the 10-foot quadrant built by Hooke for Flamsteed in 1676, the flimsy frame is separated from the wall by small wood blocks, and the Sharp mural arc may have possessed the same.[8]

[8] All of the Place engravings are reproduced in D. Howse, *Francis Place and the Early Royal Observatory* (New York, 1975). References to specific instruments depicted in the engravings are given beneath the figures.

Of all the instruments examined in this study, including engravings of instruments that no longer survive, only the Greaves and Butterfield quadrants, Nos 2 and 3, possess frames of adequate rigidity, their limbs being mounted on heavy frames of welded wrought-iron bars approximately $1\frac{3}{4} \times \frac{3}{16}$ inches in section.

The design features shown by surviving instruments are also seen paralleled in the engravings of instruments which no longer survive. With engravings we have the advantage of seeing the instrument in a complete condition, and often in the act of being used. Large astronomical instruments are depicted on three engravings by Francis Place, showing the 10-foot Hooke quadrant, Tompion's 7-foot sextant, and a quadrant of about 4-foot radius in his engravings of the interior of Flamsteed House, Greenwich

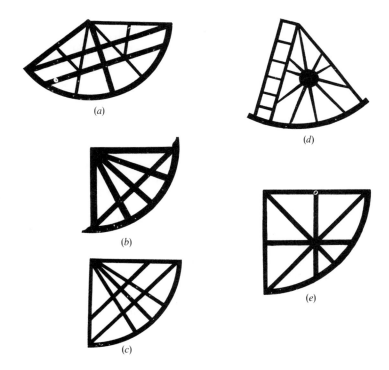

Figure 4.

Instruments known from illustrations. (*a*) Mural arc built by Abraham Sharp for John Flamsteed, 1689; radius 7-foot. Engraving by Bowen in *Historia Coelestis Britannica*, III (1725). (*b*) Screw quadrant proposed by Robert Hooke, 1674; radius about 3-foot. Hooke, *Animadversions* (1674). (*c*) Hooke's quadrant for Flamsteed, 1676; radius 10-foot. Engraving by Francis Place, *c.* 1676; reproduced in Howse, *Francis Place and the Early Royal Observatory* (1975) plate xb. (*d*) Sextant built by Thomas Tompion for Flamsteed, 1675; radius 7-foot. Engraving by Francis Place in *Historia Coelestis Britannica*, III (1725). (*e*) Quadrant of unknown make, *c.* 1676; radius about 4-foot. Included in Francis Place's engraving of the interior of the Octagon Room, Greenwich Observatory; reproduced in Howse, *Francis Place and the Early Royal Observatory* (1975), plate IX.

468

(Figures 4(d) and 4(e)). In Hooke's *Animadversions*[9] there is a fine plate showing the design of Hooke's proposed screw quadrant, while in Flamsteed's *Historia Coelestis Britannica*[10] there is Bowen's probably unfinished plate of the Sharp mural arc. This mural arc is also included in Sir James Thornhill's ceiling in the Painted Hall of the Royal Naval College, Greenwich,[11] where the limb and some of the bars are drawn a little thicker than in Bowen's engraving, but in exactly the same place in terms of design. All of these instruments seem to have been particularly thin and flimsy, considering the heavy work expected from them, and in the case of the Hooke 10-foot quadrant, the weakness of the frame soon rendered the instrument unusable.

Two features become apparent from the design of most seventeenth-century astronomical instruments, both from the pieces that survive and those shown in the engravings. Firstly, the builder aimed to support his limb on a construction of radial bars that resembled the spokes of a wheel to which the limb formed a segment. Secondly, their priority seems to have been the vertical rigidity of the arc, while paying hardly any attention whatsoever to its firmness in the lateral plane. Both these features are unique to seventeenth-century design and do not appear in eighteenth-century instruments.

The radial arrangement is apparent in the Butterfield and Sharp quadrants, Figures 3(b) and 3(c)), the two Oxford sextants, Figures 3(d) and 3(f) and (if the missing bars can be inferred from the remaining lugs), the Elias Allen quadrant, Figure 3(e). It is also clear in the engravings by Place, Hooke, and Bowen.

By 1725, however, George Graham had developed a much more efficient mounting, replacing the radial spokes with a rigid trellis of bars, meeting at right angles, and formed in the manner of Figure 5. The original instrument is still to be seen in the Quadrant House at Greenwich. This trellis design was subsequently developed by Sisson, Bird, and other eighteenth-century makers, two characteristic examples of which are to be found in the Astronomy Gallery at the Science Museum, South Kensington: the 8-foot South quadrant made by Bird for the Radcliffe Observatory, Oxford, and an 8-foot, 140° arc, built by Sisson. Both these instruments use a brass framework structure identical to that of Graham, in which thirteen great bars were used to support 90° of limb at eleven equally spaced suspension points. In this new design, the limb no longer hung from spokes, but became an integral part of the trellis, with the result that it had less freedom to move and distort.

It may be argued that the need to make an instrument rigid in the lateral plane was an integral development of the quadrant as a meridian instrument from Flamsteed's time onwards. Once Right Ascensions were being measured by timing the passage of stars across the limb of a large quadrant, it was imperative that the limb should be a perfect flat vertical plane, incapable of distorting to the left or right. With the exception of the Butterfield quadrant, No. 3, of c. 1700, none of the measured instruments in this study, nor those depicted in engravings, shows evidence of lateral bracing. The general sections of these instruments are shown in Figure 6(a), with the consequent distortion patterns.

The limb of the Butterfield quadrant has, however, a T–girdered 'back arch' to keep the limb rigid, as seen in Figure 6(b). It was an important feature of the 1725 Graham

[9] Robert Hooke, *Animadversions on the first part of the Machina Coelestis of...Johannes Hevelius* (London, 1674), fold-out plate.

[10] Flamsteed, footnote 7.

[11] A modern reproduction of Thornhill's painting can be seen in Derek Howse, *The Buildings and Instruments*, volume III of *Greenwich Observatory* (London, 1975), figure 22.

Figure 5.

Arrangement of the bars in George Graham's 8-foot quadrant, 1725. The bars are no longer positioned radially from the centre, but are formed into a rigid trellis with right-angled joints. Graham obtained lateral rigidity by using right-angled girders behind the bars, as shown in the section.

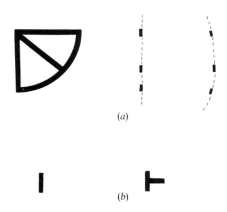

(*a*)

(*b*)

Figure 6.

(*a*) A seventeenth-century radial quadrant frame. Its simple flat bars are shown in section, and their inclination towards lateral distortion. (*b*) *Left*: Limb section of a typical seventeeth-century pattern. *Right*: The improved T-section limb on the Butterfield quadrant, No. 3.

470

instrument, that not only was its limb supported upon a trellis of bars, but the trellis itself was girdered laterally, as seen in Figure 5.

On the Elias Allen quadrant, No. 6, and the sextant, No. 5, it is clear that the limb was not fashioned from one single strip of brass, but from several pieces, brazed together with simple butt-joints. Inevitably, these limbs must have been weak, and from the frugality with which brass was used in them, one comes to appreciate how prized a metal it was in the seventeenth century. Not until the eighteenth century did brass become the general material out of which astronomical quadrants were fashioned; and one also notes that when brass plates were joined to form a smooth surface in a Bird instrument, they were locked together with a carpenter's dove-tail to give maximum strength, and not the butt-joint of Elias Allen and his contemporaries, as seen in Figure 7.

Flamsteed's best observations with the radially-mounted mural arc could achieve 10 to 15 seconds, whereas on a slightly larger radius, but with a trellis frame, Graham could reach 7 or 8 seconds. Yet after Bird had discovered the advantages of a thermally homogeneous frame, all made from one metal without the mixture of brass and iron parts, his quadrants could reliably measure down to 1 or 2 seconds. It is interesting to note that all the instruments measured in the present study contain a brass limb attached to an iron frame, although when one considers the relative crudity of these instruments—the best of which could only measure down to 27 seconds—the distortions accruing from thermal disequilibrium would have been trifling.

Figure 7.

Limb joints. *Left*: A simple seventeenth-century butt joint, as found on the Allen quadrant, No. 6. *Right*: A rigid dovetail joint as found on the Bird quadrant made for the Radcliffe Observatory at Oxford, 1772.

3. Conclusions

With proper study, careful measurement, and comparative examination of design features, it is possible for the large quadrants of three centuries ago to reveal information about the physical limitations bearing on early scientific research that is in no way inferior to that derived by the modern field archaeologist about former cultures.

But one important caveat must be borne in mind with all reconstruction work of this kind: how far can the surviving scale error of a seventeenth-century piece of brass be considered as a reliable guide to the working accuracy of the complete instrument in the hands of a contemporary scientist? Considering the possibility of damage or thermal disequilibrium between the bimetallic parts over the course of centuries, one might also enquire how far the surviving scale is an authentic guide to the instrument when new.

It must be understood, however, that in the absence of information relating to collimation, performance, and similar details, no solid claims can be advanced for the overall performance of the complete instrument in its original state. The present method of degree analysis aims, rather, to trace the skills with which artisans came to develop the difficult art of dividing, in the hope of establishing a geometrical ceiling for a particular epoch. It is primarily a technique concerned with the skill of the craftsman, rather than the perspicacity of the astronomer.

There is, moreover, no way of telling for certain the unchanged state of a scale under investigation, although certain criteria must be borne in mind: (*a*) only scales in a near-perfect condition, with no ostensible damage or buckling, are suitable for measurement; (*b*) the linear techniques does not attempt to measure graduations in direct angular units, but rather to compare the homogeneity of one part of the scale against another. By concentrating upon the homogeneity of the graduations rather than their 'absolute accuracy', local distortions caused by thermal disequilibrium or slight warping remain localized, and will not give rise to an accumulative error, as they would if measured in angular units from a fixed zero point. The error analysis allows extreme divergencies away from the mean error of the scale to become immediately identifiable.

Although it is necessary to be fully aware of the limitations that bear upon attempts to resurrect data from imperfect objects—particularly when it is of a mathematical kind—it is hoped that it may enable us to express in relatively precise terms that which has previously been the province of unverifiable conjecture.

Because accuracy was, and still is, the Alpha and Omega of science, it is of primary importance that the modern scholar should know as precisely as possible the degree of technical competence of the period with which he is dealing. Without a knowledge of the efficiency of the instruments available at a given time, he has no effective yardstick by which to compare one craftsman or scientist against another. Because relevant documents are scarce, and when they do survive are frequently misleading, it is necessary to look beyond conventional literary sources for a starting point. This starting point is to be found amongst the 'alternative documents', the physical artefacts that still survive in our observatory and museum collections.

Acknowledgments

It is a pleasure to acknowledge the facilities provided to me by Mr F. R. Maddison, Curator of the Museum of the History of Science, Oxford. I also thank the Director of the Science Museum and the Director of the National Maritime Museum for access to instruments in their charge. I am indebted to Dr G. L'E. Turner for advice on the study of scientific instruments and on the preparation of this paper.

VIII

Astronomia Practica: The Principal Instruments and their Uses at the Royal Observatory

> The great exactness with which instruments are now constructed hath enabled the astronomers of the present age to discover several changes in the positions of the heavenly bodies, which by reason of their smallness had escaped the notice of their predecessors.
>
> James Bradley[1]

Of all the mechanical arts practised in the early eighteenth century, none achieved greater delicacy of operation than those of the mathematical instrument maker, whose skills made possible the exact celestial geometry upon which the fame of the early Astronomers Royal rested. Whether one attempted to measure the motion of the Earth, or find the longitude, it was acknowledged that a solution would only be forthcoming after the making of accurate measurements of the places of the celestial objects. Because observational instruments act as extensions to natural human faculties, one may consider that the capacities of instrument makers acted as a ceiling upon the endeavours of working astronomers, and practical scientific horizons widened in a relationship that bore directly on the evolution of craft skills. The seventeenth century witnessed the beginnings of that transformation in the art of precision measurement, which was later to become the hallmark of the Observatory. By the arrival of the third quarter of the eighteenth century, the accuracy with which James Bradley could measure declinations had increased ten-fold since 1690, and over sixty-fold since the days of Tycho.[2]

Through the first century of the Observatory's existence one can trace the evolution of a concept of instrumental precision and the sequence of mechanical innovations in which it came to be embodied. In particular, these innovations centred around improved methods of graduating scales, the introduction of telescopic sights and the application of the micrometer principle to the business of delineating angles too small to be represented on a conventional scale.

William Gascoigne in 1639 had been the first to suspend a reticule wire in the focus of a Keplerian telescope and apply the same to the alidade of a sextant, although it was not until the time of the Restoration that astronomers began to develop the instrument.[3] Robert Hooke was the first to demonstrate the importance of the telescopic sight, when he calculated that, as the unaided human eye was unable to resolve angles smaller than 1' of arc, the only way to improve on the observations of Tycho was to equip measuring instruments

VIII

with telescopic sights. By counting the whole and part turns of a micrometer screw that governed the motions of a pointer it was possible to measure angular quantities to a few seconds of arc.[4] The earliest micrometer was also the brain-child of Gascoigne, who used the screw counts to move the reticule wire in the eyepiece of his telescope in an early filar micrometer.[5]

During the 1660s, Hooke devised a method of graduating quadrants and sextants that seems to have been a development from the micrometer principle. The extreme edge of his quadrant's limb was cut into a number of small teeth, like a gearwheel. These engaged a wormwheel that was attached to the sighting alidade of the instrument. To make an observation, one counted the wormwheel revolutions necessary to bring the alidade and its telescopic sight into line with the chosen star, and then converted them into degrees by means of a table. In theory at least, such a method of graduation completely dispensed with the need to equip the quadrant with conventional engraved divisions.[6]

Flamsteed's principal instrument between 1675 and 1689 was an iron equatorial sextant by Tompion that embodied both telescopic sights[7] and Hooke's method of graduation. It was fashioned as a 60° arc, with two telescopic sights.[8] One sight was fixed to the 'zero degree's' end of the arc, [142] the other being attached to the alidade, and free to move through all 60°. To measure an angular separation between two stars, an assistant first sighted one star of the pair through the fixed telescope. Flamsteed then racked the alidade across the scale, until he sighted the second. The resulting angular separation was then obtained by counting the turns applied to the wormwheel that were necessary to 'open' the telescopes to the required distance to make the sighting (Fig. 22.1).

The equatorial sextant was not restricted to meridian work, and could be made to point to any region of the sky, and track the stars in the equatorial plane. Its purpose was to survey by triangulation the place of every star in relationship to its adjacent stars, by measuring their angular separations from the Vernal Equinox. When the stellar places had been accurately determined, it was possible to chart the planetary motions against them. Flamsteed's technique was essentially the same as Tycho's,[9] though it was hoped that the optical and mechanical superiority of Flamsteed's instrument would render the problems of the longitude and stellar parallax less intractable.

Within a few months of commencing, Flamsteed discovered that his sextant was afflicted with serious errors.[10] The constant racking of the wormwheel had caused the teeth to wear to such an extent that the instrument displayed errors up to 1' of arc when tested against specimen angles laid out in Greenwich Park. Flamsteed dismantled the instrument, and with his own hands engraved on its brass limb a set of conventional divisions,[11] each degree

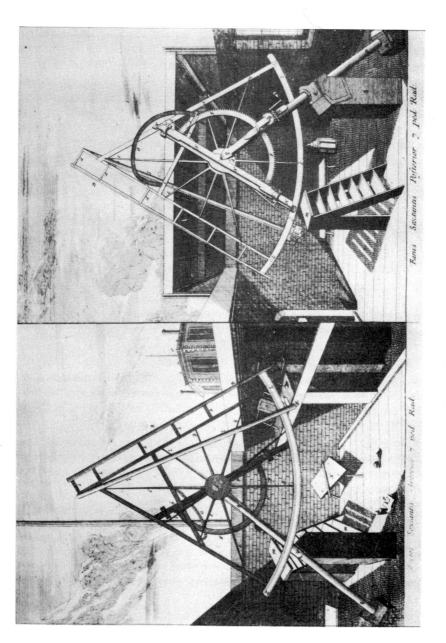

Fig. 22.1. Two views of Flamsteed's sextant.

being subdivided down to 10″ of arc.[12] The sextant was now equipped with two independent means of reading the same angle, and he soon realised the value of being able to cross-check any given observation, by either screw turns or scale degrees.[13] Furthermore, as the screw was still quite reliable over small angles he found it a useful way of reading the interval seconds between any two diagonal lines on the engraved scale.[14] This technique of being able to cross-check readings was to be of great importance, and is to be found in a refined form on all subsequent Greenwich instruments.

In his correspondence with Molyneux, Flamsteed says much about Hooke's method of screw graduation — much of it unfavourable — although it is not always easy to separate his genuine criticism from his personal dislike of Hooke as a man. He entertained little faith in the Hooke screw, and it may have been through the inventor's connivance with Sir Jonas Moore and the Royal Society that the sextant was fitted with the device at all.[15] Not even Tompion, the finest metalworker of the age, could make the gearing to the required order of perfection at this early date, whilst the lack of homogeneity of the available metals made them susceptible to uneven wear.[16] If Flamsteed seems severe in regarding the 'ingenious Mr. Hooke' as an impractical dreamer whose imagination soared beyond the skills and materials of his age, it must be remembered that Flamsteed was the only contemporary scientific man to have to use a Hooke invention in his daily work.[17] The Astronomer Royal did not even grant Hooke the credit for inventing the screw, but peevishly stated that he had 'borrowed it from the Emperor Ferdinand as you will find if you read ye Liber Prolegomeno of Tycho Brahe's Historia Celestis page 112', where a screw quadrant is described.[18] Two instruments similar to that in Tycho's description, one by Habermal, the other attributed to Jobst Bürgi, are still to be seen in the Prague Museum.[19] But neither of these instruments operated on the same principle as Hooke's, for their rackwork only served the purpose of moving or steadying the alidade, rather than measuring the actual degrees.

Regardless of the low opinion in which Flamsteed held Hooke's instruments, it is likely that they provided more inspiration for the sextant than the Astronomer Royal was willing to give credit for. The Greenwich sextant represented a substantial improvement in design and accuracy upon those of Tycho and Hevelius, in the respect that it embodied telescopic sights, screw micrometer and a geared equatorial mount. An instrument incorporating these features, with the additional refinement of having its equatorial turned by a clock rather than a hand-crank, appears as one of the plates in Hooke's Animadversions, 1674.[20]

Not until 1688 did the Observatory possess an adequate meridian instrument. So advantageous did such an instrument promise to be, that both

Hooke and Flamsteed had attempted to construct one, but without success.[21] An accurate arc, set in the plane of the meridian, was necessary to take the altitudes of the stars. Furthermore, used in conjunction with a good pendulum clock, it could greatly simplify the observing procedure. Declinations were obtained by taking the altitudes of stars as they crossed the meridian. Likewise, by timing the hours and minutes that elapsed between the meridian culminations of any two stars, their Right Ascensions were soon obtained. Both Tycho and the Landgrave of Cassel had attempted this method of fixing Right Ascensions, but with limited success.[22] It was not until the invention of the pendulum clock that the technique became feasible. [144] Measuring Right Ascensions by means of a clock was much less tedious than sextant observation, required fewer assistants and could be performed in a semi-covered building, instead of in the open.[23]

With the improvement of his financial situation after 1685,[24] Flamsteed commissioned Abraham Sharp to construct a mural arc of 7 feet radius, and covering 140° of sky, so that it would be possible to obtain the Observatory's latitude by directly observing the Pole Star, instead of having to triangulate it from the circumpolar stars with the sextant. It was strongly built of iron, and securely fixed to the wall with nails and wedges[25] (Fig. 22.2).

But the principal feature of the mural arc lay in the high quality of its degree graduations.[26] Before the graduation of the instrument could begin, it was first necessary to establish the zero point on the limb. This was achieved by first collimating the telescopic alidade, so that its index was exactly parallel to the optical axis of the telescope. The alidade was first swung from the centre of the arc, facing east, and a plumb-line made to hang, so that its wire passed across the alidade index. In this way, several transits of the star Caput Draconis were observed, and on each occasion, a mark was made at the point where the plumb-line crossed the index. Several nights later, the alidade and plumb-line were turned through 180° to face west, re-hung, and the procedure repeated. From the resulting observations, it was only necessary to bisect the two sets of marks, drawn when the alidade was hung facing east and west, to obtain the true collimation. Once this had been successfully completed, the alidade was replaced on the centre of the mural arc and so adjusted that the new-found point hung behind the instrument's plumb-line. When this was so, a small scratch was made on the polished limb of the mural arc, as a zenith starting point for the divisions.[27] Though no record remains describing the technique by means of which the 140° were laid out from this point, Abraham Sharp later stated [145] that the instrument was engraved *in situ* on the wall, and not on a horizontal table, as was to become the procedure in the eighteenth century. It is also known that the 60° point was struck from the same beam compass opening as was used to draw the

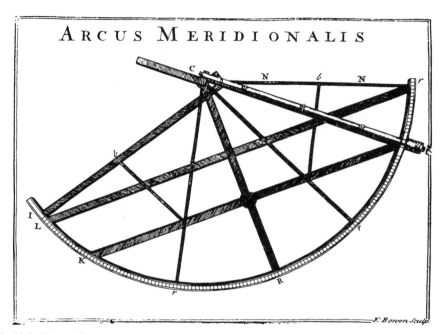

ARCUS MERIDIONALIS

Fig. 22.2. Flamsteed's mural arc. The instrument was constructed by Abraham Sharp in 1688. Its radius was just under 7 feet. After being secured to a meridional wall, it was possible to determine a star's declination by means of the telescopic sight. This engraving from *Historia Coelestis* does not show the rackwork on the limb, nor the micrometer.

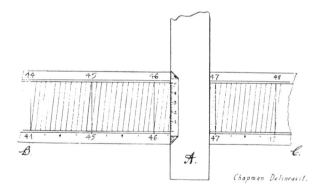

Fig. 22.3. Flamsteed's mural arc. The degree divisions (drawn full size). $B - C$, represents part of the scale of degrees. It is divided down to intervals of 5' of arc by direct subdivision. Each of these 5' spaces are joined across the face of the arc by means of a *diagonal*. *A* represents the end of the arm supporting the telescopic sight, which is joined to the centre of the arc at its opposite end. Where it passes across the scale, its edge is bevelled. This space corresponding exactly to the width of the main scale, is divided into five parts, representing minutes of arc. Each of these minutes is divided in turn into six parts, equal to 10". The scale was read at the intersection of a diagonal and the minute sub division; in the present example, $46° 16' 40''$.

radius of the 7-foot arc.[28] The remainder of the divisions were probably laid off by bisecting the 60° space to give 30° and 15° spaces, followed by trisecting and quinquesecting, to obtain the single degrees. As with the sextant, each degree was broken down by diagonals until it could be read to 10″ spaces[29] (Fig. 22.3).

This was achieved by first dividing the degrees directly to 5′ spaces on the arc. A diagonal line was then drawn across the space. It was reduced to single minutes by the 'fiducial edge' of the alidade, which carried five transverse lines drawn across it. Each of these lines was so placed that they divided the diagonal line into five parts. Whenever an observation was made, one of these lines would either intersect with the diagonal, or else come very close to it. By counting which of the five marks touched the diagonal, the scale was easily read to individual minutes. For those cases where no line made exact contact with a diagonal, it was possible to read the interval fraction by means of six intermediary marks drawn between the minute divisions, each of which represented 10″ of arc.

Sharp's use of the alidade to begin the divisions is interesting, for it also enabled the alidade to be used as a future check on the error of the instrument, as Bradley was later to do with the zenith sector. Flamsteed was thus able to discover any deviation in the zenith adjustment of the arc by first observing Caput Draconis in the zenith, and then examining how far the original zero point on the limb had strayed. He kept constant watch on the adjustment of the arc, and became convinced over the years that the meridian wall which supported the instrument was sinking. By 1715 it had moved so much that it was necessary to apply a correction of 14′20″ to all observations made with the arc.[30]

By 1688, Flamsteed had come to appreciate the advantage of having two independent means of reading any angle, so that one acted as a check upon the other.[31] No doubt this is the reason why he ordered the new arc to be fitted with a Hooke screw similar to that on the sextant. An additional refinement on the arc was a mechanism which enabled the screw to be disengaged when one desired to alter quickly the elevation of the alidade, instead of slowly having to rack it down the arc, as was the case with the sextant.[32] As both instruments were of the same radius[33] with seventeen teeth to the inch on their screws it was obviously useful to be able to compare cross-readings between them, and agreement was found to be satisfactory.[34] Not until Bradley's time, however, did the technique [146] develop of checking an instrument not merely against another of similar type, and hence likely to be afflicted with similar errors, but against one that worked on a different principle, such as the transit or zenith sector. This practice was to become the cornerstone of eighteenth-century fundamental astronomy.

VIII

Flamsteed's work had successfully revised the catalogue of Tycho, but that understanding of the lunar theory which held the key to the longitude problem still remained elusive. With the 8-foot mural quadrant, built by Graham for Halley in 1725, hopes were entertained that a solution would be forthcoming (Fig. 22.4). It embodied the best features of Sharp's arc with the additional advantage of a stronger overall construction and superior graduations. Graham's instrument was to become the prototype astronomical quadrant, to be copied and improved upon by instrument makers, until the quadrant yielded pre-eminence to the astronomical circle in the world's observatories after 1800.

To prevent flexure, the brass arc was supported by an ingeniously latticed frame of iron, built by Jonathan Sisson. In case the new masonry wall on which it was mounted should move, it was possible to re-adjust the instrument by means of set screws plugged into the wall by special cement.[35] But the principal feature of the instrument lay in Graham's graduations, which established a new standard of accuracy in instrument making. This was also the first major instrument for which a detailed description of the graduation procedure survives.[36] The dividing was performed in the 'Great Room' at the Observatory.[37]

Graham was aware that in laying off divisions with a beam compass, it was only possible to bisect angles with real accuracy. In pursuit of this end, he sought to simplify the graduating procedure, so as to locate each degree point using bisection alone. But on a 90° scale this turned out to be impossible. To proceed beyond 15°, he had to trisect and quiquesect angles in the traditional manner, reversing the compasses to cross-check each division. A solution was found by engraving a separate scale on the brass limb of the quadrant about 1½ inches removed from the 90° scale. The second [147] scale was a natural quadrant divided into 96 equal parts, which gave a sequence of numbers that could be delineated by bisection alone. After striking off the radius, and deriving division 64, Graham could bisect 64 to 32, 16, 8, 4, 2 and down to single degrees using a 'pure' geometrical technique. Conversion to conventional degree equivalents was easily accomplished by means of a table. In use it was possible to check one scale against the other, as Flamsteed had done with his arc, but to a greatly enhanced degree of accuracy. The 96-part scale was found to be more reliable, but discrepancy between the two arcs was never more than a few seconds.[38]

Both of the arcs carried degree subdivisions that read direct to 5' spaces. Graham dispensed with diagonal lines on this instrument, and read smaller quantities by a Vernier scale[39] (Fig. 22.5). Diagonal lines were tedious to draw, requiring a high degree of concentration over many hours. Each

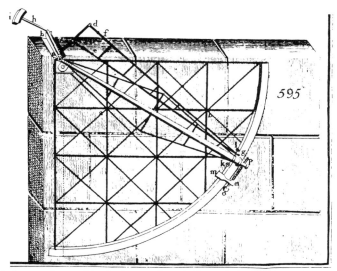

595

Fig. 22.4. The Graham quadrant 1725. The framework was built up of a lattice of wrought iron bars, constructed by Jonathan Sisson. It supported a brass limb of eight feet radius. The limb was divided by George Graham, with two scales, containing 90 and 96 parts respectively. The telescope was adjusted onto a star by means of the screw-set, shown at o, n and p. This Limb-set did not carry a micrometer head on Graham's original instrument, and was used only as a slow-motion adjustment, rather than to make measurements.

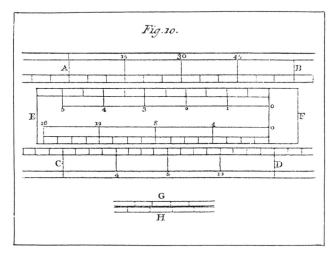

Fig. 22.5. Reading the Vernier scale. A − B, represents a part of the 90-degree scale on the quadrant's limb. E − F, represents the Vernier scale, being attached to the telescope at the point F. C − D. represents the 96-part scale on the limb. A − B, corresponds to one full degree and is divided into twelve equal parts. The Vernier adjacent to it contains ten equal parts. 'O' on the Vernier falls between 50 and 55′ on the scale. Reading along the Vernier scale, it is seen that the line between 3 and 4 corresponds exactly with a main scale division (At 15′). This 3½′ is added onto the 50′ on the main scale, thus giving 53′ 30″. The Vernier corresponding to the 96-part scale C − D works in the same way. Where there is no *exact* correspondence between any two lines as at G − H, the angle is estimated to 15″ or 7½″, on each side of the line. Stone's translation of Bion's *Instruments*

diagonal line had to bisect exactly the two dots that it connected, and a moment's lapse of attention could throw a whole sequence of divisions into error. Up to Graham's time, it had been customary to graduate an instrument by first laying off the degree points as dots on the limb, and then delineating them proper, by enlarging them into lines by means of the scribing knife and rule.[40] Degree divisions and diagonals were drawn alike by this method. Graham recognized the near impossibility of working accurately with the scriber, or 'dividing knife', and used in its stead a small beam compass, so that he could draw a controlled uniform line that would accurately bisect every primary dot.[41] To lay off the diagonals on Flamsteed's mural arc, it had been necessary to bisect 3360 dots, one at each end of the diagonal. By employing instead a Vernier scale, one effected a great economy of divisions, and automatically reduced the likelihood of error. Graham's quadrant required only 1080 points on its 90° scale. A short Vernier scale attached to the telescopic sight permitted the divisions to be read to 30″ directly, and 15″ or less by estimation.[42] Provided that the Vernier's own divisions were equally placed, there was an obvious gain in accuracy, consistency and economy of effort. Indeed, even if the Vernier did contain an error, its quantity could easily be discovered by trials, and a [148] correction applied to each reading. By rejecting diagonals, Graham simplified the whole procedure of graduating, reduced the likelihood of error accumulation, and at the same time made the errors easy to detect should they occur. On the other hand, such innovations could only hope to succeed when executed with flawless craftsmanship.[43]

With the techniques used to graduate the Greenwich quadrant, Graham commenced a 'school' of instrument graduation that was to make English craftsmen pre-eminent in this field until well into the nineteenth century.[44] Adherents of Graham's school were unanimous in their rejection of the dividing knife and diagonals. They espoused the scribing compass with its controlled accurate strokes, and realized that the best graduations were always performed with the greatest simplicity of technique and economy of effort. Sisson, Ramsden and the brothers Troughton were all of Graham's 'school', although it is agreed that the finest exponent of these techniques was John Bird.

As originally supplied in 1725, Graham's quadrant was not fitted with a micrometer, but only Verniers to read the scale degrees. The instrument was equipped, however, with an adjusting screw, by means of which the elevation of the telescope could be delicately controlled, to secure a star in the crosswires.[45] In 1745, shortly after becoming Astronomer Royal, Bradley commissioned John Bird to convert the screw-set into a micrometer by fitting it with a rod bearing 39¼ threads to the inch[46] (Fig. 22.6). One end of this rod was anchored by a bushed pivot, to a plate that was capable of being

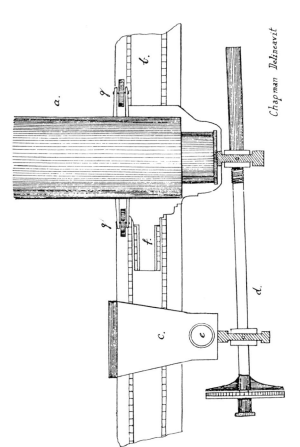

Chapman Delineavit

Fig. 22.6. Quadrant micrometer. Developed by John Bird and fitted by him to the Graham quadrant in 1745. It was fitted to all Bird's subsequent instruments.

a is the telescopic sight which transverses the limb of the quadrant *b*, with its 90 and 96-part scales. The telescope moves across the limb on the rollers *g*, and carries the Vernier *f* to read main scale divisions. *c* is an anchor plate that may be locked to any part of the quadrant's limb by means of the thumb screw *e*. Beneath this anchor plate is a bearing through which is mounted the micrometer *d*. The left-hand end of the micrometer carries a divided wheel, the head of which designates fractions of a revolution. The opposite end of the micrometer engages a screw fixed onto the sighting telescope.

After securing a star with the telescope, its position was first read to 15″ by means of the verniers. The full and part turns that it had been necessary to apply to the micrometer to bring the star to the crosswire, would then yield the individual seconds.

secured to any part of the quadrant's limb, by means of a thumb screw. The other end of the rod engaged a female thread on the telescopic alidade. By turning this rod through one revolution, the alidade was elevated exactly 53″ arc, thus enabling Bradley to read to single seconds with the instrument. To observe, one first disengaged the thumb screw on the anchor plate, allowing the telescope to move in altitude. When the star had been brought into the telescope field, the anchor plate was locked and the micrometer turned, until the culminating star was brought into the crosswire. The [149] resulting angle was obtained by first reading the Vernier to minutes, then counting the micrometer revolutions to determine the seconds. Such a micrometer represented a great improvement on the screw quadrant of Flamsteed. As it turned only during the critical moments of transit, wear and friction were reduced to a minimum. Moreover, a micrometer of such reliability as this one turned out to be an invaluable cross-check on the scale divisions, for their exact quantities could be ascertained by means of the screw.[47] Micrometers of this pattern came to be fitted to all the instruments of John Bird and remained a standard fitting until the quadrant became obsolete.[48] Although no documented ancestry survives for the limb micrometer, it is similar in form to the one with which Graham equipped Molyneux's zenith sector in 1725.[49] This in turn may have been derived from a screw adjustment that was a standard feature of most of Hevelius'[50] instruments and was used in its most rudimentary form by Tycho.[51]

Prior to setting up his quadrant in 1725, Halley's only meridian instrument had been the Graham transit telescope built in 1721 (Fig. 22.7). It consisted of an ordinary telescope, fitted with crosswires and accurately mounted with its centre wire delineating the meridian, similar to the instrument first used by Römer.[52] Though the transit had no scale to measure declinations, it could fix Right Ascension angles with great accuracy, when used in conjunction with a good clock. But after [150] the installation of the quadrant, Halley neglected the transit, and it was not until Bradley succeeded him that it was brought back into general use. The transit provided an excellent means of determining the meridian. In its passage around the pole, a circumpolar star will cross the meridian twice, at north and south culminations, scribing two arcs in the sky. If, when timing the exact periods between these culminations, an observer discovered that the star took longer to pass through the western arc than the eastern, this indicated a deviation in the instrument's axis from the meridian. Observing Capella by this method in 1742, Bradley discovered that the transit was out of true by 12″ to the west.[53] Once established, the transit's alignment was easily checked by means of terrestrial marks that corresponded to the astronomical meridian. An additional safeguard was available by reversing the instrument through 180°

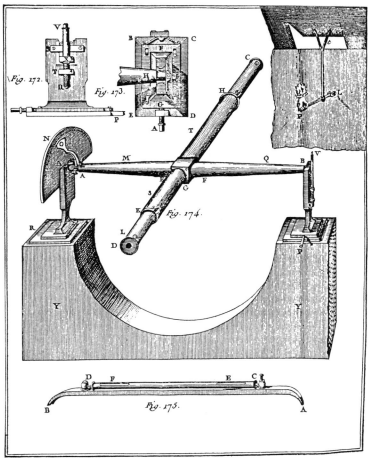

Lalande's 'Aftronomie'

Fig. 22.7. Transit Telescope, Mid-18th Century. The telescope $D-C$, rotates in the plane of the meridian on the axis $M-Q$. The inset figures 172 and 173, show the pivots that support the axis, and the screws by means of which it could be adjusted exactly to the meridian. Inset figure 175, shows the spirit level, used to test the horizontality of the axis pivots.

on its axis. If it returned to the same marks after reversal, its accuracy could be considered trustworthy.

Small deviations in the plane of the quadrant's limb were easily detected by two astronomers simultaneously observing the passage of a star with the transit and quadrant respectively. If the star transited fast or slow as seen with the quadrant, this indicated a deviation in the plane of its limb.[54] Observing some sixty stars between horizon and zenith, Bradley succeeded in plotting the limb deviations of this quadrant through every 5°.[55] This use of a simple easily-adjusted instrument as a check on a complex and more restricted one had been an advantage unknown to Flamsteed.

Halley had assiduously observed the Moon for 20 years but, as Bradley discovered upon succeeding him, had been neglectful of his instruments. In addition to being out of true both with respect to the meridian and vertical, the great quadrant[56] had become distorted so that its arc was discovered to be 16″ less than a perfect quadrant.[57] In 1742 the Admiralty had allowed £1,000 to repair the instruments, and re-equip the Observatory with new ones. Bradley ordered a set of instruments from Bird, the foremost of which was a new 8-foot quadrant, along with a transit, costing £300 and £73 respectively.[58]

Bird's new quadrant closely resembled the instrument by Graham. Its radius was supported by a frame of latticed bars. The limb carried scales of 90 and 96 parts, read by Verniers, in addition to which there was a micrometer. Yet the instrument contained as many original innovations as that upon which it was modelled. Mention has already been made that since its construction the Graham quadrant had changed shape to the extent of 16″. Bird attributed this to a distortion resulting from the different coefficients of expansion of brass and iron, which had caused the limb to buckle.[59] Graham himself had been one of the first to study the effect of heat on metals, and though he was conducting experiments on temperature compensating pendulums in 1725,[60] does not seem to have recognized the effect of temperature change on the quadrant. To preserve the shape of the new quadrant, Bird made it entirely of brass, so that all its parts would expand evenly. The success of this strategy was acknowledged by Edward Troughton when he examined the instrument in 1808.[61]

Bird's foremost innovation, upon which his future fame rested, lay in his invention of a means of dividing a 90° scale by continuous bisection alone. Using a 'scale of equal parts' that read accurately to five decimal places, he measured off a set of computed chords with the beam compasses, so that when scribed in sequence, they gave exactly the point 85° 20′.[62] This angle is capable of straight bisection down to 5′ intervals. In this way, a scale was produced by the desired 'pure' geometrical technique, every angle of which could also be tested against the set of computed chords on the scale, to guarantee their accuracy.   Bird was the first instrument maker to pay scrupulous attention to the temperature of the metals upon which he worked, and took elaborate precautions to protect his tools and materials from unequal expansion[63] during the 52 days required to graduate an instrument.[64]

After being set on the north meridian wall for three years, to observe the circumpolar stars, the Bird quadrant was set up to face south.   Graham's quadrant was re-divided by Bird and set up on the north wall in 1753.[65]   Here they remained to serve the next three Astronomers Royal, until being rendered obsolete in 1813.

Bradley's contemporary fame rested upon his discoveries of the aberration of light and the nutation of the Earth's axis, which gave the first 'proof' of its motion in space.   Both of these discoveries, made in pursuit of the elusive stellar parallax, had been made with a Graham zenith sector[66] (Fig. 22.8). This comprised a 12½-foot-focus telescope, suspended on gimbals at its object glass, so that it hung like a pendulum, with its 'swing' along the meridian. It measured zenith culminations of stars, for at this point they could be observed free from atmospheric distortion.   A delicate micrometer revealed whether a given star deviated in its zenith culminations.[67]   When the sector was moved from Wanstead to Greenwich, Bradley used the instrument to check the zenith collimation of his [151] quadrants.   To perform this operation, one first determined the exact place of a zenith star with the sector.[68]   The star was then observed with the quadrant, after bringing the telescope down to the bottom of the arc.   Any deviation between the two instruments in the resultant reading enabled the error of collimation to be deduced in the same way in which comparisons with the transit rendered possible the quadrant's adjustment to the meridian plane.   Zenith sectors and transits both relied on the principle of hanging in a simple gimbal mount, so as to delineate two regions of the sky the locations of which were essential eighteenth-century positional astronomy.   Like the transit, the sector could be checked against terrestrial standards, such as its plumb line, as well as being 'reversible'. This meant that it could be lifted out of its 'V'-shaped gimbals and turned through 180° so that any error would be immediately obvious when the observation was repeated through the reversed instrument.[69]   Spirit levels were used to guarantee the perfect horizontality of the gimbals, before either transit or sector was used to guide the quadrants.[70]

The instruments and observing techniques that had become operational at Greenwich by 1760 set a new standard in accuracy and were justly acknowledged as models of precision to be followed [152] elsewhere.[71]   It was at Oxford's Radcliffe Observatory, equipped in 1774 with a complete set of Bird's instruments, where these methods were best put to effect outside Greenwich.[72]   Bradley's secret of success lay in a relentless pursuit of

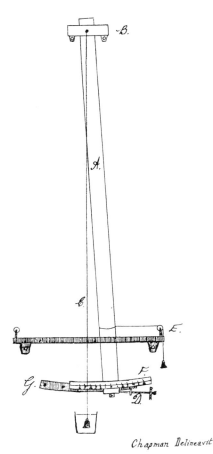

Chapman Delineavit

Fig. 22.8. Zenith Sector. *A* is the telescope, suspended on gimbals at *B*, in the meridian plane. At the lower end of the tube is attached *F*, a short scale of degrees. This scale is crossed by the plumb line *C*, so as to give the approximate zenith distance of a star. *G* is the *Back Arch*, fixed to the wall behind the sector. The micrometer *D*, can be fixed to any part of this arch by means of a thumb-screw. *E* is a weighted wire, ensuring that the telescope is kept in contact with the end of the micrometer screw. An observer first sights a selected star, and records its approximate zenith distance by means of the plumb line. Over the course of the following year he measures its culminations. The small 6-monthly deviations caused by Abberation and Nutation are then measured with the micrometer. Similarly, the instrument was used to delineate the zenith as a means of collimating the quadrants.

instrumental error that depended not only upon knowing the internal errors of every instrument,[73] but how they performed in relationship with adjacent instruments that worked on different principles. Thus the heavy mural quadrants had to accord with the gimbal-mounted transit and zenith sector, and these in turn with their levels, lines and meridian marks until eventually they were true to the heavens themselves. The whole technique was designed to ensnare any potential error in a web of safeguards before it could do mischief. Other astronomers prior to Bradley had observed great care in the adjustment of their instruments, but none before his time had done so with such thoroughness. Corrections for the smallest quantities occur in his observing books, sometimes to decimal places of a single second. It was in pursuit of his efforts to account for every disturbance that might impair an observation that he began to record temperatures and barometric pressure, so that he could account more fully for the variations in atmospheric refraction. His observing books after 1750 show each observation accompanied by temperature readings, both within and without the Quadrant House, along with the height of the barometer.[74]

From Bradley's time until 1813, the principal instruments remained essentially unchanged. After rigorously testing the instrument during the early years of his office, Maskelyne had implicit faith in the Bird quadrant. In 1781 he had unsuccessfully prevailed upon the Board of Ordnance to obtain the Ramsden equatorial for Greenwich,[75] but it was not until Pond revealed major errors in the old quadrant in 1806 that his faith in the instrument was shaken.[76]

By the arrival of the Observatory's first anniversary, the heavens had been re-charted to an unprecedented degree of accuracy and Bradley's work on aberration had rendered the first substantial proof of the Earth's motion. The longitude had been found by two methods and the lunar theory reduced for practical use in the *Nautical Almanac*. Even the observations of Mayer, which had contributed so much in this respect, had been executed at Göttingen with a 6-foot Bird quadrant. Little of this would have been accomplished but for fundamental innovations in the craft of instrument making, and the rise of a generation of astronomers who recognized that the price of perfection was eternal vigilance.

Notes and References

1. James Bradley, 'A letter to the Rt. Hon. George, Earl of Macclesfield, concerning an apparent motion in some fixed stars', *Philosophical Transactions* [hereafter referred to as *Phil. Trans.*] 485, Vol. xlv (1748), 1.

2. The accuracy of observations attained by earlier astronomers was discussed by Sir George Shuckburgh in 'Account of an Equatorial instrument', *Phil. Trans.* 1793, 75. See also the article by William Pearson, 'Circle', in Abraham Rees' *Cyclopaedia*

VIII

(1819). Flamsteed discusses a 10′ error in Tycho's observations in *Historica Narratio Vitae Meae*, Royal Greenwich Observatory (RGO) MS. Flamsteed 32B. This along with many other Flamsteed papers was printed by Francis Baily in *An Account of the Rev'd John Flamsteed* (London 1835), 38 [hereafter referred to as 'Baily']. Future references will cite both the MS. and printed Flamsteed sources. Tycho's errors were further discussed in John Flamsteed's *Historia Coelestis Britannica* (London 1725), Vol. III, 101 [hereafter referred to as *Historia*].

3.   For sources on early telescopic sights see S.J. Rigaud, *Correspondence of Scientific Men in the Seventeenth Century* (Oxford 1841) Vol. 1, 46/47: letter from Gascoigne to William Oughtred, undated but probably 1641. See also the letters printed by William Derham in *Phil. Trans.* 352, Vol. 30 (1717), 604-10. Also Flamsteed's *Historia*, Vol. III, 94. For a modern study of the telescopic sight, see R. McKeon's 'Histoire de l'acquisition des instruments d'astronomie et géodesy munis d'appareils de visée optique', *Physis. Riv. internaz. di storia della scienza* 1972, 221-42.

4.   Robert Hooke, *Some animadversions on the first part of Hevelius, his 'Machina Celestis'* (London 1674), 7 [hereafter referred to as *Animadversions*].

5.   John Flamsteed, *Historia* Vol. III, 94. See also Richard Towneley, 'An extract of a letter, written by Mr. Richard Towneley to Dr. Croon, touching the invention of dividing a foot into many thousand parts for mathematical purposes', *Phil. Trans.* 25 (1667), 457. Also Richard Towneley, 'Of an instrument for dividing a foot into many thousand parts and thereby measuring the diameters of planets to great exactness', *Phil. Trans.* 29 (1667), 541.

6.   *Animadversions*, 42-57.

7.   Flamsteed was convinced of the superiority of telescopic sights as opposed to open sights, as advocated by [153] Hevelius, saying '... Half a minute ... is so sensible a measure in the long tube I used that 'tis easier to mistake four inches in measuring a foot'. Letter to Oldenburg, 7 April 1677: Royal Society MS. (L.B.C. Sup. 3.360).

8.   *Historia* Vol. III, 106.

9.   Tycho Brahe, *Astronomiae Instauratae Mechanica* (1598), translated by H. Raeder, E. Strömgren and B. Raeder, as *Description of his Instruments and Scientific Work* (Copenhagen 1946). See the description of Tycho's sextant. Also J.L.E. Dreyer, *Tycho Brahe: a Picture of Scientific Life and Work in the Sixteenth Century* (1890), 326-8, where the observing technique with the sextant is described.

10.   *Historia* Vol. III, 103-7.

11.   Baily, 127.

12.   Flamsteed did not state the smallest quantity denoted on the limb of his sextant. On the Mural Arc it was 10″, but as both instruments were of the same radius and carried similar scales, it is also likely that the sextant read to 10″.

13.   Southampton Civic Record Office MS. (D/M, 1/1.ff.99r.v. 100r.v.) letter from Flamsteed to William Molyneux, 4 Nov. 1686. 'I find ye screw of good use in yt. proves whether I have numbered the distances noted on ye diagonals true or no.' Typescript of this letter in the National Maritime Museum (N.M.M.) Library.

14.   *Ibid.* 'Yet for a single degree they are so nearly the same [you] may securely take the halfe or third pte of ye revolves or ptes. answering to any one degree anywhere for thirty or twenty minutes in another place.'

15.   *Historia* Vol. III, 103. Sir Jonas Moore paid for the construction of the sextant.

Flamsteed cites several cases where Hooke tried to force his inventions on to Moore: Baily, 41, 42, 46, etc.

16. Baily, 41, from the MS. *Historica Narratio Vitae Meae*, RGO. MS., Flamsteed 32B, 10. William Molyneux had much difficulty with a quadrant equipped with a Hooke screw, as discussed in the letter to Flamsteed, 19 Oct. 1686, in Southampton Civic Record Office (D/M. 1/1.ff.98r.v.). Typescript in N.M.M. Library. John Smeaton discusses the limitations imposed by materials and contemporary craftsmanship upon Flamsteed's sextant in the historical part of his paper 'On the graduation of astronomical instruments', *Phil. Trans.* lxxvi (1786), 4-5 [hereafter referred to as 'Smeaton'].

17. Royal Society MS. (MSS.243.Fl 8.): letter from Flamsteed to Richard Towneley, 3 July 1675. This contains Flamsteed's best appraisal of Hooke as an impractical mechanic. Typescript in N.M.M. Library.

18. Southampton Civic Record Office (D/M.1/1ff.99r.v. 100r.v.): letter, Flamsteed to Molyneux, 4 Nov. 1686. Flamsteed also mentions the theft of the device in *Historica Narratio Vitae Meae*, see Baily, 41.

19. Zdenek Horský and Otile Skopová, *Astronomy and Gnomonics: A Catalogue of Instruments of the Fifteenth to the Nineteenth Centuries in the Collections of the National Museum of Prague* (1968), 28-30.

20. *Animadversions*, 67-8. See also Hooke's plate illustrating his quadrant's mounting.

21. Bailey, 118. Letter from Flamsteed to Sir Jonas Moore, 16 July 1676. The 10-foot quadrant devised by Hooke was so dangerous that it had 'like to have deprived Cuthbert [Flamsteed's assistant] of his fingers'. The unsuccessful quadrant is also discussed in the *Historia* Vol. III, 107.

22. H.A. Lloyd, *Some Outstanding Clocks over Seven Hundred Years, 1250-1950* (London 1958), 63. See the correspondence between the Landgrave of Hesse Cassel and Tycho, translated from Tycho's *Dani Epistolarum Astronomicarum Libri* (1601), 269.

23. The improved observing conditions made possible by the Mural Arc are discussed in Baily, 56-7, which is reprinted from Flamsteed's *Coelum Britannicum: the restitution of the places of the fixed stars, from observations made at the Observatory*, RGO MS. Flamsteed Vol. 39, 101-3.

24. In 1685 Flamsteed was presented with the living of Burstow, and in 1688, when his father died, his financial situation was eased considerably: Baily, xxviii, xxix. So wretched was his salary that he told Baron von Uffenbach when he visited the Observatory in 1710 that he could have achieved little had he not been the 'son of a rich merchant': W.H. Quarrell and Margaret Ware, *London in 1710* (1934), 22, being translated from Conrad von Uffenbach's *Merkwürdige Reisen durch Niedersachen, Holland und Engelland* (1753) [hereafter referred to as 'Uffenbach'].

25. *Historia* Vol. III, 108-9. Also Baily, 55, being from *Coelum Britannicum* [n. 23]. Sharp reduced the limb to perfect flatness with a 'peculiar contrivance', or plane, before commencing graduation.

26. The mural arc made by Sharp for Flamsteed was based on the disused framework of one of Flamsteed's unsuccessful arcs. Although it no longer survives, there are two other instruments now at Greenwich that were made and divided by Sharp. One is an equatorial instrument, equipped with diagonal scales (described by Sharp in a

VIII

letter to Flamsteed dated February 1702, quoted by Cudworth, 63/4), the other an unfinished 5-foot quadrant described by N.S. Heineken, 'Relics of the mechanical productions of Abraham Sharp', *Philosophical Magazine* xxx (1847), 25-7. Neither of these instruments were originally intended to be used at Greenwich.

27. *Historia* Vol. III, 109-10.

28. William Cudworth, *Life and Correspondence of Abraham Sharp, the Yorkshire Mathematician and Astronomer and Assistant to Flamsteed* (London 1889), 16-18. Letter, 2 Feb. 1721-2, Sharp to Joseph Crossthwait. [154]

29. *Historia* Vol. III, 110. See Fig. 22.3. The diagonals on the arc were drawn so that the line was not broken into five equal spaces, but into five spaces that compensated for the impossibility of perfectly dividing a *curved* division by means of a straight diagonal line. The place of each diagonal line was determined by calculation. The necessity of drawing *unequal* diagonal subdivisions to obtain maximum accuracy is treated by Hooke in *Animadversions*, 12. The use of curved diagonals is examined by Nicholas Bion in *The Construction and Principal Uses of Mathematical Instruments*, as translated into English by Edmond Stone (London 1758), 108-9.

30. In the *Historia* Flamsteed notes the quantities for the sinking of the wall on each page of the Observations. The movement of the wall is also mentioned by Baily, 55-6, being *Coelum Britannicum* [n. 23], 101.

31. Baily, 56.

32. Uffenbach, 25. An engraving of the mural arc, with its gearing exploded, is depicted on a plate in the original German edition, but not in the English translation.

33. The two instruments were not exactly of the same radius. The sextant's radius was 6ft. 9¼ inches, the mural arc's 6ft. 7½ inches, but both were usually considered as of 7 feet.

34. *Historia* Vol. III, 113. '... That the differences are rarely greater than those which are readily granted to arise from the difficulty of dividing and exactly placing instruments ... over the plane of the meridian.'

35. Robert Smith, *A Compleat System of Opticks* (Cambridge 1738), Book III, 337 [hereafter referred to as 'Smith's *Opticks*'].

36. Smith's *Opticks* [n. 35], 332-40. Smith's descripion of the graduation procedure of the Graham quadrant is the earliest surviving account of the graduation of a major astronomical instrument. It is also the only detailed account of Graham's working technique. All later writers who refer to Graham, including Smeaton, Pearson, Brewster and others, depended on this primary source.

37. Baily, 359, letter, Joseph Crossthwait to Sharp, 24 July 1725.

38. Smith's *Opticks* [n. 35], 334.

39. The Vernier scale was first described in Pierre Vernier's *La construction, l'usage et les propriétéz du quadrant nouveau de mathématique* (Brussels 1631). In the eighteenth century a vernier scale was commonly referred to as a Nonius, being derived, no doubt, from Pedro Nuñez (Nonius) who was one of the first mathematicians to subdivide a main degree division by means of a separate scale, in *De Crepulis* (Lisbon 1542). The sequence of developments through which the scale passed in the sixteenth and seventeenth centuries is described in E.R. Kiely's *Surveying Instruments: their History and Classroom Use* (New York 1947).

40. David Brewster, *The Edinburgh Encyclopedia* (Edinburgh 1830). See the article

'Graduation', 349.

41. Smith's *Opticks*, 335.

42. Smith's *Opticks*, 338-9.

43. James Bradley, 'A letter ... concerning an apparent motion in some of the fixed stars' (op. cit.), *Phil. Trans.* 1748. Several times in this paper, Bradley pays tribute to the skill of Graham.

44. Smeaton, 177.

45. Smith's *Opticks*, 337-8.

46. S.P. Rigaud, *Miscellaneous Works and Correspondence of the Rev. James Bradley, D.D., F.R.S.* (Oxford 1832), lv [hereafter referred to as *Bradley's Memoirs*].

47. Thomas Hornsby, *Astronomical Observations made at the Royal Observatory, Greenwich, from the year MDCCL to the year MDCCLXII, by the Rev. James Bradley, D.D.* (Oxford 1797), Vol. I. See vii, viii and xiv, headed 'The Quadrants'. [Hereafter referred to as *Bradley's Observations*.]

48. John Bird, *The method of constructing mural quadrants. Exemplified by a description of the brass mural quadrant in the Royal Observatory, Greenwich, by Mr. John Bird, mathematical instrument maker in the Strand* (London 1767), 21-2 [hereafter referred to as Bird's *Quadrants*]. For further details of Bird's micrometer see William Ludlam, *An introduction to Mr. Bird's method of dividing astronomical instruments* (1785), 64-7.

49. James Bradley, 'A Letter ... to Dr. Edmond Halley Astron. Reg. &c. giving an Account of a new discovered Motion of the Fix'd Stars', *Phil. Trans.* 406 Vol. xxxv, 637-8.

50. Johannes Hevelius, *Machina Coelestis* (Dantzig 1673). Most of the circular divided instruments represented in the plates depict screw-sets to adjust the sighting arms to the stars. See cpt. XV, and plate 'T'.

51. Tycho Brahe, *Astronomiae Instauratae Mechanica* [n. 9]. See plates 18 and 19, and description.

52. Peter Horrebow, *Basis Astronomiae, sive astronomiae pars mechanica* (Havaniae 1735).

53. S.J. Rigaud, 'Some particulars respecting the principal instruments at the Royal Observatory at Greenwich in the time of Dr. Halley', *Memoirs of the Royal Astronomical Society* IX (1836), 205-7 [hereafter referred to as *Greenwich in 1742*]. The materials from which Rigaud compiled this paper are to be found in Bradley's notebook, catalogued at Vol. 121, in the National Maritime Museum's photostat MS. of Bradley.

54. Smith's *Opticks* [n. 35], 340.

55. *Greenwich in 1742* [n. 53], 220-1. [155]

56. *Greenwich in 1742* [n. 53], 219. Bradley discovered the quadrant's vertical adjustment to be 34½" out of true. This, no doubt, resulted from the infrequency with which Halley tested the instrument with plumb-lines: see Francis Baily, 'Some account of the astronomical observations made by Dr. Edmund Halley, at the Royal Observatory at Greenwich', *Memoirs of the Royal Astronomical Society* 1835, 169-90: 182. In this paper Baily also mentions Halley's use of the transit to collimate the quadrant in 1726.

57. *Bradley's Observations* [n. 47], xiv.

58. *Bradley's Memoirs* [n. 46], lxxiv.

59. Bird's *Quadrants* [n. 48], 7. Wear on the centre pin could also have contributed towards this error, in addition to temperature distortion. In 1746 Bradley wrote that part of the error could have been caused '... by the cylinder's [i.e. centre pivot] being placed ... about 1/191th part of an inch from the true centre of the arc': *Bradley's Memoirs* [n. 46], lv.

60. George Graham, 'A contrivance to avoid the irregularities in a clock's motion occasioned by the action of heat and cold upon the rod of the pendulum, by Mr. George Graham, watchmaker, F.R.S.'. Graham read this paper to the Royal Society on 28 April 1726, it being in the Society's Library, under (R.B.C.12.739). It was also published in *Phil. Trans.* (1726).

61. Edward Troughton, 'An account of a method of dividing astronomical and other instruments by ocular inspection, in which the usual tools for graduating are not employed; and the whole operation being so contrived, that no error can occur but what is chargeable to vision, when assisted by the best optical means of viewing and measuring minute quantities', *Phil. Trans.* 1809, 138 [hereafter referred to as Troughton's *Circle Dividing*]. Making allowances for the wear that would result from sixty years' usage, Troughton found that the Bird quadrant's divisions were still good when he checked them in 1808. An interesting study of this quadrant, its working accuracy and response to temperature is derived from the reduction of the Greenwich observations by Truman Henry Safford, 'Investigations of corrections to the Greenwich planetary observations from 1762 to 1830 ...', *Astronomical papers prepared for the use of the American Ephemeris and Nautical Almanac*, Vol. 2 (1891).

62. John Bird, *The method of dividing astronomical instruments by Mr. John Bird, mathematical instrument maker in the Strand* (London 1767), 3 [hereafter referred to as Bird's *Dividing*].

63. Bird's *Dividing* [n. 62], 3.

64. Troughton's *Circle Dividing* [n. 61], 142.

65. *Bradley's Observations* [n. 47], xiv. Now that Bradley possessed two quadrants, one of which faced north, he was able, for the first time, to determine independently the latitude of Greenwich. Previously, he had derived his latitude from Lord Macclesfield's observations made at Shirburn Castle. See James Bradley, 'Letter to the Earl of Macclesfield ... concerning an apparent motion observed in some fixed stars', *Phil. Trans.* 1748, reprinted in *Bradley's Memoirs* [n. 46], 20.

66. Nevil Maskelyne, *Astronomical Observations made at the Royal Observatory, at Greenwich, from the year MDCCLXV to MDCCLXXIV* (London 1785), Vol. 1, viii-xi [hereafter referred to as Maskelyne's *Observations*]. Maskelyne gives a full description of Bradley's zenith sector of 12½ feet radius. Bradley makes references to the sector in his papers on the Aberration and Nutation (op. cit.) in *Phil. Trans.* 1728 and 1748 respectively. The first person to employ a zenith sector was Robert Hooke, who described his instrument in *An attempt to prove the motion of the Earth from observations, made by Robert Hooke, Fellow of the Royal Society* (London 1674).

67. Bradley determined that one revolution of the micrometer advanced the sector through 0.0246 inches. One 80th part of this quantity (as represented on the

divisions of the micrometer head) equalled ½" of arc at 12½ feet radius: *Bradley's Memoirs* [n. 46], 195. See also Maskelyne's *Observations* [n. 66], x, for a further analysis of this instrument.

68. *Bradley's Observations* [n. 47], vii, viii, xiv. In 1768 Maskelyne made observations of the places of eighteen zenith stars, as a means of determining the collimation of the quadrant. The error rarely exceeded 1" of arc. See Maskelyne's *Observations* [n. 66], viii.

69. *Bradley's Memoirs* [n. 46], p. L. Discusses the reversing of Dr. Pound's transit in the 1720s. This was one of the first transit telescopes in England. The technique was practised by all subsequent observers with such instruments: see *Bradley's Observations* [n. 47], ii, vi.

70. Sir George Shuckburgh, 'Account of an equatorial instrument', *Phil. Trans.* 1793 [n. 2]. See the footnote on p. 92. The level was used to check the horizontality of the Bird transit. A movement of 1/50th of an inch in its bubble corresponded to a deviation of 1" of arc in the transit's gimbals. Dr. Hornsby, at the Oxford Observatory, kept a close watch on the horizontality of the Bird transit in the Radcliffe Observatory. He discovered that the western pier of the transit frequently showed a 2" shift from the horizontal, when tested with a spirit level. See H. Knox-Shaw, J. Jackson and W.H. Robinson, *Observations of the Reverend Thomas Hornsby, D.D., Savilean Professor of Astronomy and Radcliffe Observer, made with the Transit Instrument and Quadrant at the Radcliffe Observatory, Oxford, in the Years 1774 to 1798* (1932), 20.

71. W.F. Bessel, *Fundamenta Astronomiae* (Regiomontani 1818). Bessel paid great tribute to the quality of Bradley's observations. [156]

72. E.J. Stone, 'The determination of the mean North Polar Distance … of γ Draconis from the observations made at Oxford by Dr. Hornsby', *Monthly Notices of the Royal Astronomical Society*, June 1895, LV8, 409. Stone stated 'There was no observatory in the world better equipped with meridian instruments than the Oxford observatory at its foundation.'

73. Bird's *Quadrants* 20-7, for a comprehensive description of the technique of determining the collimation and other errors of an *individual* instrument.

74. See Bradley's post-1750 observing books at the Royal Observatory, of which photostats are to be found in the National Maritime Museum. See also *Bradley's Observations* [n. 47], Vol. 1, ii, where he realised that although his earlier observations had been made with great care '… the state of the atmosphere was not determined: consequently the mean quantity for refraction can only be applied.'

75. Nevil Maskelyne to the Royal Society, letter, 4 Jan. 1781, in the Royal Society Library (Gh. 120).

76. John Pond, 'On the declinations of some of the principal fixed stars; with the description of an astronomical circle, and some remarks on the construction of circular instruments', *Phil. Trans.* 1806, 23-4. See also T.H. Safford, 'Investigations of corrections to the Greenwich planetary observations' [n. 61], 57. Safford found, in the reduction of Maskelyne's observations, that the quadrant's error increased suddenly in 1787.

Reprinted from Vistas in Astronomy 20 (1976), pp. 141-56. *Numbers in square brackets indicate original pagination.*

Reconstructing the Angle-measuring Instruments of Pierre Gassendi

In 1992, the Provençal City of Digne-les-Bains celebrated the four-hundredth anniversary of the birth of its most illustrious figure, Pierre Gassendi. As part of these scholarly celebrations, it was decided that two of Gassendi's astronomical instruments might be reconstructed, to augment the visual exhibition, and I was asked to undertake the work. These instruments have subsequently been placed on display in the Musée de Digne, Boulevard Gassendi. None of Gassendi's original instruments survive and unlike his acknowledged technical master, Tycho Brahe, he left no clear descriptions of them, but from the numerous recorded observations that were published in his *Opera omnia* (1658) it was possible to reconstruct two of Gassendi's principal pieces. They were his five [Paris] foot quadrant, and his seven-foot *radius astronomicus*.

Gassendi made astronomical observations throughout most of his adult life, and discussion of their implications formed a regular feature of his correspondence with scientists across Europe.[1] These include telescopic observations of the sun, moon, and planets, the most famous of which was his Parisian observation of the transit of Mercury across the solar disk in 1631. Virtually all of Gassendi's observations involve the measurement of angles, however, for it was from such angular measurements that the planetary and dynamic theories of Copernicus, Tycho, Kepler, and Galileo could best be evaluated within the astronomical debate of the early seventeenth century.

Renaissance astronomical instruments displayed remarkable ingenuity in their design, as scientists and craftsmen attempted to obtain the maximum practical accuracy from a set of Euclidean principles that were embodied in simple metal and wooden

---

[1] *Petri Gassendi Diniensis, Opera omnia*, 6 vols, Paris, 1658. Volumes 4 & 6 contain the main body of astronomical observations and letters. Though only the 5-foot quadrant and *radius astronomicus* were reconstructed, a shadow square, a 2-foot quadrant, and other instruments are mentioned in the observations.

From *Learning, Language and Invention: Essays presented to Francis Maddison*, ed. W.D. Hackmann and A.J. Turner. Copyright © 1994 by the contributors and the Société Internationale de l'Astrolabe. Published by Variorum, Ashgate Publishing Ltd, Gower House, Croft Road, Aldershot, Hampshire, GU11 3HR, Great Britain.

structures. These designs all hinged upon the accuracy with which the circle could be divided, and tended to employ one of two principles, both of which were evident in the instruments of Gassendi. The first of these depended upon the direct division of a circular arc using beam compasses, and the further subdivision of its degrees by means of transversals, Verniers, or micrometers. But the direct division of a quarter-circle into 90 perfectly homogeneous units is geometrically impossible, as early-seventeenth-century craftsmen well knew, so that the ensuing quadrants and sextants were recognized as carrying graduations that were not perfect.

The second graduation principle employed exploited the geometrical relationship that exists between the radius, tangent, and chord of a circle, to make measurements against an equally divided straight rod. The parallactic rulers of Ptolemy had been the earliest instruments to use these principles for astronomical measuring purposes in the ancient world, and Regiomontanus, Copernicus, and Tycho Brahe had subsequently used such rulers.[2] Tycho had also experimented with the other straight-rod variant for measuring angles, the *radius astronomicus*, and as Gassendi was later to find, commented upon its convenience as a travelling instrument.[3]

Many astronomers had used the *radius astronomicus* in its various permutations of graduation in the 150 years that led up to its valedictory employment in fundamental research in the hands of Jeremiah Horrocks in 1640. Wilhelm Schickard had described a form of the instrument that was to be influential across Europe, and which was similar to Gassendi's instrument.[4] The advantage of the *radius astronomicus* against the quadrant lay in its simplicity, and the absence of any theoretical limits upon its division such as those that applied to the arcs of quadrants. Its straight rods could be several feet in length if desired, and divided into any pre-computed number of equal divisions by continuous bisection.

When one attempts the scholarly reconstruction of an instrument from the early modern period it is always with several agendas in mind. The first arises during the process of construction, when it is possible to test the practicality of various modes of graduation from the standpoints of geometry and draughtsmanship. And when the instrument is complete and mounted, it is possible to test its practical handling qualities in the act of making angular measurements against the heavens. While a modern observer who lacks the nightly familiarity in using the *radius* and similar instruments cannot hope to reproduce the practical accuracies of Gassendi and his contemporaries,

---

[2] Tycho Brahe, *Astronomiæ instauratæ mechanica*, Wandesburgi, 1598, transl. H. Raeder, E. & R. Strömgren, Copenhagen, 1946, pp. 44–51. Tycho acquired the 4-cubit rulers of Copernicus from a Canon of Ermeland in 1584, and proceeded to build a set of his own. But he found the design of the instrument to contain too many faults to be capable of critical results.

[3] Tycho *Op. cit.* (n. 2), pp. 96–97.

[4] A. Chapman, *Dividing the Circle. The Development of Critical Angular Measurement in Astronomy, 1500–1850*, Chichester: Ellis Horwood, 1990, pp. 23–31. Gemma Frisius's *De radio astronomico*, Antwerp, 1545, was a very influential work in popularizing the *radius* amongst astronomers. The most complete modern study of the instrument is John Roche, 'The Radius Astronomicus in England', *Annals of Science*, xxxviii, 1981, pp. 1–32.

he can none the less make a good assessment of the stability and practical handling qualities of a particular design. Sighting systems can be tested, along with the ease and reliability with which a scale can be read against a fiducial edge in the dark, as a way of getting the 'measure' of early modern scientific skills.

Knowing the type and dimensions of a specific Gassendi instrument (usually given in the Paris Foot of 324.8 mm) one can extract its general mathematical features. Then, using data from specific observations and their scale digit sequences recorded in the *Opera omnia*, one can determine how the scales were divided and read, to make an accurate physical reconstruction.

The Quadrant

In August 1625, in a letter to the Hollander Willebrord Snell, Gassendi described making astronomical observations with his 5-foot (1624 mm) quadrant.[5] This large, flat instrument was used to measure angles in the vertical plane, probably hanging from a wall or stand. Most likely, its main use would have been to measure the meridian altitudes of the stars and planets for incorporation into a table or celestial map.

The frame of the replica [Figure 1] is made of walnut wood, using an arrangement of seven radial and chordal beams which was characteristic of quadrants of the period. The 'limb' is also made of wood, to provide a rigid support for the brass scale.

Figure 2

At the geometrical centre of the instrument (top left corner) is a brass peg around which rotates the measuring rule, or alidade. At the opposite, or scale-end of the alidade is a backsight with two fine horizontal slits cut into it. The separation between these two slits in millimetres is exactly the same as the diameter of the central brass peg.

To make an observation, Gassendi would have first looked through the top slit and adjusted the angle of the measuring rod until the desired star was visible *exactly* on the top edge of the central peg. He would then have repeated the procedure for the lower slit against the lower edge of the peg.

Figure 3

Because the star is at infinity, it is possible to see its light across both edges of the narrow peg. When the star appears equally bright in both positions, the astronomer can be sure that no parallax, or off-centre error has entered into the observation. The arrangement was first used by Tycho Brahe about 1575 and was standard across Europe by Gassendi's time.

The scale was divided into 90 degrees with beam compasses, in accordance with Euclid's geometrical proportions for the radius and circumference of the circle. The radius of a circle will always fit six times into its circumference. The compass used to draw the quadrant's quarter circle on the brass scale can, therefore, be used to strike

---

[5] Gassendi, *Op. cit.* (n. 1), vol. 6, Gassendi to Snell, 'Postridie Idus Augusti 1625', pp. 6–10.

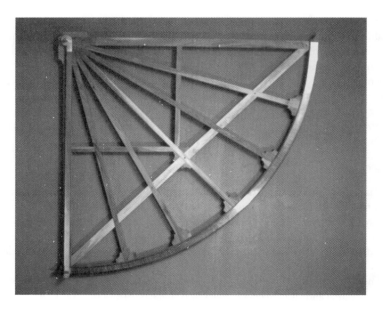

Figure 1. Replica of Gassendi's 5-foot quadrant as exhibited at the Musée de Digne in 1992

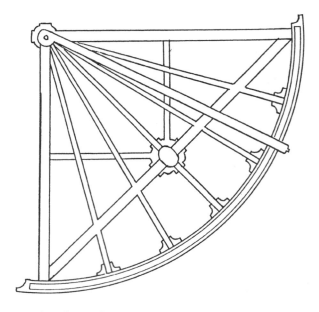

Figure 2. Replica of Gassendi's 5-foot quadrant (NB All drawings by Allan Chapman)

off its 60° point from the bottom of the brass limb. When this 60° angle is bisected to form 30°, the compass opening can be added on, to give an exact 90°. Using further bisections, trisections (threefold) and quinquesections (fivefold), the individual degrees can be drawn.

Each degree is then broken down into sixtieth parts, or minutes, by means of transversal lines. Firstly, each degree space of about 29 mm is divided into six equal parts of ten minutes. Ten equidistant lines that are concentric to it are next laid off, about 3.5 mm apart, and these are crossed by the diagonal 'transversals', six to one degree.

Figure 4

Reading an angle to a single minute then becomes simple. When Gassendi had sighted a star with the peg and slit, he would:

a)   Read off the nearest full degree against the *underside* of the alidade on the scale: 36°.
b)   Read off the full ten-minute spaces: 36° 20'.
c)   Run his eye across the scale to see where the alidade cuts *both* a transversal line and a concentric line together. By counting which of the ten concentric lines has been cut with a transversal, he could read off the individual minute which would be added onto the existing angle: 36° 20' + 4' = 36° 24'.
d)   If the alidade fell between *two* concentric lines on the transversal, then Gassendi would estimate to a half-minute, or 30 seconds: 36° 24' 30".

Though Gassendi's quadrant scale does not survive, it is clear from observations made with the instrument that it was divided into single arc-minutes, as one would expect for an instrument of large radius using Tycho's transversal scales. I have examined several transversal scales surviving from the seventeenth century, mainly in English collections, and have made micrometric measurements of six of them in an attempt to ascertain the reasonable accuracy parameters of the period.[6] The Museum of the History of Science, Oxford, has three beautifully engraved instruments from the collection of the early Savilian Professors of Astronomy. One of these instruments, a quadrant of six-foot radius, signed 'Elias Allen, Fecit Londini 1637', was used as a guide to the scale design of the Digne instrument.[7]

The five-foot Gassendi quadrant was constructed in walnut in a workshop in Digne by local carpenters, and the divisions were drawn by an engraver from Paris. The French craftsmen undertaking the manufacture worked from a detailed set of full-sized drawings which I had supplied, so that my paper templates and instructions

---

[6] A. Chapman, 'The Design and Accuracy of some Observatory Instruments of the Seventeenth Century', *Annals of Science*, xl, 1983, pp. 457–471.

[7] They include (1) Mural quadrant of 78.05 English inches radius, signed and dated 1637 by Allen, Museum of the History of Science, Accession no. 36.1; (2) Sextant, 73.26 inches radius, unsigned and undated but possibly by Allen, M. H. S. no. 36.2; (3) Iron quadrant, 25 inches radius, unsigned, possibly Allen, M. H. S. no. 36.3. See Chapman *Op. cit.* (n. 4), pp. 153–157, 197.

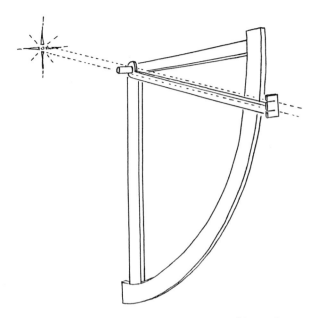

Figure 3. Sighting the star across the central peg of the quadrant

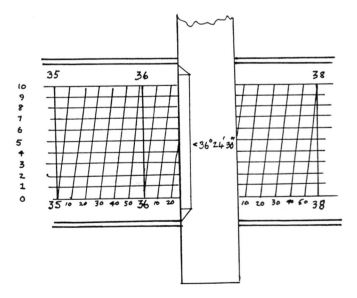

Figure 4. Reading the transversal graduations of the quadrant (full-scale)

Figure 5.   *Replica of Gassendi's* radius astronomicus *as exhibited at the*
Musée de Digne in 1992

could be used to assemble the framework of the quadrant, including joints and decorative features appropriate for an instrument of 1625. The scales likewise were copied from a fully-graduated, full-size paper scale that was part of the five-foot quadrant drawing supplied to the frame constructors. The engravers were also given precise instructions for laying off the cardinal degree points onto the scale with beam compasses, along with how to construct a seventeenth-century transversal scale reading to a single arc-minute. I fully commend the Digne craftsmen for the skill and thoroughness with which they turned my drawings into a beautiful instrument. It was decided to supply drawings, for local manufacture in Digne, to ease transportation problems.

The *Radius Astronomicus*, or Astronomical Radius

In 1619, Gassendi was making observations with an astronomical radius of 7 Paris Feet (2273 mm). This instrument was made in wood (probably beech) and was similar to that of his friend Joseph Gaultier, as described in *Opera omnia*.[8] For ease of subdivision, however, the present reconstructed instrument was slightly increased in length to the round number of 2300 mm. Unlike the quadrant, the *radius astronomicus* was made by me in Oxford and its two rods glued and jointed in Digne. It was mounted on a stand which I had designed based on a Tychonic prototype [Figure 5].

The astronomical radius was used extensively by Renaissance astronomers, being a relatively simple instrument to construct and easily dismantled for transportation. The great advantage, however, resided in its use of equidistant scales drawn upon a straight rod, which dispensed with the complexities of direct circular division.

The instrument exploited the natural trigonometrical relationship existing between the radius and tangent of a circle. All that it required was a long rod, to act as the circle's radius, and a shorter rod attached at right angles to form a 'T', and producing a tangent. Two fixed and two moveable brass sights could then be used to measure a variety of angles in the heavens once the radius was mounted onto an adjustable stand. As the quadrant was used to measure vertical (declination) angles, so the astronomical radius measured horizontal (right ascension) angles.

Figure 6

Unfortunately, Gassendi says little about his instrument beyond its rough dimensions, and there were a variety of graduation systems used by contemporary astronomers for the construction of *radii*. It has been possible to extract the characteristics of his division system, however, from the mathematical proportions used for the derivation of angles, as recorded in his observations.

The principle behind Gassendi's *radius* was the *theoretical* division of the 2300 mm rod into the 10 000 equal parts of the table of natural tangents. In reality, this rod carried no divisions at all, being a blank piece of wood, but its length was used to determine the size of the *actual* divisions drawn onto the tangent beam, and forming the 'T'.

---

[8] Gassendi *Op. cit.* (n. 1), vol. 4, pp. 75–76, 80 etc.

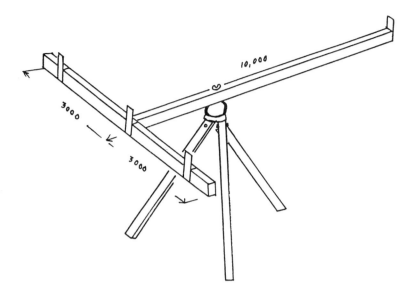

Figure 6.   *The* radius astronomicus *on its stand showing the transom and its proportions of divisions. The central axis brass sights are fixed, while the transom sights can slide*

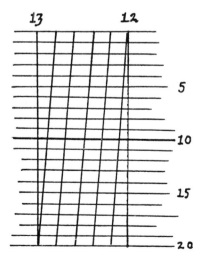

Figure 7.   *The transversal graduations of the* radius astronomicus *(full-scale)*

If the radial rod is 2300 mm long, each 1/100th division will be 23 mm. This will give the linear amplitude of each main division drawn onto the tangent, and when each space is subdivided into 100 equal parts by means of transversal lines, each single digit will, if perfectly drawn, correspond to a 0.23 mm space.

As Gassendi recorded observations on his transom amounting to 5 800 of these single, 0.23 mm equivalent divisions, his entire transom probably carried 60 main divisions, or 6 000 individual points (3 000 on each side of the central radial rod).

Though Gassendi does not indicate how the 100 individual divisions within each main space of 23 mm were laid-off, the following configuration would have been plausible, considering the dimensions of the rods.

Figure 7

Each 23 mm graduation is divided along its length into *five* equal parts, while the breadth of the tangent beam (55 mm) is divided in 20 equal parts, to produce 100 intersections by means of transversal lines. The angle would then be read in a similar manner to the transversal scale of the quadrant. The *inner* reading edge of the moving brass sight (that edge nearest to the centre of the transom) was used to locate which of the 100 intersections was crossed by a transversal line after an observation had been made. This was done by counting the whole five-digit subdivisions (20s), and adding them to the appropriate individual twentieth divisions (0.23 mm) drawn across the breadth of the transom, to obtain the total four-part figure. Thus, in the illustration, the inner edge of the sight reads:

Figure 8

– whole main divisions	=	27
– *four* whole 20ths	=	80
– single 20th divisions	=	3
TOTAL		2783

Or, for another example:

Figure 9

– whole main divisions	=	28
– *one* whole 20th	=	20
– single 20th divisions	=	13
TOTAL		2833

To make an observation of the angle between, let us say Mars and the bright star Arcturus (as Gassendi did in March 1619),[9] one would sight each object in turn by aligning it across the respective edges of the moving brass sights against the foresight

---

[9] *Ibid.*, p. 80.

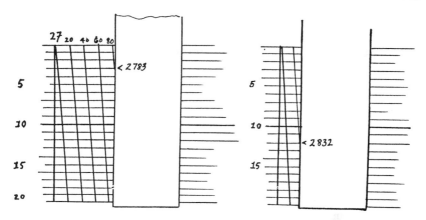

Figures 8 and 9.   Reading the transversal intersections against the edge of the brass sights on the radius transom *(full-scale)*

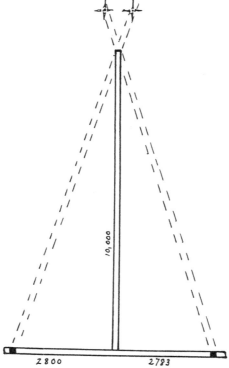

Figure 10.   Radius astronomicus seen in plan. Two observers at the extreme ends of the *transom use the brass sights to sight a pair of stars across the same tip sight*

at the far end of the radial rod. The technique would have been geometrically identical to the 'peg and slit' sight of the quadrant, and would have been parallax-free, upon compensating for the thickness of the sight on the scale.

Figure 10

Left   Right
  2800 divisions   2783 divisions

$$tangent \quad \frac{2800}{10000} = 0.28, \sqrt{} \ 15° \ 39' \qquad tangent \quad \frac{2783}{10000} = 0.27, \sqrt{} \ 15° \ 33'$$

For any angle that was beneath 16°, Gassendi could, if he desired, sight one object in the pair with the fixed central sights on the radius, and the other with either the left- or right-hand moving sight. Sometimes, he seems to have observed very small angles of one or two degrees by putting both moving sights onto the *same* transom arm, where he would have been easily able to switch his eye between the two sights.

To obtain the resulting angle, he simply took the two transom measurements and subtracted one from the other, as when in March 1619 he observed the small angle between Castor and Pollux, the two bright stars in Gemini.[10]

Figure 11

2400 divisions *minus* 1574 = 826, or tangent equivalent, 4° 32'.

It is possible (but by no means certain from his observations) that Gassendi's *radius* had a second cross-piece, or transom, at its far end, as did the instrument of his contemporary, Schickard.[11]

Figure 12

Such an instrument would have allowed angles of a greater amplitude to have been measured, though Gassendi does not seem to have recorded angles that were sufficiently large to suggest such an arrangement.

The Construction and Practical Use of Gassendi's Instruments

We possess no information about the craftsmen who constructed Gassendi's instruments. A skilled carpenter, however, could have built the wooden parts, while Gassendi himself may have drawn the divisions. Tycho Brahe and the contemporary English astronomer, Jeremiah Horrocks, were skilled graduators of their own mathematical scales. In the early seventeenth century, specialized astronomical instruments tended to be manufactured locally to suit an individual customer's specification, while the practical nature of astronomical observation itself suggests that the men who did it regularly, such as Gassendi, were probably good with their hands.

---

[10] *Ibid.*, for table of radius tangent scale numbers and their corresponding astronomical angles.
[11] Roche *Op. cit.* (n. 4), p. 29.

Figure 11. *Measuring a small angle with the* radius astronomicus, *with both brass sights at the same transom*

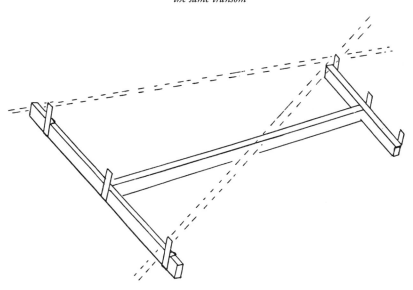

Figure 12. *Variant* radius *design. The addition of a second transom makes it possible to read a wider range of angles*

116

As Gassendi frequently travelled between Digne, Marseilles, Paris and other cities (and recorded making observations in many of them) he probably had a 'travelling kit'. In this capacity, the astronomical radius would have been ideal, because its rods could perhaps have been unpegged or unscrewed for transportation and a stand improvised. He seems, after all, to have been fond of observing the heavens from church towers. It is also possible that, having an accurately graduated transom bar and sights, (about 1400 mm long) to take on his travels, he improvised a precisely measured 7-foot (2273 mm) rod and front sight from local sources wherever he happened to be staying.

Because of the persistent cloudy weather that followed the completion of the reconstruction of Gassendi's *radius astronomicus*, and the necessity of getting it to France in good time for the opening of the Exhibition, it was not possible to test the instrument in actual astronomical observation. I did try out its scales against carefully worked-out terrestrial angles, however, and found that I could get accuracies to within a few arc-minutes. I have made astronomical observations with a replica *radius* on previous occasions and in particular, with the 36-inch instrument originally described by Jeremiah Horrocks in 1636.[12] In addition to finding the *radius* a very handy and surprisingly stable instrument to use, I found Horrocks' smaller instrument to be accurate to within well under ten arc-minutes when cross-checked against a star chart or sextant reading. This accuracy threshold could be improved, no doubt, with regular practice.[13]

The attempted reconstruction of early scientific instruments must always be undertaken with great caution and it must never be allowed to become a species of retrospective model making. This danger is always at its greatest when attempting to reconstruct instruments of conjectural design from epochs when linguistic and technical usages were different from what they are now. But when dealing with instruments of established usage, such as the quadrant and astronomical radius for which clear descriptions and sets of observations survive, one can learn a great deal from a scholarly reconstruction that is undertaken with the aim of ascertaining something about construction processes and practical handling qualities.

Acknowledgements

I wish to thank Roger Mason, FRGS, Oxford, for his assistance with seventeenth-century French linear measures, and Dr John Roche, Linacre College, Oxford, for helping me to establish a coherent system for the interpretation of Gassendi's astronomical radius values. I also thank A.V. Simcock for assistance with materials in the Museum of the History of Science, Oxford. But I take full responsibility for any mistakes.

---

[12] Jeremiah Horrocks, *Opera posthuma*, ed. John Wallis, London, 1673, 30 Aug. 1636, p. 250.

[13] A. Chapman, 'The Astronomical Art. The Reconstruction and Use of some Renaissance Astronomical Instruments', *Journal of the Br. Astron. Assoc.*, xcvi, 6, 1986, pp. 353–357. Also A. Chapman, 'Gauging Angles in the Seventeenth Century', *Sky and Telescope*, April, 1987, pp. 362–364.

X

George Graham and the Concept of Standard Accuracies in Instrumentation

The ability to make accurate measurements lies at the heart of all research in the physical sciences. Since the mid-17th century, when the Royal Society first came to systematise the "Experimental Philosophy", it was perceived that instruments formed the cutting-edge of new knowledge. Instruments enable the observer to focus his perceptions and quantify knowledge beyond the unaided senses: within sixty years, the telescope, micrometer and pendulum clock had transformed human understanding.

But if an observer was only as good as his instruments, so an instrument was only as good as its maker. When all instruments were hand-made, using craft techniques that a master had "sold" to his apprentice through the legal form of his indentures, a research tool was an idiosyncratic thing, dependent upon who had made it. How could an astronomer in Greenwich confidently collate his observations with a colleague in Paris when both men worked with quadrants made by different craftsmen, each of whom had used a "secret" technique of graduation? I believe that George Graham was the first craftsman to deal effectively with this impasse: identifying a scientific problem, inventing an instrument, analysing its performance and then applying tests and corrections to produce a design capable of technical evolution. Graham was also remarkable for combining the roles of craftsman and scientist in the same person, pursuing his own lines of scientific inquiry, and at all times describing the instruments, methods and checks required to obtain reliable results.

Rather appropriately for a man whose creative life and subsequent fame were so intimately associated with a particular institution, George Graham was the same age as the Royal Observatory, Greenwich — or perhaps a few months older. Born in either 1674 or 1675, he travelled down from his native Rigg in Cumberland to be apprenticed to the London watchmaker Henry Aske in 1688. After serving his apprenticeship — or maybe transferring part of it — he went on to work for Thomas Tompion, marrying his master's niece in 1704, entering into commercial partnership and taking over the business in 1713.[1]

While it is clear from these actions that Graham was a capable, businesslike and intelligent man, he might never have become more than a successful master tradesman. What made Graham significant was the way in which he used his business as a foundation upon which to conduct a series of researches in applied mechanics that won him both scientific and commercial renown.

One of Graham's long-lasting interests was the devising of invariable parts for his instruments, whereby distortions or inaccuracies could be made to cancel each other out. While he was not the first mechanician to think along these lines — Robert Hooke and his old master, Tompion, had entertained such ideas in the late 17th century — Graham was the first to approach the problem in a thorough-going manner. George Graham grasped the point as no one before him that no matter how skilfully a mathematical instrument was made and engraved, it would contain both random and systematic errors. Once one passed the level of accuracy where scales were required to perform to tolerances narrower than scratch marks on metal (as astronomers had by 1720) then mere craftsmanship was not enough: self-correction was essential. Much of Graham's genius resided in this recognition, and in the production of both instruments and observing procedures where systematic cross-checks could be applied at all stages.

X

Fig. 1. *George Graham, in an engraving by Hudson. On the table behind him is the open case of a clock with a mercury bob pendulum. Museum of the History of Science, Oxford.*

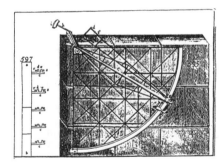

Fig. 3. *Graham's quadrant for Halley, 1725. Its frame (shown in a strangely flimsy way on this plate) was secured to a stone wall in the meridian. Note the screw adjusting mechanism that controlled the telescope's motion across the scales. Plate 597 shows the wall brackets by which it was secured and its weight evenly distributed. [From Smith's Opticks (1738) Plates 595 and 597.] It is preserved in the National Maritime Museum.* See also figure 22.4, chapter VIII, below.

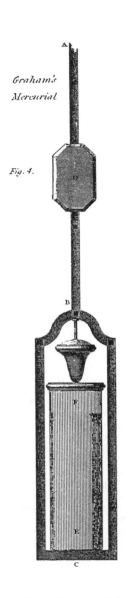

Graham's Mercurial

Fig. 4.

Fig. 2. *Graham's mercury bob pendulum. The metal stirrup hanging at the bottom of the pendulum rod carried the glass cylinder E-F in which the mercury was placed. Its upward expansion when warm counteracted the downward expansion of the rod. [A.Rees, Cyclopaedia (1819) "Horology", Plate XXXIX.]*

His first real excursion into this field was horological, as one might expect from a master clockmaker. His analysis of the behaviour of pendulum rods with changes of temperature resulted in his mercury-bob pendulum (Fig. 2), where the precise volumetric extension of the pendulum rod was carefully matched against the density of mercury, so that in warm weather the downward expansion of the rod would always be met by the equivalent upward expansion of the fluid bob (and vice versa in cold) to maintain an invariable oscillating length. The mercury pendulum, which he had perfected by 1722, worked by carefully offsetting contrary forces to maintain a homogeneous standard. Every step was capable of precise mathematical expression.[2]

An identical approach was displayed three years later when dealing with a wholly different physical problem: the graduation of a new 8 feet radius mural quadrant for Halley at the Royal Observatory. (Fig. 3) Although there were well-established procedures by which a craftsman could divide up a large arc into degrees and minutes, Graham was all too familiar with their flaws. A 90° scale could be a law unto itself, with no external standards against which it could be checked once it had been engraved, so that the astronomer became quite literally the prisoner of a potentially capricious scale. It is true that the problem of the random heterogeneity of a scale had been recognised by Hooke in 1674, though his solution of checking it against a complex worm-screw mechanism was of little practical benefit. As Flamsteed found when he tried to use his Hooke-inspired 7 feet sextant, one ended up with the frustration of trying to check a scale against the random errors of a screw in which wear was always taking place.[3]

With his astute recognition that simplicity of design lay at the heart of consistency of function, Graham cast around for a way of making two contiguous engraved scales that could act as a check upon each other. If he had merely engraved two concentric 90° scales two inches apart on the same quadrant limb he would have gained nothing, for each scale would have

contained its own random errors. Instead, he engraved an outer scale of 96 equal parts upon the same quadrant amplitude as the conventional 90° scale. He chose a scale of 96 parts because it was capable of being laid down by the 'pure' geometrical process of continuous bisection rather than the 'mixed' trisections and quinquesections necessary to produce 90°. Keeping the radius of the beam compass set to that with which he had scribed the circumferential arc, Graham transferred one of its points to the intended 0° digit of the future scale, and used the other point to strike a chord. He then utilised the Euclidean axiom that a chord of length equal to the radius of a given circle will always divide the circumference of that circle into exactly six parts(or 1 ½ parts for a quadrant) irrespective of the number of arbitrary parts into which one subsequently divided the circle.[4]

Once he had this natural chord struck off, he found that 64 was a convenient arbitrary number into which to divide the associated arc, for continuous bisection yielded 32, 16, 8, 4, 2 and 1 equal spaces. The 32 unit amplitude could then be further added to the 64 to produce a quadrant divided into 96 homogeneous parts. As the scale started and finished with the conventionally-divided arc to which it ran parallel (so that 0 = 0°, 64 = 60° and 96 = 90°) it was possible to use one scale to cross-check the other. A conversion table could easily be computed to give the exact equivalents in conventional degrees, minutes and seconds of the digits and fractions of the 96 - part scale.

For any observation made with the quadrant, therefore, it was possible to read off two different figures from the two separate verniers which, in theory, should reduce to exactly the same angle. Graham's twin scale was especially significant in that the scales were drawn in accordance with different geometrical canons (conventional degrees and continuous bisection numbers) so that congruence gave a very high certainty of accuracy. (Fig. 4)

Graham's 1725 quadrant, in addition to its double scales and superb craftsmanship,

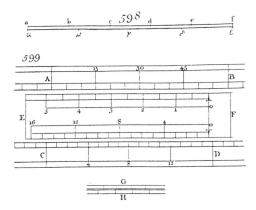

Fig. 4.   *The double vernier scale of Graham's 1725 quadrant for the Royal Observatory. The line A-B shows part of the inner 90° scale, and the line C-D the outer one of 96 parts. E-F is the vernier plate which moved across the scale attached to the sighting telescope at F. Its double verniers served both scales. G-H inset shows the method of reading an angle by a congruence of both vernier and mainscale divisions. As drawn, the scales read the angle 53'30". [From Smith's* Opticks *(1738).]*   See also figure 22.5, chapter VIII, below.

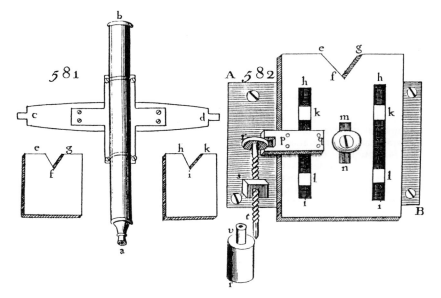

Fig. 5.   *Transit instrument from Smith's* Opticks. *Plate 581 shows the trunnions which supported the telescope a-b. Plate 582 shows one of the screw-set vee bearings in which the trunnion ends rested to produce a perfect horizontal location. The instrument shown here is a 'schematic' representation of a transit of the Graham period, as the trunnions of the Royal Observatory instrument were attached to the telescope with girder bars. (Graham's original transit is preserved in the National Maritime Museum.)*

embodied another innovation which was to be of fundamental importance to the future of mathematical instrument making. This was the provision of an adjusting screw capable of being clamped to the limb of the quadrant wherever one happened to be reading the angle, so that it could apply a delicate controlled pressure upon the sighting telescope to secure a critical alignment.[5]

In the form originally designed Graham did not supply the adjusting screw with a divided head, so that it could act as a micrometer and read between the vernier intervals to secure single seconds. Why Graham omitted this refinement is not easy to explain, for he was expert at cutting micrometer screws and equipped all his zenith sectors with them (see below). But James Bradley saw the potential [5] usefulness of such a device, and in 1745 commissioned John Bird (the 70-year old Graham was probably retired) to convert the screw on the 1725 Royal Observatory quadrant into a proper micrometer. The addition of this micrometer not only provided a useful cross-check on the verniers when making an observation, but could be used to monitor the quadrant's own internal errors as well. The micrometer had 39 ¼ threads to an inch, one turn moving the telescope through 53 arcseconds.[6]

Over the years, Bradley regularly compared the readings of the scale degrees with the micrometer equivalents, and computed the quadrant's error from the incongruity between the two sets of results. Later in the 18th century, John Bird and the Sissons (along with a host of lesser makers) also equipped their astronomical instruments with the micrometer, and the device became a basic feature of the marine sextant and surveyor's theodolite.[7]

Another prototype instrument that Graham built for Halley at Greenwich in the early 1720s was the transit telescope. Though not the inventor of the transit (Römer described his original instrument in 1675), Graham was probably the first to construct one in England, and certainly the first to design and build a transit to be used in regular research. The advantage

of the transit lay in the fact that it carried no scales or divisions, and consisted solely of a simple refracting telescope on a pair of precision trunnions. (Fig. 5) The trunnions rested in self-centring 'V' bearings, to give the instrument the appearance of a mounted cannon. When the trunnions and support bearings were oriented and levelled exactly east-west the telescope would describe a meridian arc.

In terms of mechanical efficiency it was much easier to lay the transit in the meridian plane than the quadrant.[8] The former could be made to lay down an exact meridian by pointing it to the northern sky and adjusting the orientation until a bright circumpolar star, such as Capella, took exactly half of a sidereal day to accomplish its respective eastern and western hemisphere passages around the pole, as timed by a good regulator clock. Once the meridian was established, the instrument could be reversed on its trunnions to face south, where the same clock could be used to time exact Right Ascension hour-angles for any object in the sky.[9]

The transit could also be used as a check upon the accuracy of the quadrant's vertical plane. If, in a given place, the quadrant's limb projected 1/100th inch more in one place than in another, then it would introduce an inevitable error of several seconds of arc when measuring the Right Ascensions of stars at that altitude. Because an 8-feet radius quadrant possesses a limb 13 feet long and a total surface area of around 50 square feet, with everything suspended in the vertical from a stone wall, one can understand why it was impossible to know which, if any, parts of the plane were out of true by a small fraction of an inch.

The transit instrument, however, enabled a simple yet critical test to be applied to the quadrant's plane at any time. If two astronomers, using the quadrant and transit respectively in the meridian plane a few feet apart, made a simultaneous observation of the same star, they *should* witness the R.A. culmination at the same instant. If, on the other hand, they found that stars at one altitude in the sky gave simultaneous culminations while others at different altitudes did not, then one could determine exactly where the quadrant's plane was out of the meridian,and by what number of seconds in Right Ascension. Because it was mechanically easier to adjust the transit trunnions to the meridian than the expansive plane of the mural quadrant, one had an ideal way in which one instrument could be used to check another.

While Halley possessed the facility for doing this at Greenwich with Graham's instrument, it was not until he was succeeded by Bradley in 1742 that the method came into general use. Bradley carefully "mapped" the meridian deviations of the quadrant's plane over 84°, to provide himself with a set of corrections which could be applied to any subsequent observation made with the instrument. Such procedures were central to the rationale behind Graham's instruments, with their possibility of constantly cross-checking measurements.[10]

With the possible exception of the 96 part scale and mercury pendulum, Graham invented relatively few *new* instruments. His genius lay in comprehending the functional limitations of existing designs and then improving them to create a whole new research potential. Just as the quadrant, micrometer, vernier scale and transit existed before Graham turned his attention upon them, so too the zenith sector had an ancestry dating back to the 1660s. At a time when astronomers had only an imperfect grasp of the laws of atmospheric refraction, the only way to make certain that refraction was eliminated from an observation was to make it in the zenith, where the light of a star passed through the thinnest part of the air in a straight line. As early as 1669 Robert Hooke had employed this fact in an attempt to measure the parallax of the star Gamma Draconis, which passed through the zenith in the latitude of London.[11]

Hooke's zenith sector was mounted in Gresham College, London, and consisted of a wooden tube 36 feet long hanging perpendicularly in the meridian plane from its object glass. To use the instrument

one lay on one's back, looked up into the [6] eyepiece, and noticed when Gamma Draconis culminated on the zenith meridian wire. But Hooke's sector had many defects, not least of which was the lack of any precise method of measuring the star's expected seasonal movements, for the instrument had no proper eyepiece micrometer. Others who had recognised the potential of the zenith sector included the French geodesist Jean Picard and the Astronomer Royal John Flamsteed, but the instrument remained awkward and unreliable in practice.[12]

Graham addressed himself to perfecting the zenith sector during the same decade that he was re-equipping the Royal Observatory for Halley. As far as one can tell, his first sector was for Samuel Molyneux in 1725, when Molyneux re-opened the problem of the possible parallax of Gamma Draconis. The Molyneux sector was not only the prototype for all of Graham's subsequent instruments of this class, but for most sectors made by other craftsmen thereafter. Optically it was simple, consisting of a single-element Keplerian system operating at a very low focal ratio. The brilliance of the design lay in Graham's approach as a mechanician. The cylindrical metal tube was allowed to hang dead, finding its own zenith meridian like a pendulum. With the telescope hung in the zenith, a fine-pitched micrometer screw secured to the instrument frame was brought into contact with it, and adjusted to read zero. When the screw was turned, gentle pressure was applied to the tube, moving it out of the zenith. Because these micrometer screw turns represented carefully measured fractions of the tangent of a circle centred at the top of the telescope tube, it was easy to convert thousandths of an inch on the screw pitch into seconds of arc of the whole circle. By looking at Gamma Draconis over a period of six months or a year, Molyneux hoped to be able to detect the star's slight shifts of parallax (caused by the earth's motion around the sun) by rotating the micrometer screw to move the tube, follow and measure it.[13]

Though the Molyneux instrument failed to detect a parallax, it did inspire James

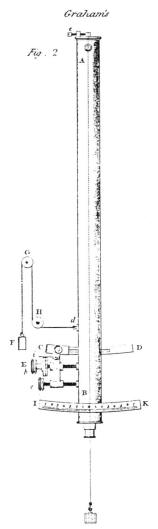

Fig. 6.   *Graham's zenith sector for Bradley,*
1727. The suspended telescope was kept in
rectification by the plumb line. The scale,
which extended 6¼° on each side of the zenith,
worked in conjunction with the micrometer
screw 'E' to track the periodic motions of
near-zenith stars. [From W. Pearson's
Practical Astronomy *(1828) Plate XXVII.]*
It is preserved in the National Maritime
Museum.

X

Bradley (the future Astronomer Royal) to commission Graham in 1727 to build him a 12½ feet instrument. This he set up at Wanstead in Essex, and left it hanging for over 20 years. Graham's 12½ feet sector for Bradley was an altogether more versatile instrument than Molyneux's. Its tube hung from the object glass in a pair of transit-like trunnions to move across a 6 ¼° scale on either side of the zenith, (Fig. 6) so that in addition to Gamma Draconis it could encompass some 200 other stars including the bright Capella. One full turn on the micrometer of this sector moved the telescope 0.0246 inches out of the zenith, while 1/80th of a revolution on its divided head moved it through 0.5 arcsecond. It was, without doubt, the most accurate astronomical instrument constructed up to that date.[14]

Though Bradley failed to detect a stellar parallax (which we now know was quite beyond the grasp of 18th century technology), he did use the instrument to make two other fundamental discoveries. In 1728 he discovered the aberration of light with it, and in 1748 the nutation, or nodding of the earth on its axis.

Bradley's discovery of the aberration was especially significant. It resulted from an unexpected observation made by him when using Molyneux's new sector: over the course of six months Gamma Draconis would appear to dip in one direction and then in the other through a total angle of 40 arcseconds. This original observation, made between November 1725 and March 1726, was perhaps Bradley's first incentive in commissioning the 12½ feet instrument from Graham, so that its wider compass of near-zenith stars could better test the universality of this phenomenon.

Bradley knew that it was not a parallax because the dip was not in the right direction. It then occurred to him, in one of those spontaneous flashes of inspiration so important in scientific discovery, that as the earth moves around the sun then for six months we will be approaching the stars in a certain direction of the sky, and for the next six receding from them.[15] In observational terms, this vector will make the apparent position of any star shift first in one direction, then in another, over the course of 12 months. Bradley was entirely correct in his 1728 interpretation, while his value of 20.00 to 20.25 arcseconds comes remarkably close to the 20.40 arcseconds accepted today.

I am not aware that Graham was especially concerned with the design and manufacture of astronomical instruments before he was commissioned to make the Greenwich pieces for Halley. Prior to that, he seems to have been concerned primarily with horological mechanisms.[16] But the success of his instruments for Halley, Molyneux and Bradley won him international attention. Graham's role in the discoveries of Bradley was fully acknowledged by that astronomer, as "depending on the accurateness of the instrument"…"of our curious member Mr George Graham, which made the Aberration and Nutation discoverable".[17] Bradley was one of the first major astronomers to fully appreciate the barriers which the quality of one's instruments imposed, and recognised in George Graham a consummate mechanician who could not only make instruments capable of working to hitherto unprecedented levels of "accurateness" but who, as a scientist in his own right, fully comprehended and shared the problems for which a solution was being sought. Nor is there any hint of Bradley being patronising to Graham, for we must remember that the latter was not only a highly successful artisan by the 1720s but a Fellow and Council Member of the [7] Royal Society in his own right.

In 1735, Louis XV of France sent an expedition to the extreme north of the Gulf of Bothnia in the Baltic. Its astronomers were to survey a line on the river Tornea, which flows north-south, to measure the amplitude of a meridional degree of arc within the polar circle. To perform this operation the expedition carried several instruments by George Graham, including a zenith sector of 9 feet radius. The sector was used to measure the exact zenith angles of stars at the northern and southern ends of the surveyed line, so that a precisely measured linear distance upon the surface of the earth (or frozen river) could be equated with a celestial amplitude. The expedition was sent to settle once and for all the dispute about the shape of the earth; was it pointed at the poles as Cassini's French meridian lines suggested, or flattened in accordance with the postulates contained in Newton's Principia? The answer was to be obtained by measuring the surface length of a degree of arc within the polar circle and comparing it with degrees measured in France and Peru.

It says a great deal about Graham's reputation in 1735 that an official French expedition carried English instruments. In their results published in 1737 Maupertuis and the French Académiciens had nothing but the highest praise for Graham's instruments, describing in minute detail the paces through which the zenith sector — as the instrument which measured the crucial zenith angles — was put. Prior to beginning the astronomical observations, the zenith sector's 5½° scale and micrometer were tested in a series of horizontal triangles. These triangles were surveyed on the flat surface of the ice with chains, to create known terrestrial proportions which the sector scale could then measure. Maupertuis and his four astronomical colleagues took turns to make independent measurements with the sector placed horizontally on wooden trestles, and when all were finished collated their results. The five men agreed that the minus error of no more than 3¾ seconds of arc could be ascribed to the contraction of the instrument in the bitter sub-zero temperatures in which they worked, rather than to any error of graduation. When further tests were applied to the homogeneity of the degree spaces one to another and to the micrometer screw, they found that no errors larger than 0.95 seconds of arc were discernible.[18]

By 1740 it had come to be recognised that Graham's instruments set an international standard of excellence, both for mathematical and horological pieces, and that this excellence derived from the principles in accordance with which they were designed and constructed, and the intimate liaison between craftsman's art and scientist's inquiry implicit in them. These principles not only enabled astronomers to quantify the errors present in any Graham instrument, and apply an appropriate correction, but to detect progressive alterations in a given instrument with the passage of time. In the same way that the French Académiciens had detected tiny systematic errors which they ascribed to the zenith sector's contraction in an Arctic climate, so James Bradley in 1746 was able to detect a 15¾ second error in the 8 feet Greenwich quadrant, which he likewise attributed to a thermal cause.[19]

The Greenwich quadrant consisted of a heavy brass limb riveted on a rigid gridiron of cross-braced iron bars. But after twenty years the differing expansion co-efficients of brass and iron had caused buckling of the brass scale. Though invisible to the naked eye, the quadrant's micrometer was able to detect the 16 arc-second error mentioned above, of which there had been no record twenty years previously when the quadrant was new. The error was found to be at its worst around the 45° point, where the distortion pressures would have been at their greatest.

When John Bird was commissioned to build a new quadrant for Greenwich in 1749 he followed Graham's design in its main principles. But Bird learned from Graham's mistakes, and eliminated thermal disequilibria by fabricating the entire instrument in brass.[20] Bird's quadrant remained in use at Greenwich until 1813. After 1753, following the completion and testing of the new Bird instrument, Graham's quadrant was removed from its original south-facing position on the meridian wall, its worn divisions re-drawn by Bird, and turned through 180° to face north where, in illustrious semi-retirement, it was to measure the culminations of circumpolar stars until 1812.[21]

George Graham represents the coming-of-age of scientific instrument making. Before him was a wealth of ingenuity, designing and devising as scientists strove for better data, and intelligent artisans did their best to make it possible. But in

X

Graham the desires of the researcher combined with the perceptive skills of the maker to produce a wholly different order of operation. This was made clear not only by the recognition which he received during his own lifetime but from the way in which his precepts were followed by later generations. In his open-handed love of technical excellence he willingly advised and made an interest-free loan to John Harrison to aid his chronometer work.[22] Jeremiah and Jonathon Sisson, John Bird, Jesse Ramsden and Edward Troughton, along with a host of clockmakers, also admired him as the inspirer of their craft. The man they looked back to was more than just a master craftsman; he was the artist and scientist who had raised their trade to the status of an intellectual discipline, and shown that the standards which it set were indispensable to the advancement of the physical sciences.

Acknowledgements

I wish to thank the National Maritime Museum for access to the Graham instruments in its collections, and Mr. A.V. Simcock, Librarian of the Museum of the History of Science, Oxford, for his assistance with illustrations.

Notes and References

1. For the major facts of Graham's life see E.G.R Taylor, *Mathematical Practitioners of Hanoverian England* (Cambridge, 1966) 120-121. Also *Dictionary of National Biography* and *Dictionary of Scientific Biography*.

2. George Graham, "A contrivance to avoid the irregularities in a clock's motion occasioned by the action of heat and cold on a pendulum rod", *Philosophical Transactions of the Royal Society of London*, (1726) 40-44.

3. Robert Hooke, *Some Animadversions on Hevelius* (London, 1674) 51-58 ff. John Flamsteed, *The 'Preface' to John Flamsteed's 'Historia Coelestis Britannica' 1725*. Edited and Introduced by Allan Chapman, National Maritime Museum Monograph No. 52. (Greenwich, 1982) 118, 204.

4. Robert Smith, *A Compleat System of Opticks* (London, 1738) 334-336. Derek Howse, *Greenwich Observatory III; the Buildings and Instruments* (London, 1975) 23-24. The instrument is still preserved at the National Maritime Museum, London.

5. Smith, *Opticks*, 337-338.

6. James Bradley, *Miscellaneous Works and Correspondence*. Edited by S.P. Rigaud (Oxford, 1832) 1v.

7. Allan Chapman, *Dividing the Circle* Ellis Horwood (Chichester, 1990) 71-81.

8. S.P. Rigaud, "Some particulars respecting the principal instruments at...Greenwich, in the time of Dr Halley", *Memoirs of the Royal Astronomical Society*, 9 (1836) 205-207.

9. Rigaud, "Some particulars.." op. cit., 210. Also, Smith's *Opticks*, 325-326. James [8] Bradley, *Astronomical Observations made at the Royal Observatory*, Edit. T. Hornsby (Oxford, 1797) ii.

10. Rigaud, "Some particulars..." op. cit., 220-221.

11. Robert Hooke, *An Attempt to Prove the Motion of the Earth* (London, 1674) 7.

12. Richard Waller, *The Measure of the Earth...by the Académie of Sciences at Paris* (London, 1688) 28. For Flamsteed's 100 foot "well telescope" see Derek Howse, *Francis Place and the Early Years of the Greenwich Observatory* (New York, 1975) 58, Plate XIIa.

13. Bradley, *Miscellaneous Works*, op. cit., 93-116.

14. Bradley, *Miscellaneous Works*, 194-195.

15. J. Bradley, "Account of a new discovered motion of the fix'd stars", *Phil. Trans.*, (1728) 637-661; 646.

16. These included his invention of the "dead beat" or non-recoil escapement for clocks, 1726, and the cylinder watch

escapement (1725); see *Britten's Old Clocks and Watches and their Makers*, Edit. Clutton, Baillie & Ilbert, 9th Edition, (London, 1982) 125-128; 149-150.

17. Bradley, "Account of a new discovered motion" op. cit., 638. Bradley also praised Graham in his "Letter to the Rt. Hon. George, Earl of Macclesfield, concerning an apparent motion of some fixed stars," *Phil. Trans.*, (1748) 1-43; 6.

18. P.L.M. Maupertuis, "La figure de la terre déterminée...mesure le degré du méridien en cercle polaire", *Mémoire de*

l'Académie Royale" (1737) 386-466.

19. Bradley, *Miscellaneous Works*, op. cit., 1v. Also, John Bird, *The Method of Constructing Mural Quadrants*, (London, 1767).

20. Bird, *Quadrants*, op. cit., 7-8.

21. Howse, *Greenwich Observatory*, op. cit., 24, 26.

22. Arthur Raistrick, *Quakers in Science and Industry*, David & Charles, (Newton Abbot, 1968) 239.

Reprinted from Bulletin of the Scientific Instruments Society *27 (1990), pp. 3-8. Numbers in square brackets in the margins indicate original pagination.*

XI

SCIENTIFIC INSTRUMENTS AND INDUSTRIAL INNOVATION: THE ACHIEVEMENT OF JESSE RAMSDEN

Like all professions, that of the scientific instrument-maker has its élite figures, for as soon as answers to many questions in natural inquiry, such as the true shape of the earth and its gravitational characteristics, were seen to hinge on the interpretation of exact physical data, then the intellectual artisan became a necessary breed. George Graham became the prototype of the virtuoso craftsman who made an international reputation, and a fortune, and contributed fundamental advances to precision instrumentation,[1] while Edward Troughton's concern with achieving precision in mechanical devices which worked with the repetition of an industrial machine took élite craftsmanship to a new level by the early nineteenth century. But the bridging figure who linked the tradition of the great 'one-off' bench artists like Tompion, Graham and Bird, with the industrial scale of Cooke, Troughton and Simms, was undoubtedly Jesse Ramsden.

Ramsden's career fits almost exactly that era which some economic historians style 'the first industrial revolution'. The hallmarks of that revolution had been the mechanization of traditional processes of hand-manufacture, by the application of specialized devices. One might also suggest that industrial manufacture embodied two further components: the elimination of variations between individual artefacts and the establishment of uniform standards.

More than any other craftsman, Jesse Ramsden turned the scientific instrument, be it an observatory circle or a draughtsman's protractor, into a cost-effective industrial artefact, but where high quality was made to cost less in real terms than ever before.

Like the products of Boulton, Watt and Arkwright, Ramsden's technological innovations were stimulated by new demands being placed upon old skills, and as with these industrial innovators, his success was based upon extensive research and development.

Several factors moulded Ramsden's career as a scientist and scientific entrepreneur from the 1760s onwards. Pre-eminent among these factors was the development of the shipboard astronomical instrumentation whereby 'lunars' (i.e. the measurement of the changing angles between the moon and certain stars)

---

[1] Arthur Raistrick, *Quakers in Science and Industry*, Newton Abbot, 1968, pp. 234–41. (Graham's Quaker credentials are not wholly proven.) Allan Chapman, 'George Graham and the Concept of Standard Accuracies in Instrumentation', *Bulletin of the Scientific Instruments Society*, 27, 1990, pp. 3–8.

could be practically determined for finding the longitude. The second was a fundamental change taking place in the research instrumentation of astronomical observatories, as the circle usurped the ancient quadrant as the fundamental angle-measuring device. The third, in order of scientific if not commercial importance, arose from the growing need for surveying and drawing instruments by architects, draughtsmen and canal engineers in the late eighteenth century. Jesse Ramsden 'industrialized' the first and third branches of instrumentation by mechanizing, in various ways, their process of manufacture, shifting reliance for accuracy from the artisan's skill to the automatic action of a machine, and greatly speeding up production. The profits generated by this high-quality 'bread and butter' trade in relatively routine instruments produced the resources whereby Ramsden could develop the astronomical circle as a fundamental research tool.

In the same way that Matthew Boulton's metal manufactory developed the wherewithal to design the condensing steam engine, so Ramsden's trade in machine-calibrated sextants and theodolites paid for the time which enabled him to devise the Roy, Palermo, Dunsink, and Shuckburgh circles which could measure angles ten times smaller than Bird's quadrants.

Devices for speeding up or regularizing specialized aspects of instrument manufacture had been in use a long time before Ramsden. The clock makers' wheel-cutting engine dated from around 1670, while opticians had long since used lathes and tools intended to give relatively uniform focal lengths to batches of lenses. Yet all of these techniques operated within relatively stable conditions of supply and demand, simply to speed up or regularize an established product. Jesse Ramsden's dividing engine in the 1770s, however, was invented to develop and manufacture what was, in practical terms, a new product, operating within new parameters of accuracy, to be used in conjunction with newly available astronomical data. While it is true that Bird had made a few large-radius hand-divided sextants with which to try out Mayer's lunar tables at sea, both the instruments and the technique had been experimental in 1760.[2]

It can be argued that the key to Ramsden's achievement, and the factor which created the problems which his particular ingenuity first came to solve, was the production of reliable lunar tables by Tobias Mayer in the 1750s and their subsequent development for use at sea by Maskelyne in the *Nautical Almanac* after 1767.[3] Before these tables could be of practical advantage in the finding of the longitude, however, it was necessary for navigators to possess a hand-held instrument which was sufficiently compact to be usable on the deck of a pitching vessel, yet accurate enough to read angles to 30 arc seconds or less: an accuracy

---

[2] John Smeaton, 'Observations on the Graduation of Astronomical Instruments', *Philosophical Transactions of the Royal Society*, 76, 1786, pp. 181–2, states that by 1760 Ramsden was thinking about the mechanical division of sextants.

[3] Derek Howse, *Nevil Maskelyne, the Seaman's Astronomer*, Cambridge, 1989, p. 14.

420

level higher than that attainable with a rigidly mounted observatory quadrant of seven feet radius a century before. Equally important, moreover, it was necessary for the instrument to be as easy to use when measuring awkward angles between the moon and a star as it was in the vertical, while its scale had to possess a sufficient amplitude to allow the navigator to shoot angles of over 100 degrees to catch stars that were behind his back. In short, the finding of the longitude at sea by lunars demanded a shipboard instrument which was smaller and more compact than an old-fashioned sea quadrant, and yet which still read to what was almost a land-based observatory standard of accuracy.[4] Though we have long admired John Harrison for his painstaking analysis of the problems of marine chronometry, little attention has been paid to the equally important problems involved in developing a viable sextant for lunars, which was Ramsden's achievement.

It was the marine octant, devised by John Hadley around 1731, which provided the principle from which the sextant developed. By using a 45-degree scale that was capable of reading to 90 degrees by reflection, the octant was a relatively compact and easily held instrument reading down to one arc minute on a 15- or 18-inch radius. But to produce an instrument which was light enough in weight to carry such a comparatively large radius at sea, the frame had to be made of thin wood and the scale of ivory. These materials inevitably limited the accuracy of the instrument, for they lacked the rigidity and thermal uniformity of brass. While a Hadley octant was sufficient in theory to take the sun's noon altitude to a minute, it was still incapable of reliably taking sightings for lunar computations which demanded much higher levels of manoeuvrability and accuracy.[5]

Not only had the 90-degree scale of the octant to be increased to 120 degrees, but the radius had to be drastically reduced to around eight or ten inches to make it more compact and easily handled. At the same time, the graduations had to be twice or three times more accurately drawn at nine inches radius than had been customary on an octant of fifteen. In short, the instrument had to be miniaturized, made more rigid and much more finely divided, equipped with a telescopic sight for more precise alignment, and capable of being supported in the right hand while the left was free to make fine adjustments. This was a tall order in terms of design, and Ramsden perceived that everything hinged on the craftsman's capacity to produce impeccably accurate divisions at ten or twelve inches radius.[6]

---

[4] A. N. Stimson, 'Some Board of Longitude Instruments in the Nineteenth Century', in *Nineteenth Century Scientific Instruments and their Makers*, ed. P. R. de Clercq, Amsterdam, 1985, p. 95.

[5] Ramsden found the centring and graduation errors of contemporary octants to be as large as 5 minutes. He claimed that his improved sextants of 15 inches radius were accurate to 6 seconds: J. Aiken, *General Biography*, 8, London, 1813, p. 451, article 'Ramsden' by Rev. L. Dutens. Dutens's essay article provides one of the major primary sources for Ramsden's ideas, motivation and character by a person who claimed to have known him: J. A. Bennett, *The Divided Circle*, London, 1987, pp. 130–34.

[6] The radius of sextants became smaller as engine division became standard. Ramsden's original

In many respects, it was a situation similar to that faced by his Greenock contemporary, James Watt, when he applied his mind to the improvement of the Newcomen steam engine, making it smaller and much more powerful by the application of a separate condenser. And in the same way that Watt's engine had only become practicable when equipped with precision high-pressure cylinders and valvework manufactured in Boulton's engineering works, so Ramsden's transformation of the octant into the sextant was only possible when uniform divisions could be speedily engraved upon small radius instruments.

One must not forget that the biggest drawback to the initial use of the chronometer at sea at this period was the scarcity and prohibitive cost of the watches, and if lunars were ever to be widely used, then an easily-manufactured, modestly priced, and wholly reliable hand-held astronomical instrument for making the necessary sightings had to be quickly developed. The hand division of ten-inch scales to the astronomical standard of accuracy required for lunars was not commercially viable in 1760. Although John Bird could probably have done it over the weeks which it supposedly took him to engrave a homogeneous scale, the finished sextants could have been as scarce and expensive as Larcum Kendall's replica of the Harrison IV chronometer.[7] The superb 20-inch sextant which Bird engraved for Captain Campbell in 1757 indicates what rarities such instruments were likely to be.

It was in Jesse Ramsden's development of the dividing engine that the problem found a solution, to bring out the industrial revolution parallels, for Ramsden was transforming an existing technology to meet new demands, greatly improving quality, and making it possible for a labourer to do in thirty minutes what an established craftsman could not do in a fortnight.[8]

The principle at the heart of Ramsden's first, 1766, and all subsequent engines was that of reduction copy division, whereby a large template, such as Hooke's 'ten-foot dividing plate', was copied on to a smaller instrument.[9] In the past, this process had been used by craftsmen to minimize the inherent errors of copy division, by working on the assumption that original errors in arc minutes when working at ten feet radius would be less than when working at ten inches, and

---

instruments had often been of around 15 inches radius, but by the mid-1790s serial number 1209 (Whipple Museum, Cambridge) was 8¼ inches while Troughton's 'snuff box' sextants were under 2 inches. Bennett (see n. 5 above), p. 137.

[7] Between 1766 and 1776, Larcum Kendall made three chronometers for the Admiralty that were effective copies of Harrison IV, to test the feasibility of duplicating Harrison's design. They cost, respectively, £450, £200, and £100: see Derek Howse, 'Captain Cook's Marine Timekeepers: Part 1, The Kendall Watches', *Antiquarian Horology*, 1969, pp. 190–92. Derek Howse, *Nevil Maskelyne, the Seaman's Astronomer*, Cambridge, 1989, p. 14.

[8] *Edinburgh Encyclopaedia*, 10, 1830, article 'Graduation', p. 357, cited from Edward Troughton. Allan Chapman, *Dividing the Circle, the Development of Critical Angular Measurement in Astronomy 1500–1850*, Chichester, 1990, p. 158.

[9] Robert Hooke, *Some Animadversions ... on Hevelius*, London, 1674, p. 14.

thereby making it more reliable to copy from a ten-foot than from a ten-inch scale. What made Ramsden's engine such an innovation, however, was his adoption of the large denticulated plate and lead screw from the horological engine of his fellow-Yorkshireman, Henry Hindley, so that conventional eye alignments could be replaced by the screw turns of what was a thirty-inch diameter precision micrometer.

Ramsden's engine thereby differed from all previous techniques of copy division on two important grounds. First, he did not use the technique merely to minimize error, for the divisions on a nine-inch radius sextant were to be no less accurate than those on the thirty-inch main wheel of the engine. Second, it was not invented merely to speed up the manufacture of traditional mathematical instruments, such as protractors, but to provide the technology which would make possible a new precision artefact at a reasonable price, in large quantities.

But Ramsden's first dividing engine of 1766 was soon found to fall short of its inventor's expectations, and while it was found to be good for the manufacture of conventional mathematical instruments, it was not good enough for sextant scales.[10] In consequence, Ramsden set about the construction of an improved model with a 45-inch plate which became operational in 1774. It was this machine that so impressed the Admiralty that they purchased it for £315, while permitting Ramsden to retain it for his own use. The Government also paid him a further £300 to write and publish a detailed description of the engine for the public benefit.[11] And though Ramsden did this, he strategically avoided describing the critical process whereby he laid off the 2160 circumference teeth on the dividing wheel, the accuracy of which made the machine so significant. Over the next quarter of a century, Ramsden's firm divided annually about forty sextants[12] on this machine—more than enough to supply each capital ship in the Royal Navy and part of the Merchant Fleet as well. The ease of manufacture of engine-divided sextants, reading down to 20 or 30 arc seconds, made possible precision instrumentation on a truly industrial scale, and one can understand why the cost-conscious Admiralty tended to favour the lunar's method of finding the longitude against that of the chronometer. It seemed cheaper, after all, to teach officers how to use a £15 sextant, solve equations, and compute from Maskelyne's tables, than it was to provide every ship with a chronometer.[13]

Crucial to the perfection of Ramsden's circular dividing engine had been the

---

[10] William Pearson, 'Graduation', in Abraham Rees, *The Cyclopaedia*, 16, London, 1819. The most complete study of the dividing engine to date is John Brooks, 'The Circular Dividing Engine; the Development in England, 1739–1843', *Annals of Science*, 49, 1992, pp. 101–35.

[11] Jesse Ramsden, *Description of an Engine for Dividing Mathematical Instruments*, London, 1777, Preface; *idem*, *Description of an Engine for Dividing Strait Lines*, London, 1779, pp. 13–16; *Edinburgh Encyclopaedia* (see n. 8 above), p. 253.

[12] Stimson (see n. 4 above), p. 98. The *DNB* states that Ramsden made 1,000 sextants by 1789.

[13] Ibid. See also Stimson's n. 26.

great improvements which he made to precision screw cutting, for while the 2160 teeth on his engine plate provided the main ten-minute division spaces, they had to be engaged by a lead screw the threads of which could measure down to ten seconds to make the engine work.[14] But once the machine was assembled, and the rachet mechanism on the lead screw adjusted to make it delineate a specific amplitude on the dividing plate, then a mere labourer could operate its treadle and scriber in alternation, to engrave superlative scales. Though Ramsden never recorded how long it required to engrave a sextant on his second machine of 1774, we do know that his rival, Edward Troughton, employed a 'young man' who could engrave one on Troughton's almost identical engine in thirty minutes. This is indeed an eminently reasonable figure, as is known from personal experience, for in 1978 the author used the same Troughton engine, preserved in the Science Museum, to engrave an octant scale. It was found that with a little practice, 36 divisions per minute of time could be engraved, working rapidly, or a whole sextant in thirty minutes with ease.[15]

An important point about Ramsden's work, however, is often misunderstood, for while his dividing engine was a superb industrial tool, it was never used to graduate the large observatory circles for which he was famous.[16] The dividing engine was reserved exclusively for graduating what might be called *secondary* observational instruments, such as sextants and conventional theodolites, which were used for making comparison observations (as from the *Nautical Almanac*), and was never used to divide the *original* scales of fundamental research tools.

The graduation of the great astronomical circles, destined as they were for some of the leading observatories of the day, was the product of a quite different technology, though they, like the dividing engine, bore important parallels to the rationale of the industrial revolution. If the dividing engine is thought of as pioneering a branch of mass production, then the great astronomical circles might be seen as a development in primary capital equipment technology. Like a canal, or a railway, a potential customer was likely to want only one great circle of the type developed and supplied to the Palermo or Dublin Observatories. And like a railway, or more particularly, like a locomotive, this brand new piece of technology would be the product of much fundamental research and development as a specification which accorded with the individual customer's needs was evolved. Though this evolution of 'watershed' technologies was not unique to the Industrial Revolution (Bird's method of dividing, developed in the 1740s, had produced a whole new generation of quadrants) circle-based astronomy, like the railway, integrated several technologies—improved optics, new bearing systems,

---

[14] Ramsden, *Description ... Mathematical Instruments* (see n. 11 above), p. 13. Rees, *Cyclopaedia* (see n. 10 above), article 'Engine', unpaginated, 'Mr Ramsden's Engine'.

[15] See Chapman (n. 8 above), pp. 158–9.

[16] A. Chapman, 'The Accuracy of Angular Measuring Instruments used in Astronomy between 1500 and 1800', *Journal for the History of Astronomy*, 16, 1983, pp. 133–7.

more elaborate micrometry, and new concepts in cross-checking—which between them possessed enormous potential for further development.

Ramsden could never have used the dividing engine to graduate circles of four or five feet diameter even had he wished to do so, for while the engine was fine when transferring its divisions on to a hand-held instrument intended to read to 20 arc seconds, an astronomical circle had to be much more precise if it was to transcend the established plus or minus one arc second of Bird's quadrants. This required a process of *original* division which, in Ramsden's hands, made the circle capable of reading to an extra decimal part of a single arc second.

Like many eighteenth-century industrial inventors who worked before the creation of a reliable patent system, Ramsden never revealed the precise techniques whereby his instruments achieved their excellence. While he was willing to publish an account of a finished product if paid to do so, or let a satisfied customer such as Sir George Shuckburgh do so for him, Ramsden never revealed how the graduations which lay at the heart of his instruments were drawn. In this respect, moreover, he was at one with his commercially-minded astronomical contemporary, William Herschel, who was quite happy to communicate detailed accounts of the *mechanics* of his reflecting telescopes to the readers of *Philosophical Transactions* while studiously saying nothing whatever about the manufacture of the *optics* which made them unique.[17]

Jesse Ramsden was not the first craftsman to construct circular instruments for astronomical observatories, though he was the first to *develop* them as pieces of fundamental capital equipment, for the astronomer's principal measuring instrument was still the quadrant when he entered the scientific scene in the 1760s. It was Ramsden who undertook the major research and development work which made possible much of the instrumentation of the nineteenth-century observatories.[18] In particular, the astronomical circle, and to a lesser extent the equatorial mount, were Ramsden's achievements, and it says something for the success of his business, as well as about his scientific foresight and ingenuity, that he could devote so much of his time and resources to developing new lines. Once again, his actions parallel those of the capitalist inventors, and in particular those of the cotton trade who were willing to disrupt the assured profits of an established technology to take time off to develop mechanised processes. Unlike Arkwright, Cartwright and other textile entrepreneurs, however, Ramsden's

---

[17] William Herschel, 'Description of a Forty-Feet Reflecting Telescope', *Philosophical Transactions of the Royal Society*, 1795, pp. 347–409. When William Lassell, in the 1820s, attempted to make his first specula, he was obliged to use John Edwards's article in the *Nautical Almanac* (1787): A. Chapman, 'William Lassell (1799–1880), Practitioner, Patron and "Grand Amateur" of Victorian Astronomy', *Vistas in Astronomy*, 32, 1989, pp. 349, 366.

[18] Chapman (see n. 8 above), pp. 112–14. George Shuckburgh, 'Description of an Equatorial Instrument', *Philosophical Transactions of the Royal Society*, 83, 1793, pp. 67–128. Jesse Ramsden, *Description of the Universal Equatorial*, London, 1791.

enterprise needed something in addition to the profit motive: a desire to advance learning. The perfection of the astronomical circle demanded sacrifice with no prospect of a financial return, though it did eventually win him a fellowship of the Royal Society, and the honoured status of member of the St Petersburg Academy.

It is not clear why Ramsden did take upon himself the development of the astronomical circle, for while he produced several famed circular instruments in the 1780s and 1790s, they all seem to have been years in building, while the original customers who commissioned them were not always the same as those who subsequently possessed them. It seems, however, that the dividing engine formed a mechanical twin with the process of original circular division of observatory instruments in Ramsden's mind, for the graduation of his engine plate into 2160 teeth posed identical problems to the division of the Palermo and other astronomical circles. All of these instruments seem to share a joint ancestry with the sophisticated horological engine of Henry Hindley (with which Ramsden became familiar by repute via John Stancliffe) and with the Duc du Chaulnes's dividing process, published in 1768.[19]

From Hindley (or perhaps from the writings of Robert Hooke) Ramsden came upon the idea of turning a precisely denticulated plate by means of an accurate tangent screw, so that the resulting dividing machine was, in effect, a large micrometer.[20] From the Duc du Chaulnes, he came to recognize the importance of the reversible properties of the circle, in which errors could be cross-checked against 180 or 60 degree marker points both to construct the circle originally and to read angles with it when complete. It was almost certainly from the Duc du Chaulnes that Ramsden also hit upon the idea of using low-power microscopes, equipped with reticules, to read the graduation points on a scale.[21]

Once more, we see in Ramsden the successful development of a set of experimental devices contrived initially by others, in the same way that Arkwright's spinning Mule brought to mechanical fruition the transitional machines of Crompton, Hargreaves, and others. On the other hand, we must not forget that practitioners in established crafts such as clock and instrument-makering had been improving on each other's ideas for well over a century by 1770. But where Ramsden differed from Tompion or Graham was not in his ingenuity, but in the wider world wherein that ingenuity could operate. Inventions, after all, are made important by their context. Watt's development of the steam engine differed from Newcomen's or Papin's in the scale of adjacent industrial operations going on in England in 1780 as opposed to 1710. Ramsden's

---

[19] John Smeaton (see n. 2 above).

[20] Hooke, *Animadversions* (see n. 9 above), pp. 55–6.

[21] Duc du Chaulnes, *Nouvelle Méthode...*, Paris, 1768; William Pearson, *An Introduction to Practical Astronomy*, 2, London, 1829, pp. 413–28.

significance as an 'industrial' figure differed from the earlier horologically-related crafts by 1775 because society was placing more diverse needs upon mechanical 'precision' than simply using it to tell the time. His range of contributions thereby extended across tool-making and the large-scale manufacture of sextants, while at the heart of his scientific (as opposed to commercial) reputation lay his development of large circular instruments intended for original astronomical research.

The experimental, or developmental, status of his large circles is clear in the many years which it took to build most of them. Having received an order for a major instrument from a private person or institution, he would mull over it for years, or in the case of the Dublin circle, for *decades*, before it was received by the customer. While Ramsden, as a perfectionist, could accept an order for an instrument of novel design and great promise, it seems that often neither customer nor craftsman knew what they were letting themselves in for when the contract was signed. Twice, for instance, did Ramsden abandon the five-foot circle ordered for the Palermo Observatory, and only after Piazzi's nagging did he take it up again in January 1788 for completion in August of the same year. The 'Shuckburgh' equatorial was possibly the instrument ordered for Greenwich in 1781, though it was a private individual who finally possessed this pioneering piece of astronomical engineering in 1793.[22] The Dublin circle was reduced from ten to nine and finally to eight-feet diameter, its design constantly changing in Ramsden's mind, until its completion ultimately fell upon Matthew Berge after the master's death, and nearly two decades after the original placing of the order.[23]

With no established precedents before him, as the quadrant makers had, Ramsden's circles grew organically, from what was often a vague initial specification. Ramsden had to solve numerous mechanical, not to mention mathematical, problems if he was to produce work that was up to his reputation. Graduating an instrument so that it would read to 0.1 arc second by microscopes was only part of the task, for the instrument had to be balanced, supported, and made proof against flexure and distortion.

Ramsden found the division of his circles painstaking but reliable, though his secret compass and microscope technique (called 'coaxing') may not have been developed in time to divide the 1774 engine.[24] What took time, however, was the devising of mounts upon which they could be used to full advantage. At the simplest level, this involved the erection of a circle which had been originally divided in a horizontal position for use in a vertical one without so much as a

---

[22] Pearson (see n. 10 above), p. 413; Chapman (see n. 8 above), p. 119; J. A. Bennett, *Church, State, and Astronomy in Ireland; 200 years of Armagh Observatory*, Belfast, 1990, pp. 24, 26.

[23] Pearson (see n. 10 above), p. 423.

[24] John Brooks (see n. 10 above) argues, p. 133, that Ramsden did not develop his 'coaxing' method in time for use on the second, 1774, engine, but I am not wholly convinced.

1/1000 inch flexure. But it became even more difficult with complex orders like the Shuckburgh equatorial, where the four-foot declination circle had to remain uniform even when turned upside down, and the polar circle had to do the same in spite of changing stresses as the declination superstructure was turned to different parts of the sky.

The most quickly executed order for a capital instrument was the theodolite for General Roy's triangulation survey. Instead of taking decades, this instrument was completed in a mere three years, between 1784 and 1787.[25] It was Ramsden's first great circular instrument to come into operation. In many ways it contained most of the technical ingredients that would be present in his subsequent large astronomical circles: the replacement of verniers with micrometers, a multiple reading facility whereby potential scale errors could be reduced by 130 and 60 degree microscopes, and reversibility—the outstanding advantage of full circle as opposed to part-circle instruments whereby the accuracy of a reading could be checked against its opposite, or 180°, point. Reversibility also made it possible to check the accuracy of every single graduation on the scale even before the instrument was used, by checking graduations against each other for internal consistency. Internal errors could be tabulated so that measured corrections could be applied when particular degrees were used to measure star positions.

As I have already mentioned, it was not the division of an instrument's scales which took Ramsden years to complete (though it could take 150 days), but the puzzling out of how to mount the same to optimum advantage.[26] With the theodolite, however, he was working within a relatively established technology, for small theodolites were already in common use. Where the great Roy theodolite was useful for Ramsden's development as an innovator was the way in which it allowed him to experiment with an enlarged version of an established design into which he could incorporate his circular, microscope-reading and reversible innovations to produce a terrestrial instrument capable of performing to an astronomical level of accuracy.

One must not forget that in the 1780s there were no engineering precedents to follow when trying to fabricate large, light-weight, rigid, and wholly accurate mechanical structures. Ramsden was no less a pioneer among instrument-makers than was his contemporary James Brindley among civil engineers or the future George Stephenson among railway designers. But one important respect in which Ramsden did differ from these engineers was in the paucity of resources at his disposal. Venturesome as it was to build a railway in 1825, Stephenson at least

[25] William Roy, 'An Account of the Trigonometrical Operations ... between the Royal Observatories of Greenwich and Paris...', *Philosophical Transactions of the Royal Society*, 80, 1790, pp. 111–270.

[26] Edward Troughton, 'An Account of Dividing Astronomical Instruments by Ocular Inspection', *Philosophical Transactions of the Royal Society*, 99, 1809, p. 114.

XI

428

had a Board of Directors behind him to maintain a regular supply of money, a site manager to see that the earth was dug, and a foundry in which to build experimental locomotives.[27] Ramsden, by contrast, had a general specification, a group of journeymen whom he was paying to manufacture profitable lines to keep the business ticking over, and whatever sum might have changed hands when the contract was signed.[28] Consequently, his undertaking was much more risky, not to mention ludicrously undercapitalized. Instead of being surprised that it often took him a decade to build an original instrument, we should be surprised that he built any at all.

In the same way, however, that Ramsden borrowed from earlier designers in dividing his engine and circle scales, so he exploited what the existing tradition had to offer when it came to mounts. He recognized the stability and excellent load-bearing properties of the cone, which had first been used in transit instruments by Bird and Sisson and also used for the beams of his 1786 balance for the Royal Society.[29] All of his great circles made use of conical supports for spokes in wheels, column supports, and bearings. He also learned from the quadrant makers such as Graham to avoid bolted joints wherever possible, and to rely instead on cast, welded or brazed fittings; though unlike quadrants, his structures had to be relatively mobile and not just fastened to a meridian wall. It is interesting to notice that most of these design features were already present in the Roy theodolite of 1787, as indeed some of them had been in the second dividing engine, thereby enabling us to trace the initial evolution of Ramsden's technical vocabulary and the way in which he applied it to new problems.

The close relationship which existed in Ramsden's mind between established practice and innovation, whereby he could produce combinations of reliable components which could be incorporated into a novel artefact, heralded a practice which was to lie at the heart of nineteenth-century industrial design. The firearms trade provides many parallels, and in particular the rapid evolution of the revolver between 1812 and 1836. Though the basic components of the chambered cylinder, barrel, and primitive ignition mechanism had existed since the sixteenth century, it was the American, Elisha Collier, who applied essential precision components to cylinder and barrel alignment after c.1812, and Samuel Colt who built upon his work, adding percussion ignition, single and double action, and industrial manufacture by the late 1830s.[30]

Gifted and innovative as Ramsden clearly was as a designer and builder of instruments intended to solve scientific problems, there are several indications that he was not a natural businessman. For one thing, he seems to have lacked

---

[27] P. J. G. Ransom, *The Victorian Railway and How it Evolved*, London, 1990.

[28] Aiken (see n. 5 above), p. 454; Ramsden employed 60 workmen.

[29] J. T. Stock, *The Development of the Chemical Balance*, London, 1969, pp. 13–14, pl. 14.

[30] A. W. F. Taylorson, 'Muzzle-Loading Revolvers' in *Pollard's History of Firearms*, Feltham, 1983, pp. 215–17; 226–32.

XI

that urge towards power and wealth which is one of the mainsprings of the breed. Even at the height of his success as a philosophical mechanician and FRS, he was still content to live above his workshops in Piccadilly, while spending his evenings talking to his apprentices in the kitchen, and eating bread and butter. No man, moreover, who could forget an invitation to attend a soirée with King George III, and absent-mindedly turn up at the palace exactly a year late, could be called a social climber.[31]

He seems to have lacked Graham's 'quaker' shrewdness in money dealing, or Troughton's close-fistedness, for as he grew older and more established, it was the grand instruments and problems in astronomy and physics which occupied his mind, rather than the relentless promotion of his business. While it is true that he was never inclined to publish his secrets without some form of reward, his methods seem to have been known to 'all the best dividers' by the early nineteenth century.[32] Yet in spite of Ramsden's pioneering work in the development of the dividing engine and sextant, it was claimed that Troughton made more money from the commercial exploitation of his improved version of Ramsden's second machine (which either John or Edward Troughton must have been allowed to examine before the Board of Longitude published the *Account* of 1777 if the first Troughton engine was operational by 1778) than Ramsden made from the original.[33]

Jesse Ramsden seems to have belonged to the same strain of industrial innovators as James Brindley, Samuel Crompton and I. K. Brunel, to whom the challenge of invention was more consistently appealing than the financial exploitation of the ensuing machine. But what made Ramsden quite remarkable in the history of science was his 'dual nationality' as a scientist of European stature in his own right, combined with his status as a technological entrepreneur and London tradesman. Considering the daring enterprises which he undertook, and the persistently under-funded character of his operation, Ramsden achieved wonders.[34] By consistently robbing Peter to pay Paul in his business dealings,

[31] Aiken (see n. 5 above), p. 456. Dutens emphasized that Ramsden was only interested in money for 'making further improvements in science', and drew attention to his lack of interest in worldly concerns. See also *DNB*.

[32] Rees's *Cyclopaedia*, 'Graduation' (see n. 8 above), unpaginated, sig. 3Y4r; Troughton (see n. 26 above), p. 113; *Edinburgh Encyclopaedia* (see n. 8 above), p. 370.

[33] Ramsden, *Description ... Mathematical Instruments* (see n. 11 above), Preface. I am also obliged to John Brooks for reminding me of Ramsden's obligation to grant access to his machine from 28 October 1775; private communication, 10 March 1992. *Edinburgh Encyclopaedia* (see n. 8 above), p. 253.

[34] At his death in November 1800, Ramsden left under £5000. I am indebted to Dr Anita McConnell for drawing Ramsden's will and related documents to my attention (Prob. 11/1355, ff. 251–2. Death duty assessment is at IR.26/49, pp. 131–2. Transcript in Science Museum Library, Court Papers, folder T). Matthew Berge continued his master's business and inherited the lease on house and premises.

XI

430

exploiting the dividing engine's varying profits to pay the research and development costs of capital instruments, and often keeping just ahead of his creditors, Ramsden revolutionized science-based technology in Georgian England, and provided both a prototype and an object lesson for the men of the industrial revolution.

XII

William Herschel and the Measurement of Space*

To give a forty-minute address to commemorate the 250th anniversary of the birth of a major man of science is one thing; to say something about him which is original, and intended for publication in the *Journal* of the honoured Society of which he was the first President, is another. William Herschel belongs to that small band of astronomers whose impact on their own time was profound and whose reputation as a fundamental innovator and creator of new perspectives has never been eclipsed. One might cite the names of Hipparchus, Copernicus, Galileo and Newton as astronomers of 'continuous currency' and it is my argument that William Herschel is one of that company. Though the original contributions of each one of these astronomers were eventually modified or superseded, they nonetheless provided marker points against which later discoveries were reckoned and, while Herschel's work on the 'construction of the heavens' is now outdated, it asked many of the fundamental questions which contemporary astrophysics is still trying to answer.

Though the most recent of the above-mentioned innovators, the overall direction of Herschel's work has never ceased to inspire scientific thought and emulation. The concepts of a dynamic as opposed to a static universe, of changing star systems bounded by mathematical laws and the identification of 'luminous fluid' in deep space, provided the rudimentary materials of modern cosmology. His deep space telescope technology gave a new excellence to the already pre-eminent British scientific instrument making trade, while his keen appreciation of research thresholds bounded by 'space penetrating power' defined the big telescope technologies of William Lassell, Lord Rosse and the great observatories of today.

His importance is also reflected in the attention which he has received from historians of science on every level, extending from reverential Victorian biographies and children's books to analytical studies published in the contemporary historical press (1). Very few aspects of his half-century long and extraordinarily fruitful career in astronomy have not been examined and assessed, including his depiction in portraiture, while the *Collected Scientific Papers* which Dreyer abstracted and published in 1912 fill up two heavy quarto volumes (2). It is with an awareness of this body of scholarship behind me that I will attempt a short survey of what is perhaps the most innovative and far reaching aspect of his researches, on the construction of the heavens,

---

* Based on a talk given at the RAS on 1988 November 11 to mark the 250th Anniversary of the birth of Sir William Herschel.

and place it within the wider context of eighteenth-century astronomical priorities.

Perhaps one reason why modern astronomers feel a particular intellectual kinship with Herschel is because he seemed to shift the ancient and traditional concerns of astronomy away from the tracing of moving lights across a black background to the fathoming of the background itself. One might say that Herschel took a static science which for millennia had operated in two-dimensional coordinates and gave to it, in his definition of 1817, 'length, breadth and depth; or latitude, longitude and profundity'(3).

Appealing as many of Herschel's ideas may be, however, we must not allow ourselves to forget that he, and his eighteenth-century colleagues, often drew radically different conclusions about the consequences of many cosmological forces from the interpretations held today. The whole of Herschel's cosmological ideas focused on one primary process which, like everything else in nature, could eventually be explained in terms of gravitation; the process of condensation. His was a condensing universe, not an expanding one, in which cosmological bodies moved to form increasingly complex, compact and symmetrical aggregations, seen at perfection in globular clusters and planetary nebulae. Under the all-pervading influence of 'attraction' and 'universal law,' individual stars moved out of the original places allotted by the Creator (like equidistant plants in a flower bed) to form clusters. Depending on the length of time during which the individual attractive forces had been in operation, the cluster could be open, globular or planetary, as all bodies moved towards compression.

The Milky Way, which Herschel considered to be a complex, irregular cluster, also partook in the universal process of condensation and appeared to be breaking down into a multitude of local clusters. Those dark patches in the starry fabric of the Milky Way, which we now ascribe to intervening dust, he attributed to local condensation spots, where all the stars in a region had come together to form a cluster, leaving a 'hole' in the stellar background around them. Herschel drew substantiation for this idea from the irresolvable nebulae and star clusters which he claimed to find in the vicinity of supposed holes in the Milky Way, citing the Messier objects 4 and 80 as examples (4). Even after Herschel's discovery of the object now known as NGC 1514 in November 1790, which finally convinced him that 'true nebulosity' existed in space (as opposed to star clusters too far away to be resolved into individual components), the nebulosity came to be explained as the medium out of which separate stars either condensed, or upon which they fed, to maintain their light output (5). At every stage, however, condensation and clustering, both between individual stars and from nebulae into stars, were the universal effects of the universal cause – gravity.

Many features of the scientific culture of Herschel's day would have inclined him to this condensing view. Mechanical technology provided some graphic evidences of the power of condensation which would have inspired a man of the late Enlightenment. Boyle's classic airpump experiments in the previous century had demonstrated the power of atmospheric pressure and compression amongst what were thought to be the atomic particles of the air. The Leiden jar after 1745 showed how the 'electric fluid' could be collected from glass by friction and concentrated in a jar, while Newcomen's steam engine gave evidence that, when steam condensed into water, mechanical powers of great intensity could be produced (**6**).

But the most compelling evidence for the power of condensation came from the careful monitoring of the force which occasioned it: gravitational attraction. Since Newton's elegant formulation of his laws in 1687, astronomers and physicists had felt it incumbent upon themselves to measure, demonstrate and quantify the action of gravity in nature. Between 1728 and 1798, Newtonian theory was shown to operate in the Solar System and on the surface of the Earth, which proved that it was indeed universal. James Bradley's discovery of aberration in 1728 and nutation twenty years later, demonstrated new phenomena that were in precise agreement with Newtonian postulates (**7**). In 1735, the French Academicians, after extensive geodetic measures in France, Lapland and Peru, declared the earth to be an oblate spheroid, in accordance with the shape already predicted in *Principia* (**8**). Having successfully demonstrated gravitation on the planetary level, it next remained to do the same for it as a terrestrial force. By 'weighing' Schiehallion in 1775 and measuring the displacement of the plumb line of a zenith sector by the mountain, Maskelyne showed that the inverse square law worked equally well between relatively small terrestrial bodies, and Cavendish obtained confirming results with his torsion balance in 1798 (**9**).

Though Herschel had always been careful not to push analogies too far, it seemed to be only a matter of time before gravity was also demonstrated in the stellar universe. As early as 1767 John Michell had pointed out that the chance of clusters such as the Pleiades occurring randomly was 500 000 to one, and that *attraction* must take place between stellar bodies (**10**). While nothing could be done at this stage to associate such phenomena with Newton's laws – and would not be possible until well into the next century – the analogical evidence was strong. Analogy, after all, constituted a powerful argument in its own right to eighteenth century minds, accustomed as they were to drawing parallels between different parts of a homogeneous Creation, and even the anonymous satirist of 1761 could conceive of God as the all-wise clockmaker whose (**11**),

> 'Sagacious Newton, lost with pondering thought
> to mathematic rules a system brought;
> God, as an Eastern Monarch, left for show,
> His Viceroy, Gravity, the God below.'

In the wake of the physical demonstration of the power of condensation in contemporary experimental physics, the elegant proofs of *measured* local gravity and the age's partiality for analogical reasoning, it is hardly surprising that William Herschel drew the conclusions that he did about the fundamental cosmological agent.

Eighteenth century astronomy was concerned, first and foremost, with precision measurement. Whether one was in pursuit of a practical solution to the longitude, like the officially appointed Astronomers Royal, or attempting to define intellectually significant quantities, such as the astronomical unit or stellar parallax, progress was seen as inextricably tied up with making better angular measurements. The instruments with which one attempted to measure the sky were already of conservative seventeenth century designs,

XII

such as the telescopic sight, micrometer and zenith sector, while the quadrant was based on geometrical prototypes which went back to classical antiquity. Improved accuracy depended not upon improved principles of design, but improved processes of manufacture, better beam compass graduations, more accurate micrometer screws and colour-free lenses (**12**). Both the instruments and the problems which they existed to solve, moreover, were relatively traditional, concerning themselves with the perfection of Solar System dynamics, and using the stars for little purpose other than to act as markers against which the 'great lights' of heaven moved. A comparison between ancient and modern measures had enabled Halley to detect the first proper motions in 1718, but the fabric, or physical construction, of the stellar universe was effectively a closed book to the astronomers of 1720 (**13**). The book was closed not because philosophers were uninterested in space, for as early as 1576 Thomas Digges had depicted the stars beyond the Copernican solar system as going on 'for ever,' but no tools of inquiry were available to enable the investigator to rise beyond pure speculation (**14**). By the early eighteenth century, the old Baconian science of the Royal Society, given a new analytical dimension by the mathematics of Newton, had defined scientific research as something capable of both mathematical proof and physical demonstration. Theories needed to be tested in nature and mathematicians needed quantities to introduce into their equations if they purported to describe the natural world, as opposed to their own intellectual confabulations.

With an instrument technology geared up to provide little more than improved R.A. and Dec. coordinates, the astronomer had no tools with which he could viably draw lines into stellar space and present even marginally acceptable solutions to the learned world.

Eighteenth century astronomy did, however, approach the problem of the stellar cosmos with a number of postulates in mind, which derived in part from the more highly developed branches of the science and in part from what the age regarded as essentially true. Foremost amongst these truths was the assumed equality of all related things in nature. Oak trees, cats and raindrops were all about the same size and displayed immediately recognizable characteristics. John Locke, whose *Essay concerning human understanding* (1690), had become the epistemological bible of eighteenth century science, had helped define nature along these lines, while Herschel, in his 1789 paper, had made the same assumption about stars as generic objects (**15**),

> I am, however, inclined to believe [that they] may not go farther than the difference in size, found in the individuals belonging to the same species of plants or animals, in their different states of health or vegetation, after they are come to a certain degree of growth.

If all stars, therefore, were about the same in size and light output, it logically followed that bright stars appeared so merely because of their proximity.

The Age of Reason's partiality for symmetry and natural order also impli. . that if the stars were generically similar physically, then the ratios between them must be mathematically ordained. As in any even distribution of homogeneous radiating objects, the ones that were further away and appeared dimmer would outnumber those that were close at hand, and bright, in accordance with a model for stellar distribution dating from the early eighteenth century (16). Though Herschel was to abandon the concept of a universe that was regularly planted with stars like flowers in a bed, he hung tenaciously on to the rule which defined stars as physically homogeneous in terms of real light output, and where apparent dimness was a function of distance. Even in his penultimate cosmological paper of 1817, by which time he was willing to recognize the presence of 'true nebulosity' in space and the tendency of stars to cluster into annular zones, he stood firm to the distance–luminosity rule (17). Had he sacrificed it, he would have lost what he considered to be the most reliable guiding principle whereby he might fathom the structure of space.

It was not for nothing that Herschel frequently spoke of a 'natural history' of the heavens in his published papers (18) for, in this respect, he was in conformity not only with some of the key guiding principles of his age but with one that was well represented in the Royal Society – taxonomy. Implicit in the Baconian creed of experimental science was the obsession with classification and the drawing up of catalogues. Lacking any fundamental understanding of *why* nature worked, many scientists hoped that by learning of *how* it did so, some sort of truth would emerge. Though the taxonomic approach to science was less important in physics and astronomy, which had been mathematicized in the seventeenth century, they were very much the order of the day in medicine, botany, geology, zoology and those sciences which were not yet susceptible to precise quantification.

Though an astronomer, Herschel's cosmological interests lay beyond the gravitationally proven Solar System and in a realm where, in 1775, there were few certainties. Nor was it obvious at the outset of his career how such forces would be detectable at those immense distances beyond the reach of Bird's quadrants or Ramsden's circles.

In consequence, Herschel developed a technique which was more in keeping with botanists and anatomists than conventional astronomers and began to comb the heavens for specimens as his Royal Society associates, Banks and Solander, had recently combed the newly discovered Botany Bay for exotic fauna. The great lists of nebulae and clusters which Herschel published in *Philosophical Transactions* were first and foremost catalogues of celestial specimens from which the natural historian of the heavens attempted to draw conclusions about the past and future states of space (19).

Herschel's work on the heavens paralleled, in many ways, contemporary techniques being employed in two other sciences that were well represented among the Fellowship of the Royal Society – physiology and fossil geology. In attempting to fathom the processes of life from cadaveric evidence, John Hunter (died 1793) had built up the most intellectually significant and experimentally determined anatomical collection to date. From the structures of dead organisms, he attempted to reconstruct their living functions (20). Similarly, in the first two decades of the nineteenth century, the English geologists William Smith and William Buckland, along with the Frenchman Georges Cuvier, tried to piece together a dynamic physical history of the earth from the apparently erratic remains of fossil animals (21). Both of these groups of men were using the taxonomic comparative techniques of the natural historian to elucidate dynamic processes from static remains and (especially with the geologists) write a *history* and indicate a *future* from scattered fragments in the here and now. It is interesting to note that when attempting to comprehend the causal processes of deep space, Herschel employed a very similar technique.

Starting from the local Solar System-based postulates of gravitation, combined with analogies drawn from other branches of physics and an over all assumption about the innate symmetry of the Creation, Herschel formulated a set of guiding principles for the universe. Stars were the fundamental building blocks of space, as were billiard ball atoms for chemists, and all possessed similar generic characteristics, fell under each other's gravitational influences and gradually moved into clusters of increasing compactness (**22**). During the 1780s, before Herschel was willing to admit to 'true nebulosity,' these basic processes could be made to account for everything visible and invisible, for even hazy patches of light could be explained away as deriving from the combined glow of many stars, none of which could be yet seen individually, but which would be when telescopes acquired a deeper space penetrating power (**23**).

This overall theory underwent some adjustment but no real modification following his observations of the 'cloudy star' (NGC 1514) in 1790. As Herschel no longer felt that it was feasible to explain away the glowing 'chevelure' around the central star of this planetary nebula in terms of either small dim stars, or very remote ones, and still preserve the homogeneity of the luminosity related to distance rule, he had to say what 'true nebulosity' might be. If a star was surrounded by flimsy matter, moreover, and that matter shone only by reflected light from the parent star, then why was it visible at all? When a star was so remote that even in a large telescope it only produced a speck of light, how could the much weaker light reflected from the adjacent nebulosity also reach us (**24**)? The only way around the problem was to posit that nebulous matter must be capable of shining by its own self-generated light, perhaps after the manner of electric discharges or aurorae, and did not therefore depend upon reflection. Both the star and the glowing cloud could thus occupy the same region of space, shine by their own independent lights, and preserve the rule that brightness was related to distance. Once recognized, the idea of a self-glowing 'shining fluid' could be extended from the planetary to account for other types of nebulous objects (**25**).

Shining fluid could now be seen as a separate cosmological material in its own right, having an independent existence from the stars while still being able to condense into them, for, as he postulated in 1791 (**26**),

Self-luminescence, it seems, [is] more fitted to produce a star by condensation than depend on the star for its existence.

Herschel now began to scour the skies for self-luminescent objects, to lay the foundation for a series of classic interpretative papers in 1811, 1814, 1817 and 1818 on the 'economy' of nebulae and their relation to the stars (**27**). He tried to classify his nebulous specimens, either as objects shining in isolation, or else with contiguous stars. Planetary nebulae, for instance, represented a shining envelope condensing into a central star; nebulae with apparent filaments attaching them to adjacent stars were being soaked up to fuel the star, while the ring shape of certain starless nebulae might indicate that they were rotating (28). Unfortunately, he failed to detect the spiral arms of the galaxies. Original and far-reaching in its implications as his economy of nebula work was, it still stood firmly in the natural history tradition, for in the absence of any firm, measurable data, Herschel was forced to fall back upon his extensive, albeit purely empirical, knowledge of celestial geography when it came to interpreting his evidence.

By 1802 (29) Herschel was drawing some remarkable conclusions about a universe which was thought to be *condensing* rather than expanding. While he was aware that Huygens in the previous century had realized that if light had a finite velocity then the rays emitted by distant stars must take many ages to reach us, and had reminded the Bath Philosophical Society of the point in 1780 (30), he brought the time-related aspect of cosmology firmly home in his 1802 paper to *Philosophical Transactions* when he said that (31),

a telescope with a power of penetrating space like my forty feet one, has also, as it may be called, a power of penetrating time past.

He had believed as early as 1802 that many nebulous objects – such as the one in Orion's sword – had undergone changes in as little as twenty-three years, and that the stellar universe was physically dynamic (32). Nor should one ever underestimate the importance of a half-century observing career and the good health which he generally enjoyed (in an age when the average life expectancy was under forty) when assessing Herschel's contribution. Being entirely dependent upon hand, eye and brain coordination for all his work, in the absence of photography, it was a singularly fortunate man who lived and preserved his faculties for long enough to notice small changes of the kind that Herschel was recording by the early nineteenth century.

Yet it would be wrong to assume that Herschel had what might be called an *evolutionary* conception of space. He saw it rather as in a permanent state of change, as nebulous fluid ultimately fuelled stars, and gravitation broke up the Milky Way into numerous island clusters. In 1814, he spoke of the disintegration of the Milky Way as somewhat like a great cosmic chronometer, measuring time from its once complete beginning, but lacking any discernible constants indicating dates or epochs (33). Herschel's universe cannot be called evolutionary, however, because the basic population of space – the stars – did not change within themselves, but only in relation to the positions which they occupied, and the nebulous matter out of which they were formed and nourished. Herschel, moreover, lived within a broad scientific culture which saw nature as complete within itself, for while spatial positions could alter within the ordained bounds of natural law, the *objects* themselves, as individuals, remained basically the same. Like contemporary naturalists, Herschel thought in terms of 'fixed species' which did not change within themselves, be they flowers, mammals or stars. It was certainly a *dynamic* and a *busy* universe, but it was not one seen as evolving into something beyond itself. Before cosmological processes could be thought of in evolutionary terms, it was not only necessary for astrophysics (like biology) to greatly increase its data base, but for the earthbound sciences to re-think their fundamental processes, and provide potential models.

Herschel's Universe was essentially a steady state one, for while generically similar stars condensed into clusters under 'attractive' influences, he always looked for potential moderating forces which prevented the whole from collapsing into a single mass. As early as his paper of 1785 he spoke of planetary clusters as the 'laboratories' of the universe, which not only contained clustering, but also 'projectile' forces, whereby matter might be somehow re-cycled back into space to forestall a final condensation (34). Tycho Brahe's new star of 1572 was cited as an example of a possible collapsed cluster of stars which, "by some waste or decay of nature, being no longer fit for their former purposes" had crashed together and disintegrated, thereby "projecting" matter back into space (35). But he did not think of the re-distribution of matter in such bodies as *exploding*, as we now think of supernovae, but as a mass of individual stars crashing together under gravity, so that their combined fragments were thrown out into space to form the raw materials out of which future clusters might develop. In Herschel's universe stars did not explode of themselves but they could crash together and shatter.

This was a universe of exact proportions, and one thing which Herschel realized at the relative outset of his career, in 1781, was the need for reliable yardsticks to give finite dimensions to what was otherwise a scale of proportions. It was a concern in no way unique to Herschel, for since the seventeenth century astronomers had argued that if they possessed the measured distance of a first magnitude star, then the rest of the universe could be drawn to scale. First magnitude stars were the natural choices for the exercise for according to the 'luminosity as a function of distance' rule, they should have been the nearest and easiest to measure. Numerous practical astronomers, such as Hooke, Flamsteed and Bradley, had searched for the parallax with zenith sectors and quadrants, and while none had found it, John Michell had pointed out in 1767 that it was most likely to be around one single arcsecond; this was a value which lay just beyond the measuring capacities of contemporary instruments (36).

Throughout his career, Herschel had worked on the assumption that once an exact parallax enabled the calculation of a stellar distance, then the distances of all other luminous bodies in space could be slotted in according to a scale of proportions based upon their magnitudes. One might suggest that Herschel's musical background added further cogency to this way of thinking for, in music theory, the pitching of one single note permits all the others to be tuned in proportion. Not for nothing had Herschel, as a young man, read Smith's *Harmonics* as well as *Optics* (37), and one might argue that he regarded standard stellar brightnesses as the visual tuning forks of the heavens. Yet in the same way as it was to remain impossible for the science of acoustics to give any precise vibrational quantities to the physics of sound ratios before the nineteenth century, so the celestial tuning forks had to wait until the Victorian technology of the spectroscope and photographic plate had been developed, but by that time the elegant dynamics of Herschel's universe no longer retained their old convincing power.

None of Herschel's work, and the remarkable cosmological ideas which he initiated, would have been possible without a new approach to instrumentation. Though he may have hunted for nebulae as a naturalist hunted for plants, Herschel's chosen part of God's Garden was not accessible to the naked eye and demanded specialized tools even before the most rudimentary interpretative schemes could be framed. Herschel was always aware of the importance of instrumentation, and never ceased to remind the readers of his papers that such results were not accessible through 'common telescopes'

In Herschel's time, mainstream astronomy was not only the most highly developed of the natural sciences, but had been the recipient of the most extensive and sustained instrumental attention. But the nature of the problems faced by mainstream eighteenth century astronomy, such as stellar cataloguing, had guided the line of instrumental innovation in one direction only – towards the measurement of increasingly accurate celestial angles. Even the telescope, which in the seventeenth century had given to astronomy a power of investigation undreamed of by the ancients, had been primarily developed for measuring purposes, to produce the telescopic sight, zenith sector and eyepiece micrometer. The instruments which lay at the forefront of research in 1770 were devices for measuring angles in the meridian plane. George Graham and John Bird had brought the quadrant to its peak of development by 1760, while Jesse Ramsden and the Troughton brothers shifted the quest for greater accuracy to the more stable and mechanically versatile shape of the astronomical circle. If the old constants were to be refined, the stellar and solar parallaxes found, and gravity shown to be universal, these were the tools with which to do the job (**38**).

To modern astronomers it is hard to understand the low priority allocated to improving the *seeing* power, as opposed to angle measuring power, of instruments by these scientists. For almost a hundred years after Newton had produced his reflecting telescope, it had failed to make any significant mark on the science although this should not be construed as meaning that the instrument was not developed in its own right (**39**). As early as 1721 John Hadley demonstrated a design for an improved Newtonian reflector with a six-inch mirror, while between 1738 and 1768, James Short established a scientific reputation and amassed a £15000 fortune as a maker of elegant Gregorian reflectors (**40**). Using the original optical system developed by Galileo and Kepler, a succession of opticians culminating in John Dollond had brought the refracting telescope to 'perfection' by 1758, with the invention of the achromatic lens (**41**).

But the optical side of instrument innovation still dragged well behind the mechanical, and did so for two good reasons, one scientific and the other commercial. On the scientific level, there was no real job which the improved telescope could do within the astronomical priorities of the day. It is true that colour free lenses were less irritating to use than those displaying chromatic aberration but the problem was a minor one in meridian astronomy where well stopped down object glasses could still give crisp images in low power telescopic sights. Astronomers generally did not need to look *at* the sky, but to delineate bright lights against it.

The telescope's real development in the mid-eighteenth century was commercial, and aimed to a large extent at the dilettante market of cultured amateurs. The kind of person who could afford twenty-five guineas for a one and a half or two foot Short Gregorian, with its exquisite brass fittings, was

XII

likely to want it as much for a conversation piece as a research tool. The Gregorian reflector's 'right-way-up' image made it ideal for terrestrial use and occasional sky gazing, and instruments of this design frequently figure as objects of domestic decoration in the sophisticated genre art of the period (**42**). The fact that so many of these telescopes have survived in excellent condition to be exhibited in modern museum collections further suggests that they were never altered or cannibalized in the way that working scientific instruments often are. Lucrative articles of commerce and *objets d'art* as these instruments may have been, their impact on the fathoming of the heavens was not great and it was not for nothing that when Herschel, as an aspiring amateur, contemplated the purchase of a London-made telescope, he found the instruments on sale to be over-priced and unsuitable for his intended work (**43**). It also says something about the contemporary state of reflecting telescope manufacture that after a few years practice, a Bath organist was producing pieces which the Astronomer Royal pronounced to be superior to any available through the London trade (**44**).

It would, on the other hand, be unjust to deny that English telescopes, of both the reflecting and refracting types, were still the best in the world as far as they went, and greatly in advance, optically speaking, of what had been available thirty or forty years before.

But what Herschel did was subject the reflecting telescope to the same research and development process that astronomers and instrument makers had been bestowing upon graduated instruments since the time of Tycho Brahe. To Herschel, the telescope was not the adjunct to a more important instrument (as was the telescope to the graduated scale) but *the* primary tool of investigation. Indeed, one of the things which makes Herschel's originality so arresting is that he saw the telescope as a research tool in its own right, and the sole key to a range of astronomical problems that went beyond the essentially cartographic activities of his contemporaries. As 'common telescopes' were not good enough, Herschel set out to fabricate uncommon ones, in which such qualities as light grasp and space penetrating power were paramount.

It must not be forgotten that William Herschel came largely to re-design the reflecting telescope, transforming it from a philosophical instrument to a device capable of crossing hitherto impassable research frontiers. His success, however, depended on first defining a scientific problem and then trying to produce a technology with which to solve it, rather than just trying to improve the operations of an instrument which still lacked a problem to exercise it. The reflecting telescope was an excellent choice for stellar work because of its greater optical surface, although it would be incorrect to assume that Herschel had been the first astronomer to comprehend the importance of light grasp; the remarkably original John Michell had drawn attention to the fact that while Robert Hooke's observation of the Pleiades in *Micrographia* had revealed seventy-eight stars through a modest refractor in 1665, a fifteen-inch diameter lens or a two-foot mirror would probably yield a thousand (**45**).

It is important to remember that Herschel never used the popular Gregorian design even for his early telescopes, preferring Newtonians, and thus avoiding the extra hazard of having to produce two curved optical surfaces (**46**). While his 7-foot focus, 6-inch mirror reflectors became a byword for excellence in their own right, the more intellectually ambitious his researches became, the correspondingly simpler were the functioning parts of his instruments. His favourite 20-foot telescope, with its 18.75-inch mirror, and his giant 40-foot instrument with its 4-foot diameter mirror, both dispensed with intermediary optical surfaces to direct the pencil of light straight into the observer's eye. Herschel's most ambitious telescopes, with which he attempted to fathom the construction of the heavens, depended on nothing more than one impeccably ground reflecting surface, working in conjunction with a simple eyepiece. This strict functionalism of design, combined with the quality of his optics, enabled Herschel to obtain not only space penetrating power, but also magnifications which his contemporaries thought incredible.

Never does Herschel fail to pay homage to the quality of his instrumentation in his major publications, although in this respect he followed precedents established by Tycho, Hevelius, Flamsteed and Bradley, who also felt it incumbent upon themselves to describe the pieces with which their results had been obtained. Convention apart, however, he left none of his readers in doubt of the new perspectives made available through his new 20-foot reflector in 1784 as he was trying to resolve the objects in Messier's Catalogue (**47**)

> by applying ourselves with all our powers to the improvement of telescopes, which I look upon as yet in their infant state ... and perhaps be able to delineate the *Interior Construction of the Universe*.

Yet one might care to argue that all his telescopes did for Herschel was to put him on an even footing, *vis-à-vis* his chosen subject of study, with the field naturalist or anatomist, who used his unaided eye or a small microscope. Herschel's instrumentation simply allowed him to see his specimens more clearly, but they did nothing else, and in the absence of the spectroscope or photographic recording media, there was little that he could do with the light which his instruments collected other than look at it with an experienced eye. Unlike his colleagues working in the more established branches of astronomy, with well developed tools and techniques of measurement at their disposal, Herschel had no proven laws or yardsticks that could be applied to deep space. He was like the anatomist who, after classifying all the parts of a body, could still provide no physical explanation of why living parts moved and dead ones did not. It was in pursuit of such measurable criteria as might give certainty to the understanding of space that Herschel worked for forty years, first attempting to measure a stellar parallax by a novel method, then 'gaging' the heavens to sample star fields statistically and, in 1817, trying to ascertain the intrinsic brightness of starlight.

Even before he came to draw his important conclusions about the structure of space, Herschel had already begun his research and development of a new instrumentation. He had never worked with conventionally graduated instruments, and one of the marks of his genius was the way in which he tried to solve old problems by new methods. In the early 1780s, he addressed himself to the centuries old problem of the parallax, which was the historical acid test for Copernican astronomy, and that was believed to be the most conclusive way of fixing the distance of a first magnitude star. Dispensing with the zenith sector based techniques of his predecessors, in which the plumb line was used to delineate a fixed point in space against which chosen star parallaxes were to be measured, Herschel opted for a class of objects that would provide their own yardsticks of comparison – contiguous pairs of what he believed to be line of sight double stars.

Measuring the angular separation of a contiguous pair of stars seemed greatly to simplify the task in hand by eliminating a range of potential variables – such as the movement of supporting walls or finely adjusted parts – which one encountered with a zenith sector or circular instrument over the months of observations required to detect a parallax. With a pair of stars, it was only necessary to monitor the angle between them, along with their positions in relation to a few nearby stars, so that all cross checks appeared within the same field of view, without the need for external coordinates.

The results of Herschel's parallax work were published in 1781 and 1782, being based on observations made with one of his seven foot Newtonian reflectors working at high magnification. So familiar had he become with the precise locations of many double stars even by this early date, that he claimed small shifts of position would be detectable to his experienced eye, while precision was to be obtained by using a micrometer (**48**).

It was to measure the positions of double stars that Herschel devised his Lamp Micrometer. The purpose of this instrument was to reproduce the double star artificially by setting two small lamps on an adjustable frame. Employing a split image principle, the lamp positions were compared with the double star, and knowing the exact disposition of the lamps in relation to the eye of the observer, an angle could be computed. The accuracy could be greatly increased, at least theoretically, by observing the stars with a very high magnification, and interpreting angles subtended as multiples of magnifications employed, so that he came to claim figures expressed to sixtieth parts of a single second. One indication of the quality of Herschel's mirrors (and eyepieces) were the powers which he claimed to employ on a 7-foot telescope, regularly cited as 460, 932 and even the incredible 6450 (**49**).

Quite apart from the Lamp Micrometer, Herschel used eyepiece micrometers of more conventional design, in which fractions of a screw turn were translated into angular fractions of a prime focus image. When one bears in mind that a single degree of arc at the prime focus of his 7-foot reflector would have measured 4.18 inches it would have been well within the micrometer screw technology of the period for him reliably to detect fractions of a single arcsecond (**50**).

Original as Herschel's technique was, it failed to produce the desired parallax, largely because the stars which he measured turned out to be too remote to display them. Yet even more important was the fact that some of the pairs measured were genuine binary systems, instead of line of sight doubles, and the probable product of the clustering tendency of stars pointed out by John Michell fifteen years before (**51**). Not until he had observed some of these pairs for over twenty years, in 1802, did Herschel realize that some of the components had moved in relation to each other, though not in parallax. Though it could not be mathematically demonstrated, or even fully recognized at the time, this motion of binary components provided the first indication that Newton's laws actually operated in stellar space (52).

Having failed to measure a parallax, Herschel attempted to apply quantitative techniques to the heavens by 'gaging' them. From 1784 onwards he devised several ways whereby he could take statistical samples of stellar densities in particular regions around the Milky Way, by breaking up the sky into measured zones of R.A. and Dec. and counting the stars within them. In 1785, he counted stars in fifteen arcmin squares, from which he was able to work out a distribution pattern indicating that the Milky Way was a 'stratum' of stars arranged in a roughly circular plane (53). While Thomas Wright had speculated upon the nature of the Milky Way in 1750, Herschel seems to have come to his conclusion quite independently and not to have read Wright's book until some time later (54). Herschel's concept of the galaxy, the principal features of which were ascertained by a combination of superior instruments, and an attempted statistical sampling, constitutes one of the most original and ingenious components of his achievement.

The accurate classification of stars, whereby they could be assigned places in the depths of space, still required a method by which the brightness of one specimen could be accurately compared with another. While the parallax of a first magnitude star would, according to the theory, provide the initial yardstick into space, the respective brightness of the lesser magnitudes would make it possible to ascribe numerical weightings to the ratios occupied and hence measure the 'profundity' of space. Various people had attempted to devise photometric techniques, such as Edward Piggott's use of graded dark glasses to compare the light output of variable stars in 1785, but in the absence of an impersonal recording medium one was still forced to depend upon eye estimations (55).

The problem had exercised Herschel for most of his career, and in 1817 he described his mature technique. This method, which he first tried in 1813–1814, depended on the presence of two optically identical 7-foot reflecting telescopes, producing stellar images of equal brightness. Standing alongside each other, the two instruments were used to view a bright star such as Arcturus (56). One of the instruments would be kept at full aperture, while the mirror of the companion telescope was masked with a series of annular rings, until the star glowed with the same light as a second magnitude star when seen at full aperture (57). In this way, it was possible to calculate the reduction necessary to make the reflecting surface of the mirror drop down one magnitude. Similar masks could be made for other magnitudes, to make it possible to express magnitude differences in terms of annular or reflecting surface proportions. By this means, any star or nebula could be compared alongside a control star of 'standard' brightness to ascertain the luminosity ratio.

While this method clearly depended upon the observing experience and eye sensitivity of the individual astronomer to make it work, it was the best that could be done until photography provided an impersonal standard by which to measure brightness. It also demonstrated two significant points in Herschel's approach to astronomy; his experimental ingenuity, experience and recognition of the need for control objects when making comparisons,

XII

412

and his ceaseless search for reliable constants in framing a natural history of the stellar universe.

In retrospect, one cannot help but feel sorry for Herschel, for it would be hard to find another major scientific figure who had already outstripped the viable research technology of his day within the first fifteen years of his career, and then proceeded to thrash around for a further thirty years in the attempt to break through the barriers which it imposed. Though Herschel had been the first astronomer to show what big telescopes could really do, he had taken these instruments to their extreme contemporary limits in a short span of years. From his first successes as a mirror grinder to the production of his 20-foot reflector in the early 1780s was but a brief span. The 1780s had been a period of intense creative activity, with major cosmological papers in 1781, 1782, 1784, 1785 and 1789, which changed man's awareness of the structure of the stellar universe. Within this decade he had determined the Sun's motion in space, and correctly computed that it was moving in the direction of the constellation Hercules (58). He had demonstrated the structure of the Milky Way from his star gages, showing it to be in a process of physical change, and enunciating his ideas on the formation of star clusters. Herschel's natural history of the heavens had soon extended the modest number of 103 nebulous specimens described in Messier's Catalogue in 1784, to thousands by the early 1790s, and made him confident that bigger mirrors would resolve the structure of these remote formations.

But with his 40-foot reflector in 1789 he had clearly hit the limit. The great telescope was a disappointment and failed either to resolve 'true nebulosity,' or to reveal the structure of distant objects. Both technically and intellectually, Herschel had gone as far as he could go by the time of his recognition of 'true nebulosity' in 1791, for no new evidences could be extracted from nature with the tools currently available.

Not until the precision engineering of the Victorian age made possible the fine large mirrors of Lassell, Rosse and their successors, and photography and spectroscopy facilitated the analysis of starlight, could the technical barricades which Herschel fruitlessly assaulted for the last thirty years of his life be eventually overrun.

Perhaps this is the reason why Herschel never developed a school, or established a new and active branch of astronomical research which continued after him. Herschel's astronomy, however, was always a one man (and with Caroline, a one woman) show, in spite of the large number of fine instruments which he was commissioned to build for others. It is true that Herschel's researches provided a vast body of classified specimens which never failed to excite the admiration of those who studied them, but every successful school must have creative prospects to offer to its graduates and by 1820 there was still nothing that a prospective deep space astronomer could do, other than faithfully duplicate the works of his master.

With the exception of John Herschel, who determined to extend his father's 'sweeps' into the southern hemisphere, the rising generation of both astronomers and instrument makers confined themselves to the more mainstream branches of the science. Bessel, Airy, Struve, Henderson, Le Verrier and Adams devoted their lives to perfecting the astronomy of the known Universe, while Tulley, Peter Dollond, Troughton, Simms and

Repsold lavished attention upon the further improvement of circular divided instruments and achromatic lenses. Though the Royal Observatory had acquired a 7-foot Herschel telescope as early as 1783, a 10-foot in 1813, and had Ramage's 25-foot reflector between 1826 and 1836, these instruments remained unused for all serious purposes (**59**). Big aperture telescopes were not needed, and even the acquisition of the Northumberland at Cambridge in 1838 and the Merz equatorial at Greenwich in 1859 (both refractors, one notes) in no way presupposed that these instruments would be used for deep sky work in the Herschel tradition.

A clear symptom of the neglect into which the large reflecting telescope fell after Herschel's innovations were the problems which the young William Lassell encountered when seeking information about building such an instrument in the mid-1820s. Inspired as Lassell had been by the objects described in Herschel's catalogues, he was still obliged to go to John Edwards's account of how to make a reflecting telescope, which had been published as a supplement to the *Nautical Almanac* for 1787, to find instructions (**60**). Lassell scrupulously copied the complete article in longhand, along with several Herschel pieces – a mute testimony to the patience that one required in the days before photographic methods of duplicating documents were invented.

The career of William Herschel teaches us many things. It indicates the impact which a mind of extraordinary and unconventional genius could exert in an age when Newtonian physics had proved the lawlikeness of nature, and those laws might perhaps be shown to extend to the furthest depths of space. It also shows how analogies could be drawn from one body of scientific knowledge, and applied to another, to form a 'natural history of the heavens.' But perhaps most of all, it shows how a relatively neglected piece of contemporary instrument technology could undergo an almost whirlwind development in the hands of a man who combined the roles of craftsman and scientist, and, in the course of fifteen years, be pushed as far as the capacities of the age would allow. Then having pushed the instrumental techniques, and with them the public standards of proof upon which science depends, to the limit, Herschel was to spend the next three decades trying, metaphorically speaking, to squeeze cosmological blood out of speculum stone.

Much of the formidable reputation which Herschel enjoyed in his lifetime stemmed from the new astronomical possibilities which he had shown to his contemporaries. In this respect, he belongs in the same league as Hipparchus, Copernicus, Galileo and Newton, and like them, he left more questions hanging than he could reasonably answer. But then again, it is the ability to delineate boundaries between what can be proven and what might actually be the whole story which lies at the heart of so much creative science, and in this respect, the debt owed to Herschel by the subsequent history of astrophysics is immense.

ACKNOWLEDGMENTS

I wish to thank Mr Peter Hingley, Librarian of the Royal Astronomical Society, for the trouble he took to make several of the sources used in this paper easy to obtain. I also thank Mr A.V.Simcock, Librarian of the

XII

414

Museum of the History of Science, Oxford, for similar assistance. My particular thanks go to Dr D.W.Dewhirst, of the Institute of Astronomy, Cambridge, for his correspondence, suggestions and advice during my research in preparing this paper.

NOTES AND REFERENCES

(1) For a good critical biography of Herschel see Angus Armitage, *William Herschel*, Thomas Nelson (London, 1962). See also M.A.Hoskin, *William Herschel, Pioneer of Sidereal Astronomy* (London, 1959). An excellent analysis of his cosmological writings, which also re-prints sections of many of the key papers is to be found in M.A.Hoskin, *William Herschel and the Construction of the Heavens*, Oldbourne (London, 1963). An interesting study of the wider culture in which Herschel moved at the outset of his astronomical career is to be found in A.J.Turner, *Science and Music in 18th Century Bath* (Bath, 1977)

(2) A.J.Turner, 'Portraits of William Herschel,' *Vistas in Astronomy*, **32** (1988), 65–94. *The Scientific Papers of Sir William Herschel*, ed. Dreyer, J.L.E., 2 vols., Royal Society and Royal Astronomical Society (London, 1912). All further citation and pagination to Herschel's papers will be made from Dreyer's edition under *Scientific Papers*.

(3) Herschel, 'Astronomical observations, and experiments tending to investigate the local arrangement of the celestial bodies in space,' *Phil. Trans. Roy. Soc.* **107** (1817), 302–331; see *Scientific Papers*, II, 575–591; 575.

(4) Herschel, 'On the construction of the heavens,' *Phil. Trans.*, **75** (1785) 213–266; *Scientific Papers*, I, 223–259; 253.

(5) Herschel, 'On nebulous stars, properly so called,' *Phil. Trans.*, **81** (1791) 71–88; *Scientific Papers*, I, 415–425.

(6) J.T.Desaguliers, in *A Course of Experimental Philosophy*, 3rd edn (London, 1758), **II**, demonstrates experimentally the power of air condensation in Lecture XI, 'Of the air-pump, condensing engine and wind-gun,' 375–411. Also, in Lecture XII, 'Of Engines,' 412–537, he discusses the power of steam condensation, and the work of Savery and Newcomen, see 466–467. John Harris, in his *Lexicon Technicum* (London, 1736), speaks of condensation and contraction as essential powers of the steam engine, in which water 'gravitates' to form 'spring and elasticity.' See vol. **I**, article 'Engine,' section 'Description of a Fire Engine' (no pagination). The Leiden jar was sometimes referred to as a 'Condenser' in an age which still thought of electricity as a 'fluid'; see W.D.Hackmann, *Electricity from Glass ... 1600–1850*, Noordhoff, (Netherlands, 1978), 90–103.

(7) James Bradley, 'A letter to Dr. Halley giving an account of a new discovered motion of the fixed stars,' *Phil. Trans.*, **35** (1728), 637–661; and Bradley, 'A letter to the ... Earl of Macclesfield concerning an apparent motion ... of the fixed stars,' *Phil. Trans.*, **45** (1748), 1–43. P.L.M.Maupertuis, 'La figure de la terre ... mesure le degré du Méridien ... Cercle Polaire,' *Mémoires de l'Académie Royale* (1737), 386–466.

(8) I.Newton, *Principia* (1687), Bk. **III**, Prop. 20, Problem IV; see Andrew Motte's translation (1729), revised by Florian Cajori, University of California (Berkeley, 1966), 428–433. Nevil Maskelyne, 'An account of observations made at the Mountain Schehallion for finding its attraction,' *Phil. Trans.*, **65** (1775), 500–542.

(9) Henry Cavendish, 'Experiments to determine the density of the Earth,' *Phil. Trans.*, (1798) 496–526. It is important to remember that John Michell was the first person to discuss a torsion balance apparatus before the Royal Society, though he died before he was able to draw any conclusions. Cavendish began his 1798 paper by paying full credit to the initial work done by Michell.

(10) John Michell, 'An inquiry into the probable Parallax and Magnitude of the fixed stars, *Phil. Trans.*, **57** (1767), 234–264; 246.

(11) From the anonymous *Vanity of Philosophical Systems* (1761), cited in R.W.Harris's *Reason and Nature in 18th Century Thought*, Blandford Press (London, 1968), 232.

(12) The importance of improved angle measuring instruments to the advance of astronomy was made clear by Flamsteed in The *'Preface' to John Flamsteed's 'Historia Coelestis Britannica' 1725*, edited and introduced by Allan Chapman, based on a translation by Alison Dione Johnson, National Maritime Museum Monograph No. 52 (1982), 111 ff.

(13) Halley determined the proper motions of three stars, Sirius, Arcturus and the 'eye of the

Bull' in 1718 by comparing classical observations some 1800 years old with those of his own day. He also pointed out that Tycho's position for Sirius was two minutes away from its correct place, but was uncertain as to whether this was due in part to a refraction error. E.Halley, 'Consideration of the change of the Latitudes of some of the principal fix't stars.' *Phil. Trans.*, **30** (1718), 736–738.

(14) Thomas Digges, *A Prognostication euerlastinge*, (London, 1576), see fol. 43, 'A Perfit description of the Caelestiall Orbes.' See also, Antonia McLean, *Humanism and the Rise of Science in Tudor England* (London, 1972) 146–147.

(15) Herschel, 'Catalogue of a second thousand of new nebulae and clusters of stars; with a few introductory remarks on the construction of the heavens,' *Phil. Trans.*, **79** (1789), 212–255; *Scientific Papers*, **I**, 336.

(16) In 1720, Halley published two brief but far-reaching papers in the same volume of *Phil. Trans.*; 'Of the infinity of the Sphere of the Fix'd stars,' **31** (1720), 22–23, argued that the stars, through mutual attraction, should eventually, 'in process of time coalesce and unite' into one mass. By being regularly planted in space, however, their mutual attractions would cancel each other out and maintain an equilibrium. In his following paper, 'Of the number, order and light of the Fix'd stars,' *Phil. Trans.*, **31** (1720), 24–26, he developed the equal distribution argument, saying that dimness must be a function of distance, for as the magnitudes diminish, apparent star numbers increase. Both of these concepts were to form touchstones for Herschel's early work.

(17) Herschel, 'Astronomical observations and experiments tending to investigate the local arrangement of the celestial bodies in Space to determine the extent and condition of the Milky Way,' *Phil. Trans.*, (1817), 302–331. *Scientific Papers*, **II**, see 576–579.

(18) The natural history analogy was brought home quite forcefully early in his 1789 paper, *op. cit.*, when he compared himself with a natural philosopher attempting to understand 'an inconsiderable number of specimens of a plant or animal,' and investigate 'the history of its rise, progress and decay,' *Scientific Papers*, **I**, 330.

(19) Herschel's great taxonomic lists from which so many of his conclusions were drawn, included 'Catalogue of double stars,' *Phil. Trans.*, **72** (1782), 112–162; *Scientific Papers*, **I** 58–90. '[Second] Catalogue of double stars,' *Phil. Trans.*, **75** (1785), 40–126; *Scientific Papers*, **I** 167–222. 'Catalogue of One Thousand new nebulae and clusters of stars,' *Phil. Trans.*, **76** (1786), 457–499; *Scientific Papers*, **I**, 260–303. 'Catalogue of a second thousand new nebulae and clusters of stars,' *Phil. Trans.*, **79** (1789), 212–255; *Scientific Papers*, **I**, 329–369. 'Catalogue of 500 new nebulae … and clusters,' *Phil. Trans.*, (1802), 477–528; *Scientific Papers*, **II**, 199–237.

(20) John Hunter's Museum, now part of the Royal College of Surgeon's Museum, is perhaps the greatest monument to 18th century taxonomy. The intellectual principles which underlay the collection run through all of his writings. See, *The Works of John Hunter*, *F.R.S.*, 4 vols. (London 1835). In his 'Life' of Hunter, Drewry Ottley emphasised '…Hunter made considerable advances towards the natural classification of the animal world which Cuvier has so admirably effected in modern times' (vol. **I**, 136–137).

(21) William Buckland, in *Geology and Mineralogy Considered with Reference to Natural Theology* (London, 1835), **I**, vii–viii, clearly describes this task of the early geologists. See also, A.N. Rupke, *The Great Chain of Nature*, O.U.P. (Oxford, 1983).

(22) Herschel describes this basic theoretical mechanism in *Phil. Trans.*, (1785), *op. cit.*, *Scientific Papers*, **I**, 224–225.

(23) Ibid., *Scientific Papers*, **I**, 226.

(24) Herschel, *Phil. Trans.*, (1791), *op. cit.*; *Scientific Papers*, **I**, 422.

(25) Halley had speculated on the existence of an independent, non-stellar, glowing medium in space some 70 years before: 'An account of several nebulae or lucid spots like clouds, lately discovered among the Fixt Stars,' *Phil. Trans.*, **29** (1716), 390–392.

(26) Herschel, *Phil. Trans.*, (1791), *op. cit.*; *Scientific Papers*, **I**, 423.

(27) Herschel, 'Astronomical observations relating to the construction of the heavens, arranged for the purpose of a critical examination, the results of which appear to throw some new light upon the organisation of the celestial bodies,' *Phil. Trans.*, (1811), 269–336; *Scientific Papers*, **II**, 459–497. See also, 'Astronomical observations relating to the sidereal part of the heavens and its connection with the nebulous part; arranged for the purpose of a critical examination,' *Phil. Trans.*, (1814), 248–284; *Scientific Papers*, **II**, 520–541. Also, *Phil. Trans.* (1817) *op. cit.*; *Scientific Papers*, **II**, 575–591. Also, 'Astronomical observations and experiments, selected for the purpose of ascertaining the relative distances of clusters

XII

and stars, and investigating how far the power of our telescopes may be expected to reach into Space, when directed to ambiguous celestial objects,' *Phil. Trans.*, (1818), 429–470; *Scientific Papers*, **II**, 592–613.

(28) Herschel examines the possible mechanisms for these nebulae in his paper *Phil. Trans.*, (1814), *op. cit.*; *Scientific Papers*, **II**, 520–541.

(29) Herschel, *Phil. Trans.*, (1802), *op. cit.*; *Scientific Papers*, **II**, 212. The gravitational 'condensing' process is emphasised again in *Phil. Trans.*, (1814), *op. cit.*; *Scientific Papers*, **II**, 533.

(30) Herschel, 'On the existence of Space' (Read to the Bath Philosophical Society, 12 May 1780), *Scientific Papers*, **I** lxxxiv–lxxxvii.

(31) Herschel, *Phil. Trans.*, (1802), *op. cit.*; *Scientific Papers*, **II**, 213.

(32) *op. cit.*; *Scientific Papers*, **II**, 213. Herschel claimed to have detected changes in the Orion Nebula as early as 4 March, 1774, at the very outset of his scientific career, when he noted that the nebula looked different from its depiction in Smith's *Opticks* (1738); see M.A.Hoskin, 'William Herschel's early investigations of nebulae; a reassessment,' *J.H.A.* **X**, (1979), 165–176; 167. Herschel also referred to the changes which he believed (incorrectly) were visible in the Orion Nebula in *Phil. Trans.*, (1811), *op. cit.*; *Scientific Papers*, **II**, 489–491: and *Phil. Trans.*, (1814), *op. cit.*; *Scientific Papers*, **II**, 525–526. This last paper conveys a good sense of the diverse, dynamic, yet essentially bounded concept of cosmological change which Herschel envisaged.

(33) Herschel, *Phil. Trans.*, (1814), *op. cit.*; *Scientific Papers*, **II**, 541.

(34) Herschel, *Phil. Trans.*, (1785), *op. cit.*; *Scientific Papers*, **I**, 225. Herschel was very much concerned with the way in which the universe retained its stability, and argued (as a speculation) that *repulsive* as well as *attractive* forces must exist. He alludes to these forces, briefly, in his *Phil. Trans.*, (1789) paper, (*Scientific Papers*, **I** 334), and claims to have discussed them as early as 1780 in Bath. Perhaps as a supplement to the 1789 paper, he left an undated manuscript 'On Central Powers,' where he discussed a possible repulsive force which could counteract Newtonian attraction, having its own inverse law, to maintain stability. See, *Scientific Papers*, **I**, cxi–cxiv.

(35) Herschel, *Phil. Trans.*, (1785), *op. cit.*; *Scientific Papers*, **I**, 259.

(36) Michell, *Phil. Trans.*, (1767), *op. cit.*, 241.

(37) The quote in which Herschel stated that he went to bed 'with a basin of milk or glass of water and Smith's *Harmonics* or *Opticks* ...' is cited without reference by Dreyer in *Scientific Papers*, **I**, xxiv. In the entry following 7 March 1766, he mentions reading Smith's *Harmonics*; *Scientific Papers*, **I**, xx. He said in 1766 that it had been the study of Smith's *Harmonics* 'by which I was drawn on from one branch of mathematics to another,' *Scientific Papers*, **I**, xix.

(38) A.Chapman, *Dividing the circle*, unpublished *Oxford University D.Phil. thesis*, (1978), 83–141 & 197–233.

(39) John Hadley, 'An account of a catadioptrick telescope ... , *Phil. Trans.*, **32**, (1723), 303–312.

(40) David Bryden, *James Short and his telescopes* (Edinburgh, 1968) 32.

(41) H.C.King, *History of the telescope*, Charles Griffin (London, 1955), 144–158ff.

(42) Several such representations occur to my knowledge. An excellent example is the Drake-Brockman family group by Highmore (?) *c.* 1745, reproduced in Marc Girouard's *Life in the English Country House*, Yale University Press (1979), Plate xix.

(43) Herschel made several references to hiring telescopes, along with the purchase of tools to commence making them in 1773; *Scientific Papers*, **I**, xxii, xxiv.

(44) In June 1782, one of Herschel's telescopes was found by Maskelyne and others 'to exceed in distinctness and magnifying power all they had seen before'; *Scientific Papers*, **I**, xxxv and footnote‡. For the best modern study of Herschel's telescopes, see James Bennett, '"On the power of penetrating into space," the telescopes of William Herschel,' *J.H.A.* **7** (1976), 75–108.

(45) Michell, *Phil. Trans.*, (1767), 260–61, *op. cit.* Hooke describes the Pleiades in *Micrographia* (1665), Observation 59.

(46) Herschel's very first attempts at telescope making, in the autumn of 1773, were directed towards making a $5\frac{1}{2}$-foot focus Gregorian, but he found the mirror very troublesome to maintain. From 1774 January, he directed his efforts towards the production of Newtonians; *Scientific Papers*, **I**, xxiv–xxv. One of the main mechanical faults with the Gregorian reflector derived from the practice of using springs to hold the primary mirror

in place. When the springs underwent thermal or position changes, they caused an unsteadiness of the image. The problem was analysed by John Edwards in 1787, who recommended the removal of such springs, which seem to have been standard fixtures on commercially made Gregorians. Herschel's telescopes, said Edwards, owed part of their excellence to his original abandonment of such springs; John Edwards, 'An account of the cause and cure of the tremors particularly affecting Reflecting Telescopes,' published as part of the *Nautical Almanac* (1787), 49–60; 60.

(47) Herschel, 'Account of some observations tending to investigate the construction of the heavens,' *Phil. Trans.*, **74** (1784), 437–451; *Scientific Papers*, 157–166; 166.

(48) Herschel's 'On the Parallax of the Fixed Stars' was read to the Royal Society on 6 December 1781, and published in *Phil. Trans.*, **72**, (1782), 82–111; *Scientific Papers*, **I** 39–57. It was followed soon after, on 10 January 1782, with his extensive 'Catalogue of Double Stars,' *Phil. Trans.*, **72** (1782), 112–162; *Scientific Papers*, **I**, 58–90; See 58–59 for Herschel's claim that he could 'estimate by diameter' an angle of 0·25 arcseconds, whereas conventional micrometers were only good to one arcsecond. See also, C.A.Murray, 'The distance of the stars,' *The Observatory*, **108**, no. 1087, December (1988), 199–217.

(49) Herschel, 'Description of a lamp-micrometer, and the method of using it' *Phil. Trans.*, **72** (1782), 163–172; *Scientific Papers*, **I**, 91–96. The high magnifications claimed by Herschel clearly aroused comment, for he published a letter to Sir Joseph Banks entitled 'A paper to obviate some doubts concerning the great magnifying powers used,' *Phil. Trans.*, **72** (1782), 173–178, *Scientific Papers*, **I**, 97–99. The letter describes the techniques by which Herschel computed and used his telescope magnifications.

(50) Herschel used several conventional micrometers, some of which are in the Science Museum, London (inv. nos. 1925–468 & 469), and the National Maritime Museum. He found conventional filar micrometers to be unsatisfactory, however, and enumerated five defects, including irregularity of screw threads and the necessity of illuminating the field to make the wires visible, thereby destroying the telescope's space penetrating power; *Phil. Trans.*, (1782), *op. cit.*; *Scientific Papers*, **I**, 91–92.

(51) In his 'Catalogue of 500 new nebulae…' *Phil. Trans.*, (1802), *op. cit.*; *Scientific Papers*, **II**, 199–237, he acknowledges the existence of genuine binaries, 201. He goes on to admit that '*many of them have actually changed their situation with regard to each other*' over the years he had observed them, 204. Herschel's italics.

(52) Though he could not demonstrate mathematically that such binaries did move in accordance with Newton's laws, he clearly considered it to be likely, being 'united by the bond of their own mutual gravitation towards each other,' *Phil. Trans.*, (1802), *op. cit.*; *Scientific Papers*, **II**, 201.

(53) Herschel, *Phil. Trans.*, (1784), *op. cit.*; *Scientific Papers*, **I**, 157–171. The newly-operational 20-foot reflector was used for these 'gages,' the 15 arcminute field of which seemed to determine their amplitude, 158. The stars were counted, or 'gaged' in bands 15 degrees long by 2 degrees across. The *Phil. Trans.*, (1785), *op. cit.*; *Scientific Papers*, **I**, 233–259, study made use of this technique.

(54) Thomas Wright, *An Original Theory or New Hypothesis of the Universe* (1750). Michael Hoskin, in *Stellar Astronomy; Historical Studies*, Science History Publications, (1982) has indicated that it was Immanuel Kant who proposed a disk-shaped galaxy, though he was in some degree influenced by a summary of Wright's work printed in Hamburg; see *ibid.*, Section B3, 'The cosmological thought of Thomas Wright,' 101–116. Hoskin also argues that while Herschel subsequently acquired a copy of Wright's book, he was not aware of either his or Kant's speculations in 1785, and drew his own independent conclusions from observational evidence; *ibid.*, 114–115.

(55) Edward Piggott, 'Observations of a Variable Star,' *Phil. Trans.*, **75** (1785) 127–136. Piggott discusses three methods, including tinted glasses, to ascertain stellar brightness changes, 135. None were found to be satisfactory.

(56) Herschel, *Phil. Trans.*, (1817), *op. cit.*; *Scientific Papers*, **II**, 579–585. The problem of measuring the remote stellar distances with telescopes, which had been with him throughout his career, formed the subject of his last major paper, when he was in his 80th year; see *Phil. Trans.*, (1818), *op. cit.*, *Scientific Papers*, **II**, 592–613.

(57) A series of 'gaging' ring sizes are to be seen in the R.A.S. Herschel Mss. 4/31.2.

(58) Herschel's constant search for changes that were seen to be taking place in space comes over clearly in the sub-title to his paper, 'On the proper motion of the sun and solar system; with an account of several changes that have happened among the Fixed Stars since the

418

time of Mr. Flamsteed,' *Phil. Trans.*, **73** (1783), 247–283; *Scientific Papers*, **I**, 108–130. For the treatment of Herschel on the Solar motion, see Hoskin, *William Herschel and the Construction of the Heavens, op. cit.*, 54–59.

(59) Derek Howse, *Greenwich Observatory*, **III** (1975), 116–117. Johann Schröter of Lilienthal in Germany was perhaps the only man to use a Herschel telescope for significant original research, other than Herschel himself. Schröter bought a 27-foot focus reflector in 1793, although he used it for lunar and Solar System work, rather than deep space; King *History of the telescope, op. cit.*, 135.

(60) William Lassell copied out this article onto 23 r. & v. leaves of a notebook around 1821; see R.A.S. Lassell Mss. 1 :4.

XIII

AN OCCUPATION FOR AN INDEPENDENT GENTLEMAN: ASTRONOMY IN THE LIFE OF JOHN HERSCHEL

Abstract

Sir John Frederick William Herschel occupies a pivotal position in the history of British astronomy (1). He formed the living link between two styles or traditions of science by being the last major specimen of one breed, and the inspiration and intellectual rôle model for the generation that would follow. For John Herschel was perhaps the last significant figure to devote himself wholly and full-time to fundamental research in astronomy and its related sciences on the strength of a private fortune. And while the stature that he enjoyed did much to stimulate the concept of the 'professional' astronomer in Britain, so many of those men of the rising generation who admired his thorough-going dedication to science were themselves more obviously professional in the respect that they earned their livings through academic science. One sees in him, therefore, an eclectic blend of attitudes towards what science was, how it should be pursued, and how it should be paid for.

1. Private Astronomy and the Scientific Establishment

For most of his active life, John Herschel epitomised science and the scientific establishment to the nation at large, and it is all the more strange, therefore, that two hundred years after his birth and 121 years after his death, he still awaits extensive biographical treatment, a published letter series, and an academic 'industry' devoted to the study of his career and its context which so many of his contemporaries have already received (2). Perhaps it is because he made no single discovery that changed the way in which people thought — as did Darwin, Faraday and even his own father, Sir William — or deliberately provoked contemporary debate. Or perhaps it is because he was simply too prolific, spending his life in a ceaseless stream of research, detailed analysis and publication, so that the modern student of John Herschel feels swamped by a deluge of printed tables and finely honed statements on a range of subjects that extend from the elements of binary stars to University reform.

On the other hand, the historian might argue that John Herschel has failed to excite major interest, until his bicentennial celebration in 1992, because he only added to the volume of astronomical data without being able to interpret it towards radical new conclusions. The telescope technology of the day, for instance, never enabled him to settle definitively the problem of nebula

Reprinted from *Vistas in Astronomy* 36 (1994), pp. 1–25. © 1993 with kind permission from Elsevier Science Ltd, The Boulevard, Langford Lane, Kiddlington OX5 1GB, UK.

XIII

composition, though he massively extended the 'database' initiated by his father. And while his work on double and triple stars likewise completed that of his father, and he, Savary, Encke and others were able to demonstrate the Newtonian elements of these systems, it only *proved* what Sir William and others had long since suspected.

He was in the tradition of those 'gentlemen free' out of which the English scientific movement had been born in the early Royal Society: men of ample fortune who could afford to provide not only the leisure required for original research out of their own private resources, but the necessary physical hardware as well (3). And in John Herschel's case, these resources had been happily preceded by the finest scientific education that the world could offer, culminating in the winning of a First Wranglership from Cambridge University (4). Combining the training of the professional mathematician with the resources of a rich private gentleman who never needed to earn his living by taking paid employment, he was an *amateur* in the noblest tradition; a person who (as he was to style William Lassell on the occasion when Sir John, as President of the Royal Astronomical Society, presented the Society's Gold Medal), belongs "to that class of observer who have created their own instrumental means, who have felt their own wants, and supplied them in their own ways" (5).

Yet John Herschel was by no means the last 'Grand Amateur' of British astronomy. Friends like John Lee and W.H. Smyth, Lord Rosse, William Lassell and James Nasmyth, not to mention many more, were to perform front-rank research out of their own pockets. Yet of that company of distinguished Victorian astronomers of private means, John Herschel was unique in so far as he was formally educated to science. He had been active in its various branches from youth onwards and was sufficiently distinguished to become an F.R.S. at twenty one, even allowing for the greater ease whereby gentlemen of philosophical turn and long pocket could join the Society in those days (6). Unlike Francis Baily and Stephen Groombridge whose fortunes, made in adroit City dealings, made possible astronomical careers from early middle age, or men like Lassell and Nasmyth who retired early on large industrial fortunes, John Herschel never needed to *make* his money as a preliminary to astronomy (7). Perhaps the nearest approach to Herschel as an independent social type was Lord Rosse, though Rosse's particular status gave him obligations to the people of Ireland that restricted his participation in full-time astronomy.

But as Sir John neither earned his living as an astronomer nor came to astronomy after another career, he can be considered neither as a professional nor as an amateur in any real sense. Perhaps he is best considered as an *occupied* astronomer, in so far as he occupied himself with it and the other sciences, in the same way that other independent wealthy gentlemen occupied themselves with antiquities, art, or field sports.

If Herschel represented the end of the 'gentleman free' tradition of astronomy, he did so because of reasons that were rapidly professionalising the science. While Britain only employed a handful of professional astronomers in 1825, such as the Astronomer Royal and a few University professors, the new professional breed was most clearly to be seen in Europe, where the great German Universities not only appointed salaried professors in the science, but publicly funded the observatories at their disposal in which to advance astrometry and the most abstruse branches of the science (8).

The circumstances which gave John Herschel his financial independence are not easy to define in detail. His father, a former performing musician and Royal pensioner, enjoyed a modest middle-class income of £200 per annum which he supplemented by the sale of telescopes (9). On the other hand, whatever fortune Sir William Herschel could have acquired by these means would have been wholly insufficient to maintain his son in high gentlemanly style for eighty years (10). Though it has not been possible at this stage to locate exact facts and figures, the fortune must have come from Mary Pitt, John's mother and the widow of a City merchant, whom Sir William had married in 1788 (11). As a young man, Sir John had clearly not felt it necessary to enter a paying profession, and after abandoning a half-hearted attempt to qualify for the Bar in 1815, felt free to pursue his scientific passions.

Nor is there any evidence that Sir John married money (12), for while Margaret Stewart came from a distinguished family of Scottish academics and clergymen, she had plenty of brothers and sisters with whom to share her deceased father's modest fortune. On the other hand, money was present in abundance. When John took his family and entourage to the Cape of Good Hope in 1833 for instance, he experienced no difficulty in paying the £500 one-way fare in the Indiaman, *Mountstewart Elphinstone* (13). Similarly, he bought 'Feldhausen' for £3,000 within a few months of arrival at the Cape (after first leasing it for £225 per annum), spent a considerable sum on improving it, and put the £500 profit which he made when he sold the property in 1838 towards defraying the return fare home in the *Windsor Castle* (14).

The growing Herschel family found 'Observatory House' at Slough rather cramped after their return to England and by August 1839 he was negotiating the purchase of what he would rename 'Collingwood', at Hawkhurst, Kent (15).

According to John Herschel's private diary, the asking price was £10,500 with an estimated £1,400 which needed to be spent on improving the property (16). Yet visits to the City of London were all that seem to have been needed to secure the house and estate. The sale was agreed on August 29th 1839 as John Herschel "Went into City & sold stock 5500 3[per cent] Cons[ols] for which netted £5005." Other sums were liquidated soon after, added to which was the sum made from the sale of 'Feldhausen', culminating in the diary entry for October 29th 1839 : 'Sold stock & bought 4000 Exchequer Bills. Our 7th child and 4th daughter was born at 11pm.' (17) As Herschel still retained the old family house, 'Observatory House' at Slough, none of the costs incurred in the purchase of 'Collingwood' could have been met by its sale, though it might have been leased (18).

Within a period of six years, 1833–39, John Herschel's diaries record some £14,000 in fresh capital transactions alone, to say nothing of the enormous running costs of his on-going scientific activities. The Cape expedition itself consumed £6237 in ordinary running costs (19), though he was still able to graciously *decline* all offers of recompense from the government upon his return. When one bears in mind that the Astronomer Royal's salary for the same period was £800 a year and a good professorial chair might only bring in £500, one begins to put the magnitude of the resources at Herschel's disposal into perspective (20).

These comfortable resources exerted a fundamental influence upon John Herschel's attitude towards science as a human activity. On the most obvious level, they enabled his extraordinary mind to range across the sciences in the eclectic way that it did, to make original contributions not only to astronomy but also chemistry, optics, botany and meteorology. The independence of action (if not leisure) which he enjoyed enabled him to devote time to the invention of novel photographic processes, conduct a vast correspondence across Europe and America, and keep the general public informed as the principal scientific pundit of the age.

But very importantly, his independent wealth made him see science as an essentially gentlemanly activity, in the way that Robert Boyle had seen it a century and a half before. While he recognised that contributions to science could be made by persons in a variety of social circumstances, it was the men of education and leisure who were the Baconian 'Lamps' that drew it all together and made it cohere. It was in this sense that John Herschel saw science as a professional activity, as the full-time *occupation* of independent gentlemen, who were (or should have been) above the clamours of lucre or office. It was an increasingly antediluvian view of science by the 1850s, and while he had every respect for paid scientists, such as his friend Airy, John Herschel saw independence of means as intimately connected with independence of mind.

This may seem strange for a man who, in so many ways, epitomised the rising professionalism of science, and was closely associated with vociferous professionalisers like Charles Babbage and Airy, and who stood as a 'professional' candidate against a 'dilettante' in the disputed Royal Society Presidential election of 1830 (21). Yet his deep-seated gentlemanly attitude towards science was clearly demonstrated in three sets of circumstances. The first of these was his dislike of being paid for any of his astronomical work, as seen above. The second was his anathema of imposed routine or time-consuming administrative duties; while the third was his instinctive dislike of being told what to do or how to spend his time.

As a young mathematics tutor at St John's, he soon grew tired of the drudgery of 'pupillizing' though at the outset of his Fellowship his friend Babbage warned him of the tedium involved in "examining 60 or 70 blockheads, not one of whom knows his right hand from his left, and not one in 10 of *them* knows anything but what is in the book" (22). Herschel could not stand teaching for long and soon quit his tutorial duties in the University to seek intellectual pastures new, though he retained his College Fellowship until his marriage in 1829. And over thirty-five years later, when he decided for reasons which are not entirely clear to accept the Mastership of the Mint in 1850, he found the duties and mode of operation of that, his only paid job, so hateful that he resigned the Office after only five years, with his nerves in tatters (23).

Having his creative labours organised by others was profoundly distasteful to Herschel. When Vernon Harcourt was trying to enlist his support in forming the professionalising British Association for the Advancement of Science in 1831, Herschel made it crystal clear that to his mind, creativity and independence of thought were inseparable. No committee could 'chalk out districts for individuals or combined diligence to explore,' especially in England, 'where freedom of action and independence of thought are highly prized and so energetically asserted on all occasions' (24).

He hated bureaucracy and believed that trying to 'manage' science from a centralised body could only pervert research through 'an overwhelming mediocrity' (25). For a man who happily moved among reforming politicians and Benthamite philosophers on an academic level, he was none the less adamant that no experiments in social engineering should touch him personally. Nor should it touch science, for as he asserted to his friend William Whewell in 1831, 'Perfect spontaneous freedom of thought is the essence of scientific progress' (26).

Though Herschel's vision of the power of scientific knowledge to bring about necessary social change inspired many people in Victorian England, he was quite incapable of belonging to that fundamental unit of modern science: the research team. All of his Herculean labours in half a dozen different sciences over sixty years were performed as a one-man show, for like the artist (but unlike the subsequent generations of scientists) he saw his work as a singular enterprise inextricably bound up with his individual vision, style of investigation, and personal resources.

Yet while it is true that John Herschel hated routine drudgery in the context of undergraduate teaching or office work at the Mint, this dislike seems to have been confined to that form of tedium which involved social relationships or teamwork. Ironically, he possessed enormous stamina for the routine work when operating on his own. One wonders at the countless thousands of hours of back-breaking observation that he put into his nebulae and double star "sweeps" in both hemispheres, while his diary records the painstaking reduction work that was necessary to produce the great lists of objects published in his catalogues between 1822 and 1865 (27).

2. John Herschel's Early Astronomical Career

With a father like Sir William behind him, one can understand why John Herschel did not choose astronomy as his first avenue of scientific investigation. Pure mathematics, optics and chemistry were his early choices, and the subjects in which he won his first scientific laurels. As a mathematician of brilliance, he was able to perceive many branches of science which could benefit from the new techniques of analysis being developed on the Continent and in Cambridge.

But in many respects, his father's work on the nebulae had been a product of a different style of science: a taxonomic rather than a mathematical style whereby the astronomer hunted for specimens possessing characteristic features and tried to infer developmental sequences therein. It was what Sir William himself frequently styled a "natural history" of the heavens (28). On the other hand, the nebulae and other recently discovered deep-sky objects were bodies which presented great challenges towards understanding the nature of the Universe, though to complete the picture, Sir William's surveys of the Galaxy had to be refined, reduced and interpreted.

It was this responsibility which Sir John shouldered, rather reluctantly, in the late summer of 1816, when Caroline Herschel recorded that on August 16th in that year, Sir William left Slough for a recuperative holiday at Dawlish in Devon, and that John accompanied him (29). It was in Dawlish that the seventy-eight-year-old astronomer prevailed upon his son to complete his programme of observation and classification of nebulae, clusters and double stars, and one suspects that John accepted the task more from filial duty than from enthusiasm (30). Almost by way of valediction, John relinquished his rooms in St. John's in preparation for what would turn out to be a lifetime of self-imposed drudgery, and by October 1816, he was hard at work in Slough.

Though John Herschel might have been reluctant back in 1816 to shoulder his father's astronomical mantle, he rapidly became not merely a superlative observational astronomer, but an extremely enthusiastic one. One has only to peruse his diaries to sense the regular pleasure that he received from looking at the sky, for in addition to the back-breaking reviews of the heavens which he undertook, the diaries record a fascination at watching the processes of nature: spectacular aurorae, blue moons, glories and meteors (31).

Perhaps the thing which most challenged his imagination about becoming an astronomer, however, was the science's poor state of cultivation 'in its more exact departments' in England in the early 19th century. Though astronomy was distinguished on its observational side—and nowhere more so than in the work of his father—'Mathematics were at the last gasp, and astronomy nearly so' as 'the chilling torpor of routine had begun to spread itself over all those branches of science which wanted the excitement of experimental research', or so Sir John later bemoaned (32).

In this context, one feels that Sir John is referring to the veritable obsession with routine meridian astronomy which had dominated both Greenwich and the leading private observatories in Britain for over half a century, the fruits of which had frequently been published in *Philosophical Transactions*. It had, moreover, frequently been meridian astronomy pursued with no wider apparent goal than the pleasure of being able to measure progressively tiny angles, as one gentleman after another had informed the Royal Society of the triumphs of his latest transit circle.

As John Herschel was aware, it was a different story on the Continent, where Bessel, the elder Struve, Encke, and others were active in researches which combined both the observational and mathematical branches of the science (33). These professors were using their officially provided observatories to undertake complex astrometrical projects for the determination of the elements of the proper motions of the stars, and the search for parallax changes (34). Armed with excellent refractors, many of them by Fraunhofer, they were undertaking a class of work by the early 1820s for which there was no parallel in England. This work, moreover, while based on the cataloguing of stellar coordinates, was much more inquisitive and direction orientated than much of the English meridian work which often accumulated data without a specific end in view. The Continental astronomers, and especially the Germans (including Germans working in Russia), were leading the field in these researches into the development of celestial mechanics and the refinement of gravitation theory.

This kind of high-level astrometry naturally appealed to a mathematician like John Herschel, and in his twenty-odd-year career as an active observer (from 1816 to about 1840) he undertook a good deal of astrometric work. Yet unlike that of his European colleagues, most of Sir John's work was undertaken with reflecting telescopes, supplemented with an equatorially mounted refractor.

The instrument with which he performed the greatest part of his deep-sky work in both hemispheres was the same "20-feet Reflecting Telescope" referred to in the title of the seven monumental papers published between 1826 and 1836 (35). It was also used to make the most of the Cape observations, published as they were in a heavy quarto volume in 1847. This remarkable telescope (probably the only single instrument to be used by the same observer to complete a review of both hemispheres) was made by John in the late 1810s under the supervision of the aged Sir William. It was to replace Sir William's original and now worn-out instrument of the same dimensions (with an 18.25 inch aperture mirror). John tells us that every effort was made to re-capture the exquisite figure of his father's original speculum, and that for optimum clarity, if not for light-grasp, it was best stopped down to 12 inches aperture (36).

When Sir William had commenced his survey with the original 20-foot in the early 1780s, he had used an observing technique whereby he made a series of horizontal 'sweeps' in which he could view a 12–14° zone of sky in about five minutes of time. This method had been replaced by a more convenient one by 1786 in which the telescope was moved through a series of 2° *vertical* sweeps, the upper and lower extremities of which could be defined by the rings of a bell (37). When Sir John commenced work with his own 20-foot, he used an almost identical method, though as he lacked an assistant, in the way Sir William had enjoyed the assistance of his sister, the work was greatly slowed down. This delay was occasioned not only by the necessity of his breaking off sustained observation at the eyepiece to record what he saw, as Sir John complained, but more significantly because the candle-light to which he was obliged to expose himself in writing down the observations inevitably spoiled his night vision (38).

He also encountered the common problem of maintaining a large speculum in a stable position in its cell, especially when working at different altitudes, though by 1833 he believed that he had devised a mirror support which eliminated this mischief (39).

It should be pointed out, however, that John Herschel never attempted to re-furbish his father's great 40-foot telescope, built at a cost of £4,000 on the strength of a Royal grant in 1789 (40). But this instrument was never as good in practice as had been hoped, and by 1820, it was already in an advanced state of neglect. Before it was demolished in 1839, however, it made a parting contribution to science by becoming a photographic subject. Sir John made the world's first glass-based negative of the telescope's timber framing, immediately prior to the instrument being dismantled by workmen (41).

It is interesting to note, moreover, that Sir John had completed the observational side of his northern and southern hemisphere surveys on the eve of the reflecting telescope being given a new lease of life following the invention of steam-driven machines for figuring mirrors. Both Lord Rosse and William Lassell were steam-machining very high quality mirrors of two, three and six-foot diameter within seven years of Herschel's return from the Cape (42). But John Herschel made no attempt to join in the new aperture-race, preferring to devote himself to the reduction of his massive Cape Observations rather than commencing a new assault upon the northern heavens.

3. Double Stars

In many respects, double-star astronomy was part of John Herschel's ancestral property, for his father had been the first observer to show attention to this class of object with a specific end in view. Sir William had begun to collect very close doubles, preferably only a few arc-seconds apart and only visible at high magnifications, in the hope of using them to derive parallax angles. But by 1803, he had come to realise that many pairs of stars were physically related systems that had shown movement over twenty years, and published a list in *Philosophical Transactions* (43). Wilhelm Struve at Dorpat had also begun to do first-class astrometric work on some of Sir William's binaries, to which he added several of his own, following the installation of the world's largest refractor, the Fraunhofer 9.9 inch, at his Observatory in 1824 (44).

Between 1821 and 1823, John Herschel and James (later Sir James) South began to re-observe Sir William Herschel's double and triple stars with South's 5-foot equatorial refractor at his London observatory. The result was a monumental analysis of 380 sets of stars that occupied 412 pages of *Philosophical Transactions* (45). This study brought double star astronomy into the limelight of the science, as it were, for it was a field of great potential. It combined the highest branches of both mathematical and observational astronomy, held the prospect of demonstrating the action of Newtonian laws in the stellar universe, while providing a major spur towards the improvement of telescopic optics and mechanics. In the 19th century, however, John Herschel was the only major double star astronomer to use a reflector as his fundamental instrument, for while the handling of large aperture, long focal length reflecting telescopes was something of a Herschel family technology, this astronomical specialism naturally favoured the equatorially mounted refractor.

The 20-foot reflector first came into its own for double star work in the *Memoir* of 1826, wherein John Herschel provided 'Descriptions and approximate places of 321 new Double and Triple Stars' (46). This was followed up by regular additions reaching and passing the 'first thousand of these objects' by 1828. Over two thousand had been clocked up by 1833 (47), and on the list went, eventually culminating in 3347 for the northern hemisphere and 2102 for the southern by 1847 (48). One should note, however, that when Herschel spoke of the *approximate* places of these objects in his catalogue headings, he was working with an altazimuth mounted reflector lacking exact setting circles. The price which he was forced to pay for the enormous light-grasp, and hence magnification, of the 20-foot, was the absence of equatorial coordinate aids. Sir John, therefore, like his father before him, was forced to rely upon his intimate knowledge of the sky in ascertaining the places of new, dim objects, by placing them within the field in relation to brighter stars of established Right Ascension and Declination. The positions, therefore, would be obtained by a mixture of micrometric measurement and practical estimation, depending on the presence of familiar and well-established markers in the field of view. On the other hand, he would have had no difficulty in measuring the precise angular relationships of the components of binary pairs to each other, for this could be achieved by placing a micrometer into the field of view.

If Sir John was as handy with the 20-foot as his father had been with his, it would appear that the absence of setting circles was only a minor disadvantage if one possessed a thorough mastery of the dim stars in the main constellations. In a letter to *The Times* of 1838, Sir James South claimed that Sir William Herschel was able, from a cold start, to find any object in the sky in under five minutes with the 20-foot (49). In addition to the 20-foot altazimuth reflector, however, it must not be forgotten that Sir John Herschel had a fine 7-foot equatorial refractor with a five-inch object glass at his disposal both in England and at the Cape. This instrument, mentioned in his catalogues of 1833 and 1835, was further employed to make 2194 micrometrical measures of stars recorded in the Cape *Results*, where it is described along with a detailed account of its mode of operation (50).

By the 1820s, several double stars, first recorded by Sir William around 1782, had moved sufficiently far from their original places that it was clear that they were binary pairs. Indeed, the star η Coronae Borealis had completed an entire revolution (51), while many others had not only moved appreciably, but also changed in colour and magnitude. It was still not proven, however, what agency had moved them, though gravity was seen as an obvious candidate. John Herschel had suggested in 1825 that constant observation of ξ Ursae Majoris should be undertaken, arguing that if the elements of the binary orbit could be measured with accuracy, then it could be established whether they conformed to Newtonian criteria (52). This is a far more complex problem than might appear at first sight, however, because the observer had no prior knowledge, and one had to establish the three-dimensional orientation of the system to the observer's line of sight, before orbital characteristics could be completed.

It was Savary, in 1830, who computed John Herschel's earmarked star ξ Ursae Majoris, to demonstrate its elliptical orbit from four consecutive chords. Encke, in Berlin, did the same soon after for the binary, 70 Ophiuchi, while in 1832, John Herschel published an 'elegant graphical construction and numerical calculation' of several stellar orbits, which had the advantage of being simpler and 'well adapted for the amateur' (53).

By the time of publishing his highly influential *A Treatise on Astronomy* (1833), Sir John was already able to provide a list of nine binaries with their known periods and elements (54). When

XIII

this work was extended into his even more influential *Outlines of Astronomy* (1847) the detailed elements of fifteen separate binaries were provided, including the closely agreeing elements for several others which had been determined up to four times by separate astronomers (55). The elements of the much-observed ξ Ursae Majoris, for instance, were published as determined by Savary, Mädler, Villarceaux and John Herschel. By the 1840s, however, so many binary and triple star elements had been determined as to enable Captain W. H. Smyth to publish a list of thirty-four "of the most remarkable of those to which an *annus magnus* can be assigned" (56).

4. John Herschel and the English 'Amateur' Astronomical Community

It is interesting to note that John Herschel's method of determining double star elements should be described by Smyth as 'well adapted for the amateur', for this is precisely what Smyth and many other English double star observers were (57). Not trained to academic mathematics like Herschel, but gentlemen of means and culture, these individuals came to serious astronomy after making their fortunes and original reputations in non-scientific professions before turning to astronomy from a love of looking at the sky.

In England, double star astronomy was largely in the hands of private individuals rather than publicly funded observatories, and it is interesting to see how John Herschel related to this community of observers. His early friendship and collaboration with Sir James South was of the greatest importance in this respect, for it had been at South's observatory in West London that his initial work on the joint catalogue of 1824 was performed. South was a surgeon, whose strategic marriage to a wealthy heiress in 1816 had enabled the abandonment of his practice so as to devote himself fully to astronomy, performed with the best instruments that Troughton, Tulley and others could build for him. South was not an academic mathematician, but he was a dedicated and highly-skilled practical observer who had achieved a great deal until his quarrelsome disposition led to the squandering of his energies in acrimonious lawsuits, egged on by his arch-enemy, the Reverend Richard Sheepshanks (58).

As a result of these lawsuits in the early 1830s South lost many friends, including John Herschel. In one of his frequent, raging letters to *The Times* in 1838, Sir James South looked back upon his former creative friendship with Herschel, "with whom, during some years, whilst a frequent inmate of my house, I passed some of the pleasantest parts of my life", before affirming that such intimacy would never be resumed (59). As late as 1846, the increasingly alienated and rather crazed Sir James was still fuming to the press, for on February 1st of that year, John Herschel made a long diary entry which expressed his exasperation at the latest public revelations from his former scientific friend. South, it seems, was virtually claiming to have been Sir John's teacher of practical observational astronomy by recalling his first observation of an occultation of a star by the moon, but Sir James merely "proves himself an *unsafe companion* as I have long considered him" and a person whom prudent and honourable men avoid (60).

Yet before and during the 1820s, Sir James had been a major figure in British astronomy, and a great encourager and teacher of rising astronomers, including Herschel and G. B. Airy, before he was almost baited to madness by the uncharitable Sheepshanks (61). And if Sir William Herschel was the first discoverer of binaries and Sir John was the next person to extensively re-examine this exciting class of object, then Sir James was the man who (after the death of his father) had taught John the trade of the practical astronomer, gave him access to an excellent set of instruments (including a five and a seven-foot equatorial) and provided expertise. It is tempting to speculate what further work in double star astronomy, beyond his great Herschel co-authored paper of 1824 and a further independent study of 1826 (62), might have been performed by this gifted amateur astronomer, had Sir James South not broken up his Campden Hill observatory to sell it for *scrap* in a ludicrous fit of pique in July 1839 (63).

By this time, however, the active centres of English double star astronomy outside Slough had become Aylesbury and Bedford. Dr John Lee, a wealthy ecclesiastical lawyer and landed gentleman, had established a major observatory in the grounds of his mansion, Hartwell House, Aylesbury, in 1827, and even when keeping law terms in Doctors' Commons, employed professional assistants such as James Epps to undertake courses of observations on his behalf (64). And at Bedford, Captain (later Vice-Admiral) W.H. Smyth set up his private observatory with an expressed intention to study double and variable stars.

Indeed, Lee and Smyth formed the nucleus of a group of south-midlands astronomers, including the brewer and landowner Samuel Whitbread of Cardington near Bedford, all of whom shared an interest in this important branch of astronomy (65). These gentlemen, moreover, were all active in the Royal Astronomical Society and enjoyed affable relations with James Glaisher of the Royal Observatory not to mention with Smyth's son, Charles Piazzi, who was working as an assistant at the Cape of Good Hope in the late 1830s before becoming Astronomer Royal for Scotland (66).

Nor must one forget that Piazzi Smyth's Director at the Cape Observatory was Thomas Maclear, a former surgeon at the Bedford Infirmary and a member of that group of local astronomical friends, before turning professional and becoming His Majesty's Astronomer at the Cape in 1834 (67). And the one common link which all of these men shared in the social demography of early Victorian astronomy was the friendship of Sir John Herschel. In addition to references to each of these men in his diary and correspondence in England (especially Admiral Smyth), Thomas Maclear had been Sir John's neighbour and close scientific collaborator during his four years at the Cape. Indeed, not only the astronomers but also the Herschel and Maclear wives and families forged friendships at the Cape that endured for decades (68).

The enduring monuments to the Bedford group, however, were two massive publications that were devoted to double and variable star astronomy. They were W.H. Smyth's celebrated *Cycle of Celestial Objects* (1844) and its supplements, and two of the three volumes of Smyth's *Aedes Hartwellianae* (1851). Volume two of Smyth's *Cycle*, entitled 'The Bedford Catalogue', showed the Victorian amateur astronomer at his best (69), and Sir John Herschel's friendship and pre-eminent status as a stellar astronomer is apparent throughout. Indeed, in the Index to the 'Bedford Catalogue' it is stated that references to Sir John's researches were 'quoted throughout', were too numerous to itemise individually, and that the reader was likely to encounter his name at every turn!

By the time of publication of Smyth's *Cycle* in 1844, however, double star astronomy had already passed into the realm of acknowledged regular research. What it, and the *Aedes Hartwellanae* made clear, however, was the pre-eminent rôle of Sir John Herschel in the initiation and continuation of this work whereby the laws of gravitation were shown to be truly universal and to apply with equal force even in the depths of the stellar universe.

But what the Aylesbury and Bedford connection also made clear was Sir John's intimate relationships with leading amateur astronomers who were engaged in front-rank research. And nowhere is this more aptly shown than in the humour which obviously existed in their relationship, as was made manifest in Piazzi Smyth's delightful watercolour cartoon of 1850. This little painting, for Lee's Hartwell House "Album", shows his father (Admiral W.H. Smyth), John Lee, and John Herschel, each carrying a dark lantern and wearing his characteristic 'club' observing cap, shaking hands at the top of the Hartwell House staircase at the conclusion of what seems to have been an observing session. Sir John's presence is made unmistakeable by his observing cap, which is embroidered with the word 'Slough' (70).

XIII

5. Nebulae and Star Clusters

One sometimes wonders whether John Herschel felt haunted by the nebulae, as objects which he had inherited from his father and about which he could never really make up his mind. Like many other inductive scientists, he was cautious when dealing with them, for as he wrote in 1833, they "furnish, in every point of view, an inexhaustible field for speculation and conjecture"(71). At this relatively early stage in his astronomical career, he hoped that after sufficient investigation with large telescopes their nature might be revealed just "as the double stars have yielded to this style of questioning"(72).

Yet that promise was never fulfilled in his lifetime, and in the introductory pages of his definitive "Catalogue of Nebulae and Clusters of Stars"(1863) one senses the presence of a problem long since recognised but not yet solved. In spite of the enormous improvement in telescope technology which he acknowledged had taken place since the 1830s in the hands of men like Rosse, Lassell, and others in Europe and America, "The brighter nebulae cannot be viewed to any advantage, and the fainter cannot be viewed at all, except by the aid of telescopes of large aperture" (73). In this valedictory utterance on the subject which had occupied a major part of his life, Sir John presented coordinate and descriptive data for 5063 nebulous objects in both hemispheres, which, unlike the double stars, had none the less refused to reveal their "natures" to inductive inquiry.

Like his father before him, Sir John Herschel saw condensation as the causal force through which all cosmological processes took place. And like his father, he regarded stars as the fundamental components of cosmological processes, though he exercised a much greater caution than his father when it came to discussing the rôle played by non-stellar nebulosity within these processes.

It is not clear, especially from his early writings, whether John Herschel believed that all stars possessed a generic size (as his father had done), though after his photometric researches at the Cape he realised that they could vary greatly in terms of absolute magnitude. Sir William, in fact, had borrowed an analogy from natural history when discussing star sizes (as he frequently did in his "natural history of the heavens") and had argued in the 1780s that all objects known to science possess an approximate generic size (74). Men, oak-trees and raindrops in their mature condition all fall within respective size bands, and while there could be individual growth differences, variation was contained within fairly strict limits. And if objects in nature followed the generic rule, then it seemed likely that stars, which were also part of God's Great Design, would do the same. The point that the stars varied in their respective sizes by a factor no greater than the size differences between a tall man and a short one, formed an axiom in Sir William's mechanism for change in the universe as well as his "distance–luminosity" ratio. In this ratio, a slightly larger star would exert an "attraction" upon a smaller one, and their combined attraction would attract others, until a rent or hole had been made in a hitherto uniform stellar distribution, and a cluster was formed (75).

By 1830, with Savary's publication of the elements for ξ Ursae Majoris (76), it had been shown that stars *were* gravitationally related bodies, though it was still not certain which star in a given pair had captured which, and to what extent masses were related to luminosity. But it now seemed very likely that the Inverse Square Law was at the heart of cluster formation. From the types of cluster collected by his father, and added to by his own observations, Sir John could reasonably argue that clusters of increasing compactness or condensity were formed, after passing through various open types before culminating in the Globular form. Though he was aware of the interpretative hazard of saying that a given type of open cluster would one day condense into a Globular one when he knew that no human being would ever witness such a maturation, it seemed a natural corollary to the laws of gravitation. As early as 1826, however, he had drawn attention to the hazards of using the naturalists' practice of demonstrating lines of growth or generic affinity in organic species in cosmology, for while the naturalist could see individual living creatures growing to maturity in time, no such empirical substantiation was possible with star clusters and nebulae

(77).

The answer could only lie in a developing telescope technology, for as Sir William's and Sir John's reflectors had already resolved many nebulous bodies into individual stellar components, so it seemed that yet bigger telescopes would resolve the tightly packed nuclei of the densest clusters, or at least demonstrate that the individual stars were too densely packed to be seen on their own (78). And once this has been done, a very powerful argument would be at hand to substantiate the clustering mechanism, by revealing the existence of individual stars, predicted by theory, yet hitherto invisible.

John Herschel came to be of the opinion that nebulae were of a similar composition to star clusters and composed of stars, or at least, of individual pieces of matter. While in 1833 he recognised that some nebulae such as θ Orionis, or 'Robur Caroli', in the southern hemisphere are "formed of little flocky masses like wisps of cloud" (79) (like the glowing "chevalures" of light acknowledged by his father after the 1790s) he was moving against the idea by the late 1840s, especially in the wake of the structural details revealed inside nebulae by Lord Rosse's telescope. For, "It may very reasonably be doubted whether there be really any essential physical distinction between nebulae and clusters of stars, at least in the nature of the matter of which they consist" (80). But his arguments against the existence of "true nebulosity" or "luminous fluid" in the way that it had been spoken of by Halley, and, to some extent, his father, depended on more than just the triumphalism of bigger telescopes (81). It also stemmed from his concern with the plausible dynamics of such materials.

Though John Herschel was willing to accept the existence of what we might now call interstellar dust, he came down against "luminous chevalures" and similar tenuosities on gravitational grounds. Before any physical system could operate in the universe, it needed *discrete particles* moving in orbital relationships. Gravity needed masses, and this in turn predicated the existence of discrete physical entities and not vague luminous streaks. The logical necessity for such particles—be they as big as stars or as small as dust grains or even atoms—became more apparent after Lord Rosse claimed to have detected stars in the Orion Nebula, and spiral structures in nebulae such as M51, for the dynamics of rotation demanded something substantial on which to act (82).

Sir John also agreed that no "flocky" nebulosity could actually glow and emit light if it did not contain discrete, *projectile* particles. As he was at pains to point out, a gas or mixture of gases such as air cannot become luminous simply through heating, at least as far as such substances were understood in the 1840s. Only when "carbonaceous particles" were mixed in with air could the whole give the illusion of glowing, as these paricles themselves began either to reflect light, or else, if hot enough, to radiate it. Indeed only when solid *projectile* particles were mixed with a gaseous vehicle could any cloudiness ever be seen, as in the case of common water-based clouds, or airborne dust in a shaft of sunlight. The gaseous medium itself, however, never becomes visible. In consequence any nebulosity seen in space must, by definition, be particulate in character, to reflect or generate light, or move into organised structures. Within the limits of the physics and chemistry of the day, Sir John's inductions were most elegant and fitted in beautifully with his predictions of triumphalism, implying that all nebulae must eventually succumb to resolution, at least in theory (83).

This particulate matter could then be acted upon by gravity to provide a fundamental cosmological mechanism. John Herschel also argued that this matter was quite probably distributed through many regions of space, especially in the peripheral branches of the Milky Way, and was present in certain star clusters. Very significantly, he argued that vestiges of this material were still present in the solar system among the inner planets close to the sun. The glow of the zodiacal light was seen as an indicator of the presence of such material, for at certain seasons of the year after the sun had set, a long, lenticular projection could be seen along the plane of the ecliptic which appeared to be of a similar stuff to the flimsiest nebulosities of the Milky Way branches (84).

XIII

This particulate stream in the ecliptic plane was not only capable of glowing by reflecting sunlight (like dust in a sunlit room) but also acting upon comets as they approached the sun (85). Whatever the comets were made of, it appeared that once they entered the dusty zone of the sun, at perihelion, they encountered mechanical pressure from its particles. As the comet inevitably accelerated under the influence of the Inverse Square Law, the result was a heating-up process which resulted in the comet's tail. John Herschel always maintained a lively interest in comets, for not only his published works but also his diary make regular reference to viewing them whenever possible (86). He published several drawings of cometary heads, and what he interpreted as solid nuclei reacting with the solar zodiacal envelope, to produce layered streams in the head and tail. He further argued that the particulate nature of the potentially glowing matter of space was substantiated from "meteorolites" from the combustions which they made in the air, and the stony forms that fell to earth (87).

In 1847, Sir John was at pains to point out the differences between the supposed formative processes of nebulae and clusters derived from Laplace's Nebula Hypothesis, and what he called the "theory of stellar aggregation" (88). Though he admitted that the Laplacian model had appeal, he argued that its status was entirely hypothetical, for no causal processes were known whereby vague nebulous matter (the composition of which was unknown) formed into physical systems under gravity. This very clearly touched upon the mature Herschel's caution at having any truck with "luminous fluid" and its variants. The theory of stellar aggregation, on the other hand, by confining itself to discrete particles, not only stayed within the plausible bounds of Newtonian dynamics but seemed to have the big telescope technology on its side, as well as meteoritic phenomena.

Early in his career, however, in 1826, he had been willing at least to countenance (in the footsteps of his father in the 1790s) the possibility that the Orion Nebula might be made of a "self-luminous or phosphorescent material substance in a highly dilated gaseous state" (89), though he admitted that such a turn of phrase only highlighted the mysterious nature of nebulous matter and qualified the statement by warning against hypothesising. Nor was he willing to go too far when talking about phosphoresence itself, for while he pointed out that phosphoresence was well known in nature, such as when the sea glowed on a dark night, all of these terrestial cases were the products of known chemical substances or organic beings (90).

Of all the nebulae visible in the northern hemisphere, the Great Nebula in Orion had long been the principal object of its class to interest astronomers, and not inappropriately, it became the subject of a major critical review by John Herschel in 1825 that was published in 1826. His report, which appeared in the *Memoirs of the [Royal] Astronomical Society*, reviewed previous studies recorded in the literature before going on to present Sir John's meticulous zonal examination of the Nebula made with his 20-feet telescope. And in this paper, John Herschel carefully followed his own oft-repeated precept of eschewing speculation about formative processes and keeping as far as possible to reporting the observed facts (91).

One of his intentions in the study of 1826 was to see if measurable change was visible in the Orion Nebula, for several observers had claimed to detect changes, including Sir William Herschel himself. The unequivocal detection of such change would naturally play a significant part in interpretations of the Nebula's composition by demonstrating that some sort of dynamical agencies were at work to produce them. Unlike previous observers who had simply drawn the Nebula, however, John Herschel broke it up into a series of topographic zones, each of which he examined for matter-distribution, to produce both written and visual records. Though he was of the impression that no change was visible in the Nebula, he none the less argued that a study such as he had made might provide a bench mark for anything that was done in the future.

Sir John returned to the Orion Nebula thirteen years later, when at the Cape, and being impressed by the quality of the skies and the Nebula's ideal position for observation, he commenced a second zonal and micrometric examination. In South Africa, he was immediately struck by the appearance of "a multitude of nebulous branches, convolutions and other details, of whose existence I had never before had the least suspicion" (92). On the other hand, Herschel was only confirmed in his earlier belief that no change had taken place, for the previously noted points of the topography of the Nebula were as they had been in 1825, while the faint new material derived from the superior transparency of the Cape skies (93).

An extremely skilled observer, Sir John was aware of how different objects such as nebulae can appear under different local conditions or speculum reflectivity, and provides us with one of the several instances where his cosmological views differed from those of his father. Sir John was of the opinion that deep-sky objects were in themselves too vast, remote and ancient to produce visible change over the brief time-scale of human observation.

By the time of publication of his *Outlines of Astronomy* in 1847, his old prediction that the nebulae would be resolved into stars was appearing to come true, for he pointed out that both Lord Rosse with his 72-inch and W.C. Bond at Harvard (working with a 15-inch Merz refractor) had broken down some of the "curdling" of the Orion Nebula into stars (94).

In his 1826 paper, John Herschel had also provided a detailed description of the Nebula in the girdle of Andromeda, Messier 31. He had discussed this elongated and highly condensed "elliptical" nebula as an object which differed fundamentally from the Orion Nebula, for as the Orion object was diffuse with a great deal of structure, that in Andromeda was highly compacted, did not seem to contain visible stars, and apart from a brightening towards the Centre, appeared uniform in its luminosity. A little over twenty years later, however, the Harvard and Parsonstown telescopes had come into operation and were revealing things invisible in the 20-foot, and pointing towards the Andromeda Nebula's resolvability (95). The new superior telescopes showed an appreciable brightening towards the centre which substantiated the idea of stellar condensity, while Bond claimed to have seen the Andromeda Nebula "thickly sown over with visible minute stars, so numerous as to allow 200 being counted within a field of 20 [arc minutes] diameter in the richest parts" (96).

Likewise, the vague projecting "laminae" that one could just manage to see emerging out of Messier 51 (the "Whirlpool") with the 20-foot were resolved into spiral arms by Lord Rosse's 72-inch, and from the wealth of detailed structure revealed by the enormous instrument, combined with the increase of faint stars in the field, further convinced Herschel in his belief that the nebulae were resolvable into particulate stellar objects, and were much more than indeterminate nebulosities (97).

6. At the Cape of Good Hope

Though most of Herschel's fundamental observational work had already been accomplished and his astronomical postulates formulated by 1833, the four years that he spent at the Cape between January 1834 and March 1838 sealed his achievement. Not only did he complete his own and his father's surveys of the heavens, but acquired observational and statistical data which modified or fine-tuned earlier ideas, especially on the "Construction of the Heavens". He also came to think of the Galaxy as a ring or annulus of stars (perhaps bearing some similarity to the annular nebulae β and γ Lyrae mentioned in his 1833 catalogue of which "The Powerful Telescope of Lord Rosse resolve[s].... into excessively minute stars, and shows filaments of stars adhering to its edges") and of which five more specimens had been discovered by the 1840s (98).

His work at the Cape also came to give a more finely delineated view of the distribution of nebulous and stellar matter in space. Herschel found that one third of the nebulous contents of the entire heavens, for instance, is condensed into a zone of one eighth of the total sky area around the northern galactic pole, in the constellations of Leo, Leo Minor, Ursae Major, Coma Berenices, Canis Venatica, and especially, Virgo. In the southern hemisphere, "a much greater uniformity of distribution prevails" (if one excepts the densely packed nebula fields in the Magellanic Clouds),

XIII

as the nebulae appear less localised into particular regions (99). Yet if the nebulae fields appear to be at their densest around the galactic poles, then the opposite applies to the "aggregations" of clusters and globular clusters which tend to be located around the peripheral regions of the Milky Way and hence are relatively close to the galactic equator (100). In his section on "The Law of Distribution of Nebulae and Clusters of Stars" in the Cape *Results*, he went so far as to suggest that the "nebulous system is distinct from the sidereal", though to some degree connected with it (101). It was his work at the Cape that made such conclusions possible, along with the suggestion that the uneven nebulous distribution indicated that the solar system was not at the centre.

The fundamental work which Herschel undertook on the two Magellanic Clouds—or "Nebulae"—could only have been possible at the Cape, and was of great significance to his understanding of stellar aggregation. As with the Orion Nebula, he made detailed surveys of the clouds and of the diverse "Stars, Nebulae and Clusters" which they contained with 226 objects logged in the small cloud and 919 in the large one (102). Herschel was also impressed by the enormous angles subtended by the Magellanic Clouds, especially the larger, and while eschewing speculation, was awed by the nature of a configuration of complex objects that could be so vastly remote while still illuminating a 3-degree area of sky.

On April 9th 1834, Lady Herschel wrote to friends at home mentioning that "Herschel" (as she always called him) had discovered two planetary nebulae in his first few weeks at the Cape and already felt "richly rewarded" by their trip to Africa (103). These objects, "exactly like planets and one of a fine blue colour", had been found by John almost before regular "sweeping" had begun, so he went on to inform his Aunt Caroline in June 1834 (104). Soon after these discoveries Thomas Maclear believed that he had detected motion in one of these Planetary Nebulae, though Herschel was sceptical (105). Only half a dozen Planetary Nebulae had been discovered prior to 1834, and Sir William Herschel had discovered four or five of them. Then they grew thick and fast, for by July 1835, Lady Herschel was growing quite blasé about such discoveries, as she mentioned to her sister-in-law that "A few more planetary nebulae are added to the list," and by 1847, Sir John was stating that some twenty-four or five Planetary Nebulae were known, some three- quarters of which were in the southern hemisphere (106). With characteristic modesty, however, Sir John made no claim for himself in the discovery of most of the objects included in this reseach.

While both Sir William and Sir John had recognised half a dozen basic types of nebulous objects, the Planetary were the most puzzling, not so much from their rarity as specimens, as from their paradoxical appearance and inexplicable character. They had been called *planetary* by Sir William because they looked very similar to normal planets, generally possessing sharply defined shapes and even surface luminosity, though they were clearly in deep space (107). When he discussed them in 1833 and in 1847, he showed his familiar wariness in speculating about their physical natures, for such little evidence was available that "it would be a waste of time to conjecture" (108).

Yet he *was* willing to conjecture within clearly prescribed limits, for his particulate conception of nebulous bodies provided some guidelines. Throughout the fourteen years that he was writing about Planetary Nebulae he was willing to conceive of them "in the nature of a hollow spherical shell", but whether this shell was filled with solid or gaseous material he was reluctant to speculate. While he was very cautious in talking about gas in space in a free state (presumably because he could not explain its gravitational behaviour from known data) he did not seem to mind it filling a volume terminated by some kind of solid shell. The Planetary Nebulae, moreover, appeared to betray no surface detail in spite of the 20 and 12 arc second surface areas which many, such as ν Aquarii and Andromeda, displayed. There was no central condensation, though Herschel and Arago in Paris conceived that these shells may be illuminated internally by central stars the point sources of which were invisible because of diffusion through matter contained in the shells (109). (One might suggest that Herschel saw them glowing by a similar process by which a frosted glass globe glows when a light is placed inside it.)

1. John Frederick William Herschel, 1845, Mezzotint, after Pickersgill.Museum of the History of Science, Oxford.

2. "Dr John Herschels neuste Entdeckungen in der Mondwelt" *c.* 1835. The scene depicts Richard Adams Locke's farcical stories printed in the New York *Sun* in August, 1835, whereby Sir John had supposedly discovered winged humanoids on the moon. According to Locke's story, Herschel obtained the colossal magnifications that were necessary by augmenting the light which came into the telescope with special lamps. This no doubt explains the presence of the two men operating a lantern and bellows. The image was then cast onto a screen. Engraving , covered with a layer of cracked and discoloured shellac, on the lid of a circular laquered box, $3\frac{5}{8}$ inches [92m.m.]. Royal Astronomical Society, on loan to the Museum of the History of Science, Oxford. See reference 123.

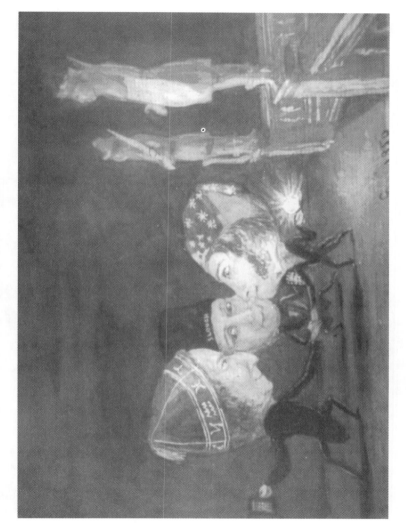

3. Charles Piazzi Smyth's small watercolour "A Three-Fold Adieu or a Tri-Partite Good Morning", 9 October 1850. John Lee's "Album" II, Museum of the History of Science, Oxford, 147. Left to right; Dr John Lee, "Slough" (Sir John Herschel ?), Admiral W.H.Smyth. See reference 70.

4. One of the many photographs taken by Sir John of his father's derelict 40-foot telescope shortly before it was demolished in December 1839. The Museum of the History of Science , Oxford, has several paper prints of this instrument, including one from the first glass negative mentioned in reference 41. Diameter, 3.7inches [95 m.m.].

But it was the sheer vastness suggested by the geometry of the Planetary Nebulae which awed John Herschel. In 1833 he argued that from their angular diameters in the sky, these objects must be at least as large in spherical volume as the entire orbit of Uranus (110). By 1847, however, fresh discoveries of Planetary Nebulae at the Cape and elsewhere, the discovery of Neptune, and the measurement of several stellar parallaxes, had widened the range of analogies. The Planetary Nebula β Ursae Majoris with its incredible 2′ 40″ arc diameter, if placed at the same distance as the recently parallax-measured 61 Cygni would be *seven times greater* than the orbit of Neptune (111). The source of luminosity inside the Planetary Nebulae, however, must be quite feeble, and "infinitely inferior" to the power of the sun. Herschel calculated that a 20 arc second (the diameter of the Planetary Nebula ν Aquarii) piece of the sun's surface would still glow as brightly as one hundred full moons, whereas ν Aquarii was invisible to the naked eye (112). All the evidence pointed, therefore, to these bodies possessing some form of diffused, feeble incandescence rather than being intense sun-type objects, which provided the reasoning behind Herschel's shell model.

By 1850, however, when Lord Rosse published his re-examination of John Herschel's 1833 Catalogue of Nebulae with the greatly enhanced light-grasp and magnification of the 72-inch telescope, he found that no less than five of the objects which had originally appeared as Planetary Nebulae in the 18-inch speculum were indeed ring nebulae (113). But this fitted in nicely with John Herschel's predicted optical triumphalism as diffuse objects came to reveal internal structure in progressively large instruments and that structure *implied* a particle-aggregate basis rather than a mass of "shining fluid".

Not until the middle and late 1860s was the true nature of the Planetary Nebulae resolved, when William Huggins discovered by means of the spectroscope that they were gaseous objects, though by that time Sir John Herschel was effectively retired from active astronomy and had not been involved in regular observational research for over two decades.

One of the things which had most consistently frustrated Sir William Herschel in his attempts to construct a plausible model for the three-dimensional arrangement of the stellar universe in the late 18th century was some sort of mensurational yardstick. Lacking any stellar parallaxes, he was forced back onto schemes that hinged upon interpretations of visible magnitudes, and his axiom that a star's apparent magnitude as seen by a terrestrial observer was in inverse proportion to its distance. His star distribution models were forced to depend, therefore, upon experience-based estimates of brightness, or at best, upon the use of comparison stars in the same field, or the insertion of annular rings into the telescope. These rings were employed to step-down the reflectivity of the mirror by a known amount in the hope of equating changing square inches of reflecting surface with genuine actinic differences between stars (114).

Sir John inherited the problem, and still lacking any knowledge of stellar parallaxes up to the time of leaving South Africa in March 1838, he also used comparative apparent magnitudes as his starting point. One wonders whether it was the novelty of the southern skies, and the diverse classes of objects which they contained that spurred him on in his magnitude comparison work. It should be noted, moreover, that one of the objects which aroused his curiosity in this context was α Centauri, for which one of the earliest parallax measurements was made by Henderson at the Cape Observatory and completed by Maclear in 1839–40 (115).

Sir John was aware of the measurements made earlier in the 19th century, whereby William Wollaston determined the sun to be brighter than the full moon by a factor of 1:800,000. On the 5th and 6th of March 1836, his Cape diary records experiments with a device which he named the "Astrometer" whereby Sir John applied the same principle of luminosity comparison to the stars. To read his account of the perfected instrument, in the Cape *Results*, however, one is struck by the singularity of the device. With that eclectic blend of firm physical principles, the seasoned observer's heightened sensitivity to changes in a visual image, and Heath-Robinsonian ingenuity, the Astrometer was Herschel cleverness at its best (116).

XIII

16

The operational heart of the Astrometer was a reflecting prism that sent moonlight through a simple eye lens to produce a point source. Sir John Herschel would then use this standard (which could itself be related to solar brightness) as a source of comparison for any given star. The reflecting prism was moved radially by means of cords, to and from the observer's eye, until its point source was of exactly the same brightness as the star under examination. From the measured positions of the eye lens and prism, a ratio could be computed that would express the degree whereby the moon's light had to be 'dimmed' to make it equal to that of the star.

John Herschel's technique, derived at the Cape, made it possible to arrange stellar brightness in accordance with a series of what he called "sequences" based upon stars such as α Centauri. It made it possible for him to reduce individual stars to the brightness they would exhibit at a standard distance and arrange them in a continuous sequence. He determined the luminosity of 191 suitably placed stars by this method, and when these stars in turn were placed into sequence alongside the sun some extraordinary differences in absolute magnitude emerged: Vega, for instance, was *forty* times brighter than the sun, and Arcturus *two hundred* (117).

This implied that stars not only differed greatly from each other in terms of brightness, but probably did so in terms of size and mass as well. Though it is not clear whether Sir John ever fully abandoned his father's distance–luminosity ratio discussed above, it is hard to imagine it surviving intact in his mind after his photometric work at the Cape.

At the Cape of Good Hope, John Herschel did much more than merely extend his father's work into the southern hemisphere. For while the largest part of his work remained taxonomic in character, it none the less produced unexpected revelations about the 'Construction of the Heavens', accomplished a great deal of research that defined the study of coloured and variable star astronomy (such as η [Eta] Carinae) and established the concept of absolute magnitude comparisons.

And viewed as a social enterprise, the Cape expedition confirmed John Herschel's status as an independent gentleman of science, for its four-year duration placed as much new and challenging data before the astronomical world of the day as the state-founded enterprise of Tycho Brahe had in the early 17th century. And while that great patron of astronomy, and dedicatee of the Cape *Results*, the Duke of Northumberland, provided funds whereby the work could be more sumptuously published than had originally been envisaged, Sir John Herschel's monument to astronomical free enterprise lay in the observations and conclusions that it contained (118).

7. The Philosopher of Nature

Significant as were the taxonomic and mensurational contributions of John Herschel to the science of his time, he produced no singular revolution in scientific ideas as did his contemporaries Faraday and Darwin. Yet his reputation, both within the scientific community and beyond into the worlds of general culture and popular opinion, was at least equal to if not greater than theirs. One might suggest that he had come, by his fortieth birthday, to represent something greater than the mere making of discoveries: he embodied the power of reasoned intellect and true knowledge. In an age which generally adopted thinking men as its heroes, Sir John Herschel filled an important space in the romantic pantheon. Beethoven and Walter Scott epitomised the heightened romantic power of feeling, whereas Sir John Herschel provided the rôle-model for the questioning intellect of prodigious power. He was the age's philosopher in the noblest sense of the word.

In some respects, this reputation also formed part of his inheritance, like the nebulae and double stars, for in *his* lifetime, Sir William had come to enjoy sage-like status as the man who had given "length, breadth, depth and profundity" to space, and demonstrated its orderly vastness (119). Interpreting the universe, after all, was an endeavour grand enough to satisfy the hungriest romantic imagination.

But even if John Herschel came into the field with some family advantages behind him, he would never have stayed the course for sixty years had his own creativity not been fully up to the task. Although it is true that Sir John was deeply respectful to his father, he was, none the less, his own man as a scientist. It was not so much that Sir John lived under his father's shadow as is frequently assumed, but that relatively few new techniques of inquiry into the 'Construction of the Heavens' had become available when he concluded his own career as an observational astronomer around 1840, than had already been there in 1790. Very large telescopes, spectroscopy, and astronomical photography had still not made their appearance, and so Sir John could do little more than broaden or verify his father's old data-base with almost identical instrumentation.

Wherever his findings, especially at the Cape, led him to conclusions different from those of his father, however, he saw no objection to their acceptance, and one can cite four examples where Sir John produced fundamental re-appraisals of evidence that were different from those of his father.

Firstly, he was much more cautious than his father in countenancing the presence and cosmological importance of "shining fluids" in space. One might suggest that John Herschel's rigorous training in gravitational dynamics made him suspicious of the rôle played by unknown forms of matter in potentially gravitational situations.

Secondly, he disagreed with his father about the existence of observable change in nebulae.

Thirdly, he had, in all practical respects, abandoned his father's 'distance–luminosity' axiom for explaining stellar distribution, for his very significant Astrometer work on absolute magnitudes in 1836 and 1837 had undermined the concept of stars possessing generic sizes and luminosities.

Fourthly, his work in both hemispheres, but especially his work on the Milky Way, Clusters, and nebula-distribution at the Cape, had made him modify his father's flat, planar model of the galaxy into a ring.

But if John Herschel's actions as an inquirer into the most abstruse regions of nature qualified him as the supreme natural philosopher of his age, then the qualification was only strengthened by his conscious actions as an author on natural philosophy. His *A Preliminary Discourse on the Study of Natural Philosophy* (1831) provided the supreme statement of a great scientist writing about what great science was and how it should be conducted. It had a formative influence on many scientists of the rising generation, and was acknowledged as such by Charles Darwin (120). In the *Preliminary Discourse* one saw the ethos of Baconian science in action, how rules of verification must be observed and speculation avoided. It embodied an affirmation of intellectual triumphalism, showing how a rational study of physical nature from Bacon and Galileo onwards had not only re-defined the idea of scientific knowledge away from the Aristotelian qualitative tradition, but produced those very canons of common sense to which every educated Victorian subscribed.

Heavy going as Herschel's classical prose and un-erring logic might seem today, the 'educated general reader' to whom the *Preliminary Discourse* (and its companion volumes in Lardner's *Cabinet Cyclopaedia*) was aimed, found it accessible, and the book was immediately influential. On the other hand, it would have seemed less formidable to a leisured readership that liked its literature tough and read the *Quarterly Review, Paradise Lost*, and the *Waverley Novels* for light entertainment!

It was also at this same readership that Sir John Herschel's *Treatise on Astronomy* (1833) (also published in Lardner's *Cyclopaedia*) and *Outlines of Astronomy* (1847) were aimed. But one reason why John Herschel's "popular" astronomical books were so influential was because of a lack of competitors. They filled a great gap in the literature, for there were no equivalent books that set out to explain *the entire* state of contemporary astronomy without equations in language that a seasoned reader of *The Times* could understand. And in addition to his books, he wrote a wealth of magazine articles, and definitive encyclopaedia sections that took his reputation into every drawing room to address not only natural philosophy in all of its branches, but subjects as diverse as earthquakes and the probabilities involved in accurate target shooting (121).

XIII

He exerted a romantic fascination upon the intelligent public of the day, was the *beau idéal* of the all-knowing English Christian Gentleman and by definition, an object of admiration. Indeed, nowhere is this image of John Herschel the profound philosopher better captured than in the superb, mannered photographic studies by Julia Margaret Cameron, made in 1867 (122). One might also argue that Sir John's international status as the universal astronomer had already been securely established by 1835, for it needed a singularly famous astronomer to discover men on the moon if the "Lunar Hoax" story of that year was going to boost the circulation figures of an obscure American newspaper (123).

But if John Herschel's reputation was so great in his lifetime, why did it not carry on after 1871 in the way that the reputations of other illustrious scientists did, and why, after getting his tomb in Westminster Abbey, did he not even receive that standard accolade of an eminent Victorian, a three-volume "Life and Letters"?

Some of the reasons for the eclipse of his reputation have been mentioned already, and perhaps the most important is the fact that he had made no fundamental discoveries that changed the way in which people think. Accumulating vast quantities of data and fine-tuning his father's cosmological system might well inspire awe in his own lifetime, but it is not the stuff to ring through the decades. This was especially so by the 1870s, when the increasingly professionalised sciences were coming to be associated in the public mind with the making of such discoveries. Likewise, Herschel's status as a natural philosopher and scientific pundit was essentially a contemporary one, for he largely addressed himself to issues and interpretations that would be out of date and forgotten fifty years hence, while "natural philosophy" itself was fast becoming an archaism by the 1860s. Perhaps his most enduring contribution to astronomy was his work on photometry and the idea of absolute magnitudes, which does, after all, lie at the heart of so much modern astrophysics.

That eclecticism through which his independent talents found their natural expression was also rather dated for a major scientist by 1870. The very professionalism in science for which Sir John had stood in the late 1820s had left his own reputation somewhat stranded fifty years later, for the new breed of mid- and late-Victorian "scientists" (a self-consciously professionalising term coined by his friend Whewell in 1840) (124) were much more focussed in their interests, and did not operate in half a dozen sciences as he had done. Nor would these new men have considered themselves to be "philosophers", for men who owned to that academic title by 1875 were dealing with different types of problems to the men of science.

John Herschel's eclecticism was clearly demonstrated in his approach to photography. One might assume that the inventor of the chemical fixation of photographic images, devisor of so much early photographic nomenclature and inspired systematic photochemical researcher in the period following his return from South Africa, would at least have gone on to use and promote photography after 1840. Yet once the *discovery* phase was over, his interests moved elsewhere, and within a short period of time he had returned to the artistic simplicity of the Camera Lucida when it came to taking pictures (125).

In the case of any person other than Sir John Herschel, ironically, such eclecticism would itself be a defining characteristic for the new and more pejorative way in which words like *amateur* and *dilettante* were coming to be used by the later 19th century. By the time of his death, Sir John had become one of the last surviving specimens of a breed which for two centuries had given British science such pre-eminence, and he was the last astronomer of that breed. While one might argue that eminent scientists of independent means were still around, such as Charles Darwin, Lord Rayleigh and William Huggins, their circumstances and attitudes were different in many ways from Herschel's. Most importantly, each of these men, irrespective of funding, worked only in one science, and perhaps with the exception of Darwin, within a highly specialised part of that science. William Huggins, whose spectroscopic researches took up astrophysics from the visual *impasse* where Sir John had left it, confined himself to one intensive branch of astronomy, and because his independent financial circumstances were nowhere near so opulent as Herschel's (126), he was not above receiving material assistance when it was available.

Yet even if the attitudes and circumstances of astronomy had changed by the late 19th century, John Herschel's life-enterprise was still an object of admiration. And it is interesting to note that when Thomas Hardy created the character of Swithun St. Cleeve as the ambitious young private gentleman astronomer in his romantic novel *Two on a Tower* (1882), it was the nebulae and clusters that fired his curiosity, and it was to the Cape of Good Hope that this fictional "occupied" astronomer took his great telescope in the conscious footsteps of Sir John Herschel (127).

8. Acknowledgements

I wish to thank the many persons and institutions that have assisted me in the research and preparation of this article, though I personally take responsibility for any error of fact or interpretation that it may contain. I thank Sheila Edwards and her staff at the Royal Society Library, and Peter Hingley and Mary Chibnall at the Royal Astronomical Society. I also thank Patrick Moore and the William Herschel Society. I also thank Miss Lynn Norman of the M.H.S. for her assistance with photographs. My particular thanks, however, go to A.V. Simcock of the Museum of the History of Science, Oxford, for his unfailing assistance with archival matters and his willingness to make the Library available outside the usual hours.

9. Notes and References

1. J.F.W. Herschel, born 7 March 1792; died 11 May 1871.
2. The only modern biography is Gunther Buttmann, *In the Shadow of the Telescope: A Biography of John Herschel*, transl. B.E.J. Pagel (Lutterworth, Guildford & London, 1974). Also, Constance Lubbock (ed.) *The Herschel Chronicle; the Life-story of Sir William Herschel and his Sister Caroline* (Cambridge, 1933). Desmond G. King-Hele (ed.) *John Herschel 1792–1872; A Bicentennial Commemoration; Proceedings of the Royal Society held on the 13th May 1992* (Royal Society, 1992). Also, Ivor Grattan Guinness, "Dark star in a new light", *The Times Higher*, 4 Sept 1992. Patrick Moore, *John Herschel, explorer of the southern skies* (Herschel Soc. Bath, 1991).
3. Thomas Sprat, A History of the Royal Society (London, 1667) 67 for "*Gentlemen*, free and unconfined".
4. *Memoirs and Correspondence of Caroline Herschel by Mrs John Herschel*, (John Murray, London, 1879) 120. Caroline's "Diary", 25 Jan. 1813 and 17 March 1813.
5. John Herschel, "Address", *Monthly Notices of the Royal Astronomical Society*, xvii (1849) 192.
6. The election procedure of the un-reformed Royal Society (but *not* in relation to the election of Sir John Herschel) was attacked by Sir Charles Babbage, *Reflections on the decline of science in England* (London, 1830) 53–56. The Royal Society, furthermore, was only accessible to men who could afford the £50 fee. The R.A.S. fee was only half that amount, *Reflections*, 43.
7. The commercial and financial bases of so many early F.R.A.S.' were discussed by William John Ashworth in his "A Stellar Economy : Baily, Herschel, Babbage and the Business of Astronomy". Unpublished paper given to the conference "Astronomy in Nineteenth-century Britain", National Maritime Museum, 14 November 1992. I thank Mr Ashworth for kindly giving me a copy of this paper.
8. Allan Chapman, "The astronomical revolution"(in early 19th-century Germany), in *Möbius and his Band*, J. Fauvel, R. Flood & R. Wilson (eds.) (O.U.P. 1993) Chapter 3.
9. Angus Armitage, *William Herschel* (Thomas Nelson & Sons, London, 1962) 25. The puzzling source of John Herschel's money is mentioned in *Herschel at the Cape; Diaries and Correspondence of Sir John Herschel*, D.S. Evans, T.J. Deeming, B.H. Evans & S. Goldfarb(eds.) (Univ. of Texas, Austin & London, 1969) xx.

XIII

20

10. The Museum of the History of Science, Oxford, possesses a sheet of manuscript accounts which itemises the returns on a group of 3% Consols between 1800 and 1810. The capital was for £2305 and the dividend only £34-11-6 which can only have represented a small part of Sir William's assets. Mss Museum 58.

11. Lubbock, *Herschel Chronicle* (Ref. 2) 173–175.

12. *Lady Herschel: Letters from the Cape 1834–1838*, Brian Warner (ed.) (Friends of the South African Library, Cape Town, 1991) 9.

13. *Herschel at the Cape* (Ref. 9) xx. Herschel paid £500 for the return fare in 1838, which was the same sum as he had paid from England to Cape Town. Also, *Lady Herschel* (Ref. 12) 56.

14. For the purchase of the "Feldhausen" estate see *Herschel at the Cape* (Ref. 9) 18 Nov. 1834 and 27 Feb. 1835; xx, 107 & 148. *Lady Herschel* (Ref. 12) 24.fn.

15. John Herschel, Manuscript "Diary", Royal Society Library; 29 August 1839.

16. Herschel "Diary" 20 July 1839 for "Collingwood" repair costs; 29 Aug. 1839 for purchase price.

17. Herschel "Diary" (Ref. 15) 21 Oct. 1839.

18. The Herschel family retained the "Observatory House", Slough, until it was demolished in 1963. John Herschel's "Diary" makes several post–1840 references to its occupants who presumably rented it from him. I also thank Patrick Moore for other information about "Observatory House".

19. *Lady Herschel* (Ref. 12) 156–157 for accounts. It is also stated (page 156) that the Herschels' expenditure for 1832 had been £829. It is also sobering to realise that Sir John was also troubled by the tax man; "Attended Appeal Income Tax at Cranbrook expecting (like a block-head) plain common justice. Got none"."Diary" (Ref. 15) 10 Jan. 1843.

20. G.B. Airy received £800 p.a. as Astronomer Royal; Airy to Lord Auckland, 10 Oct. 1834, Cambridge University Library Mss, RGO 6 1/ 153. As Plumian Professor of Astronomy in 1828 Airy had received £500 but only after having had it increased from £300; Airy, *Autobiography*, (ed.) W. Airy (C.U.P. 1896) 86. Charles Piazzi Smyth recieved £250 p.a. as Assistant Astronomer at the Cape Observatory under Thomas Maclear, and £300 p.a. as Professor of Practical Astronomy at Edinburgh University in the 1850s; H.A. & M.T. Bruck, *The Peripatetic Astronomer, the life of Charles Piazzi Smyth* (Adam Hilger, Bristol and Philadelphia 1988) 4.

21. Charles Babbage to John Herschel, 18 December 1815, Royal Society Mss. RS: 20;30. See also, *Collections of the Royal Society; Letters and Papers of Sir John Herschel, a guide to the manuscripts and microfilm*, Project editor Paul Kersaris (Royal Society, 1990) xxix.

22. Buttman (Ref. 2) 177–178. Herschel's "Diary" (Ref. 15) for 1850 indicates that he did not relish taking up his appointment. See also, Graham Dyer, "One of the best men of the business'; a critical examination of Herschel's Mastership of the Royal Mint 1850–1855", *John Herschel..Bi-Centennial* (Ref. 2) 105–113, which deals in detail with Herschel's traumatic years at the Mint. Herschel's health does not seem to have been good by middle age, especially his "weak lungs". According to his "Diary" (Ref. 15) he suffered from sciatica (8–11 Jan 1839), rheumatism (21 Aug. 1844) and various digestive complaints. At the Mint, he suffered from vertigo (early April 1852) and a general collapse in late 1854 according to *Collection of the Royal Society...Sir John Herschel*(ref.22) xxxvii. The trauma of his work at the Mint is also discussed by Larry J.Schaaf, *Out of the Shadows, Herschel, Talbot, and the Invention of Photography* (Yale Univ. Press, New Haven & London, 1992) 158.

24. Herschel to Vernon Harcourt, 5 September 1831, in J.Morrell & A.Thackray (eds) *Gentlemen of Science; Early Correspondence* (Royal Historical Society, 1984) letter 37, 55–57.

25. Herschel to Whewell, 20 September 1831, in *Gentlemen of Science* (Ref. 24) letter 45, 66–68.

26. Herschel to Whewell, 20 Sept. 1831 (ibid.).

27. *Catalogue of Scientific Papers (1800–1863) Compiled and Published by the Royal Society of London*, III (1867). See 322–328 for list of Hershel's papers.

28. For the most complete study of Sir William Herschel's cosmological ideas, see M.A.Hoskin, *William Herschel and the construction of the heavens* (Oldborne, London, 1963). Also, Allan

Chapman, "William Herschel and the measurement of space", *Quarterly Journal of the Royal Astronomical Society*, 30 (1989) 399–418;403.

29. *Herschel Chronicle* (Ref. 2) 346.

30. *Herschel Chronicle*, 352.

31. Though he did not undertake systematic observation after returning to England from the Cape, his Mss. "Diary" (Ref. 15) frequently mentions many detailed studies, especially of Comets, made with a 7-foot reflector, as of 11 June 1845, when he studies "Gibson's Comet", or 24 Feb. 1846, "Beila's", and several more.

32. J.L.E. Dreyer & H.H. Turner, *History of the Royal Astronomical Society 1820–1920* (R.A.S. 1923) 16, from Augustus de Morgan's *Memoirs* Section III.41.

33. Robert Grant, *History of Physical Astronomy* (London, 1852) 561–562.

34. Grant, *History*, 560–563.

35. These occupied 8 *Memoirs of the R.A.S.* each of which specifies the 20-foot reflecting telescope in the title. See Catalogue of Scientific Papers (Ref. 27) 324–25.

36. John Herschel, "Account of some observations made with a 20-feet Reflecting Telescope" *Memiors of the R.A.S.*, II (1826) 459–497; 460.

37. William Herschel, "Catalogue of one Thousand new Nebulae and Clusters of Stars", *Philosophical Transactions*, 76 (1786) 457–499, re-printed in The Scientific Papers of Sir William Herschel, I (Royal Society &. R.A.S. 1912) 261.

38. John Herschel, "Account...with a 20-feet.." (1826) (Ref. 36) 461. Herschel had three mirrors for the 20-foot, so that one freshly polished one was always ready. The instrument worked especially well at the Cape;JFWH to Caroline H., Feldhausen, 2 Feb. 1835, *Memoirs..of..Caroline Herschel* (ref.4) 370-271.

39. John Herschel, "Observations of Nebulae and Clusters of stars made at Slough, with a Twenty-feet Reflecting Telescope between 1825 and 1833" *Phil. Trans.*, 123 (1833) 359–505. See "Appendix" 489.

40. The best modern account of Sir William's telescopes is given by James Bennett, "'On the power of penetrating space'; the telescopes of William Herschel", *Journal for the History of Astronomy*, vii (1976) 75–108.

41. The original glass negative is in the National Museum of Photography Film and Television, Bradford, and is reproduced by Schaaf, *Out of the Shadows* (ref.23) 87. The Museum of the History of Science, Oxford, has a presentation print made from this negative entitled *The Forty-Foot Telescope at Slough* dated August 1890, along with several of Sir John's negatives and prints of the 40-foot.

42. Allan Chapman, "William Lassell(1799–1880): Practitioner, Patron and 'Grand Amateur' of Victorian Astronomy", *Vistas in Astronomy*, 32 (1988) 341–370. H.C.King, *The history of the telescope* (Griffin, London, 1955) 206–217.

43. William Herschel, "Account of the changes that have happened, during the last Twenty-five years, in the relative Situation of Double Stars; with an investigation of the causes to which they are owing", *Phil. Trans.*, 93(1803) 339–382, in *Scientific Papers*, II (Ref. 37) 277–296.

44. Grant, *History* (Ref. 33) 561.

45. John Herschel & James South, "Observations of the apparent distances and positions of 380 double and triple stars, made in the years 1821, 1822 and 1823, and compared with those of other astronomers" *Phil. Trans.*, 114 Part III (1824) 1–412.

46. See Ref. 38.

47. *Memoirs of the R.A.S*, II-IX; see Ref. 27 324–325. John Herschel, "Fifth Catalogue of Double Stars observed at Slough with a 20-feet Reflector; containing the places, descriptions and measured angles of positions of 2,007 of those objects, of which 1,304 have not been found described by any previous collection..." *Memoirs of the R.A.S.*, VI (1833) 1–74.

XIII

48. John Herschel, *Results of Astronomical Observations made during the years 1834,5,6,7,8, the Cape of Good Hope* (Smith, Elder & Co. 1847). In his numerical catalogue of double stars, ǃ 171–242, numbers 3347 to 5449 were made at the Cape.

49. James South, Letter to *The Times*, 20 November 1838. Cutting of article in John Lee "Astronomical Scrapbook", Museum of the History of Science, Oxford, Ms. Gunther 36, fol. 26

50. Herschel, Cape *Results* (Ref. 48) 247–264.

51. William Henry Smyth, *A Cycle of Celestial Objects*, I (London, 1844) 291.

52. W.H. Smyth, *The Cycle of Celestial Objects continued at the Hartwell Observatory to 1859 ... including details from 'Aedes Hartwellianae'* (London, 1860) 211.

53. Smyth, *Cycle*(Ref. 51) 292.

54. John Herschel, *A Treatise on Astronomy*, (London, 1833) 392.

55. John Herschel, *Outlines of Astronomy*, (London, 1847) 573, Art 843.

56. Smyth, *Cycle* (Ref. 51) 300.

57. Smyth, *Cycle*, 292.

58. Michael Hoskin, "Astronomers at War; South v. Sheepshanks", *J.H.A.* xx (1989) 175–212, and "More on South v. Sheepshanks", *J.H.A.* xxii (1991) 174–179. The enmity between South and Sheepshanks was intense, and neither could miss a chance to injure the other. Herschel recorded an R.A.S. meeting which he chaired 10 January 1840: "Sheepshanks gave a lecture in very indifferent taste about the Ring Micrometer—abusing Sir J.S. (not of course by name)—wanted to have stopped him but Galloway advised against it on the ground it would only be making bad worse". "Diary" (Ref. 15) 10 Jan. 1840.

59. South, letter to *The Times*, 20 Nov.1838 (Ref.49).

60. J.Herschel, "Diary" (Ref. 15) 1 February 1846. Herschel's debt to South was none the less great, and *D.N.B* "Herschel" states that Sir James had been one of his major influences as a practical astronomer.

61. E. Walter Maunder, *The Royal Observatory, Greenwich. A glance at its history and work*,(Religious Tract Society, London, 1900) 105, 114, for South as Airy's teacher.

62. James South, "Observations of the Apparent Distances and Positions of 458 Double and Triple Stars, made in the years 1823, 1824, 1825..." *Phil. Trans.*, 116 Part 1 (1826) 1–391.

63. Hoskin, "Astronomers at War" (Ref. 58) 198.

64. See Refs 52 and 53. Dr John Lee's papers, including his "Astronomical Scrapbooks" (Ref. 29) and "Albums" (Ref. 70) are preserved in the Museum of the History of Science, Oxford, Mss Gunther 9 & 10 (Albums) and Gunther 36,37, & 38, ("Scrapbooks"), James Epps's professional card is preserved in Gunther 37 fol.45, and his tombstone inscription (1773–1839) is recorded in "Album" I, Gunther 9, fol.17.

65. W.H. Smyth's papers are preserved in the R.A.S. and Bedford County Record Office. Also, Keith Sugden, "An eclectic astronomer", *Sky & Telescope*, January (1982) 27–29. Part of Samuel Whitbread's papers are also in the Bedford Record Office. Whitbread's private observatory at Cardington,Bedford, is virtually unique as the surviving observatory of a private Victorian amateur. I am indebted to Humphrey Whitbread Esq., Samuel's descendant, for granting me access.

66. H.A. & M.T. Bruck, *Peripatetic astronomer* (Ref. 20) 4–5.

67. Brian Warner, *Astronomers at the Cape of Good Hope. A History with emphasis on the nineteenth century* (A.A. Balkema, Cape Town 1979) 37.

68. *Lady Herschel* (Ref. 12). On many occasions in her Cape letters, Lady Herschel mentioned her friendship with Mary Maclear and her family, while they corresponded long after 1838.

69. Smyth's *Cycle of Celestial Objects* and *Bedford Catalogue* (which formed vol.II of the *Cycle*), along with his *Aedes Hartwellianae* (London, 1860) (Refs 51 & 51), were extremely influential publications and show the quality of serious amateur work, their instruments, and relations with John Herschel. See also Ref. 67.

70. Charles Piazzi Smyth's watercolour, "A three-fold Adieu or a Tri-Partite Good Morning" dated 9 October 1850, in John Lee's "Album" II, M.H.S. Oxford, Ms. Gunther 10. One must assume that this cartoon was somewhat retrospective, for in October 1850, John Herschel was on the verge of becoming Master of the Mint, and had not lived at Slough for a decade. See also A.A. Hanley, *Dr John Lee of Hartwell 1783-1866* (Buckinghamshire Record Office, undated,c.1985) plate 6.

71. Herschel, *Treatise* (Ref. 54.) 406.

72. Herschel, *Treatise*, 407.

73. Herschel, "Catalogue of Nebulae and Clusters of Stars", *Phil.Trans.*, 154 (1864) 1-137; 1. (N.B. The paper was received 16 October 1863, though the *Phil. Trans.* volume bears the title page date 1865).

74. William Herschel, "Catalogue of a Second Thousand of new Nebulae and Clusters of Stars.." *Phil. Trans.*, 79 (1789) 212-255, in *Scientific Papers of W.H.*, I (Ref. 37) 336. Chapman, "William Herschel and the measurement of space" (Ref. 28) 402-403.

75. William Herschel, " On the Construction of the Heavens",*Phil.Trans.*, 75 (1785) 213-266; *Scientific Papers of W.H.*, I (Ref. 37) 223-259; 253. Hoskin, *W.H.and the Construction of the Heavens*, (Ref. 28) 83, 86.

76. Felix Savary, "Sur la détermination des orbites que décrivent autour de leur centre de gravité deux étoiles très rapprochées l'une de l'autre", *Connaissance de Temps*, (1830) 56-69, 63-172.

77. J. Herschel, "Account of some Observations..." 1826, (ref.36) 488.

78. This faith in optical triumphalism and resolution of nebulae was a consistent feature of his work, i.e. *Treatise*, 1833 (Ref 54) 403; *Outlines*, 1847, (Ref. 56) Art 871.

79. Herschel, *Treatise*, 1833 (Ref. 54) 403-404.

80. Herschel, *Outlines*, 1847 (Ref. 55) Art. 871.

81. Edmond Halley had spoken of a "Medium" in space which could glow without the presence of illuminating stars; Halley, "An Account of several Nebulae or Lucid Spots..." *Phil. Trans.*, 29 (1716) 390-392.

82. Herschel discussed his particulate theory and "projectile forces" in "Observations of Nebulae and Clusters..." 1833 (Ref. 39) 501, and in *Outlines*, 1847, (Ref. 55) Arts. 871, 872, 882. Lord Rosse discusses the resolvability of Nebulae, especially Messier 51, in "Observations on the Nebulae", *Phil. Trans.*, 140 (1850) 499-514; 504-5.

83. It was in the Cape *Results* (Ref. 48) 138-139 that John Herschel provided his most tightly-argued case for particulate nebulosity and telescopic resolvability.

84. Herschel, *Treatise*, 1833 (Ref. 54) 407. *Outlines*, 1847 (Ref.55) Art 897.

85. *Outlines*, (Ref.55) Art 897.

86. He wrote an extensive essay "On Comets" in *Familiar Lectures on Scientific Subjects* (London & New York, 1866) 91-141. In his "Diary" (Ref. 15) are several references to visible comets [see also Ref. 31 above] while a major section on Halley's Comet was included in the Cape *Results* (Ref. 48) 393-413. He also wrote on them to the press, as indicated by the undated newspaper cutting which discusses comets and nebulae in John Lee's "Astronomical Scrapbook" (Ref. 64) Ms Gunther 37, fol.95v.

87. Herschel, *Outlines*, 1847 (Ref. 56) Art 898.

88. *Outlines*, (Ref. 56) Art 872. John Pringle Nichol, however, was the leading advocate of the Laplacian hypothesis of true nebulosity and its "condensing principle" in the 1850s. Nichol's *Views of the Architecture of the Heavens* (Edinburgh, 1853) 115-138 was very influential. On page 135, Nichol discusses the naturalist-parallel of nebulae collecting, though differing from Herschel (who was cautious about predicating developmental connections between static specimens) by arguing that gravity formed the mechanism whereby condensation took place.

89. Herschel, "Account of Observations..." 1826 (Ref. 36) 487. *Treatise*, 1833 (Ref. 56) 406.

90. Herschel, Cape *Results*, (Ref. 48) 138, Art.110.

91. Herschel, "Account of Observations..." (Ref. 36) 487.

XIII

92. Cape *Results*, (Ref. 48) 25–32, see Plate VIII.

93. Cape *Results*, 31, Art 67. The definitive pre-photographic study of the Orion Nebula, which placed great weight upon the work of Herschel, though not agreeing with him about the absence of physical change, was Otto Struve, "Observations de la Grande Nebuleuse d'Orion", *Mémoires de l' Académie Impériale de St Pétersbourg*, VII series, Vol.V, No. 4 (St.Pétersbourg, 1862), 97. Struve spoke of the Nebula's "d'agitation continuelle", 115.

94. Herschel, *Outlines*, (Ref. 55) Art 885. On 8 April 1844, Herschel was busy copying for Lord Adare "my picture of the Nebula in Orion that he may re-examine & compare it with the object in Lord Rosse's telescope", "Diary" (Ref. 15). Otto Struve in his "Observations de la Grande Nebuleuse..." (Ref. 93) was often highly critical of contemporary observers such as Bond.

95. Herschel, *Outlines*, (Ref. 55) Art. 874.

96. *Outlines*, Art. 874.

97. Herschel, *Outlines*, Arts 881–2. Rosse, "Observations on the Nebulae" (Ref. 82) 504.

98. Herschel suggested to Sir William Hamilton in June 1836 that the "Milky Way is not a stratum, but a ring"; cited (without source) by Agnes Clerke, *The Herschels and Modern Astronomy*, (Cassell, London, 1895) 173. Herschel, however, was always cautious when speculating in print, and in his books preferred to keep to detailed descriptions of the Milky Way; *Outlines*, (Ref. 56) Arts., 875, 786–788. Rosse, "Observations on the Nebulae", (Ref. 82) 506. Perhaps one might consider Herschel's definitive statement on the Milky Way to be that in Chapter IV of the Cape *Results* (Ref. 48) 373–392.

99. Herschel, Cape *Results*, (Ref. 48) 134–136. Arts., 97, 103.

100. Herschel, *Outlines*, (Ref. 55) Art. 869. Cape *Results*, (Ref. 48) 389, Art. 335.

101. Herschel, Cape *Results* (Ref. 48) Art. 104.

102. Herschel, Cape *Results*, 155–163.

103. *Lady Herschel* (Ref. 12) 33, 9 April 1834.

104. *Herschel at the Cape* (Ref. 9) 71–72, 6 June 1834.

105. *Herschel at the Cape*, 61, 8 April, 1834.

106. *Lady Herschel* (Ref. 12) 79, 15 July 1835. Herschel, *Outlines* (Ref. 55) Art.875.

107. Herschel, "Observations of Nebulae...", 1833 (Ref. 39) 500–601. Herschel, *Outlines*, 1847 (Ref. 55) Arts 876–77.

108. Herschel, *Treatise*, 1833 (Ref. 54) 405.

109. Herschel, *Outlines*, 876.

110. Herschel, *Treatise* (Ref. 54) 405.

111. Herschel, *Outlines* (Ref. 55) 876.

112. Herschel, *Treatise* (Ref. 54) 405.

113. Rosse, "Observations on the Nebulae..." (Ref. 82) 506.

114. William Herschel, "Astronomical Observations and Experiments tending to investigate the local arrangement of Celestial Bodies in space...." *Phil. Trans.*, 107 (1817) 302–331; in *Scientific Papers*, II (Ref. 37) 575–591; 580. A.Chapman, "William Herschel and the measurement of space" (Ref. 28) 411.

115. Grant, *History of Physical Astronomy* (Ref. 38) 551–52. As Henderson's original observations of α Centauri went back to 1832–33, John Herschel recognised his priority as a parallax measurer, *Outlines* (Ref. 55) Art.807 Also, C.A. Murray, "The distance of the stars", *The Observatory*, 108 (Dec. 1988) 199–217.

116. Herschel, Cape *Results* (Ref. 48) 353–372 for an account of the "Astrometer".

117. For "Sequences" and Astrometer technique, see Cape *Results*, 353–372, and *Outlines* (Ref. 56) Arts 779–785. Buttman, *Shadow of the Telescope* (Ref. 2) 95.

118. The Cape *Results* were dedicated "To the Memory of the Late Hugh, Duke of Northumberland". The Duke clearly took an active part in Herschel's work and the production of the published volume: on 14 June 1845, for instance, Herschel recorded visiting the Duke to show him the nebulae plates; "Diary"(Ref. 15).

119. Sir William Herschel's public status as a philosopher of nature is aptly summed up in an anonymous and undated (almost certainly 1822) poem lamenting his recent death cut out of an unspecified newspaper and pasted into John Lee's "Astronomical Scrapbook", M.H.S. Oxford Ms. Gunther 37, fol. 189; see (refs. 49 & 64). John Herschel's friend, the geologist Adam Sedgwick, also reminded Lady Herschel of Sir William's fame as a natural philosopher, 19 December 1846, by reciting an old rhyme "Oh Herschel! Oh Herschel! where do you fly? To sweep the cobwebs out of the sky", and asked when Sir John would complete and publish his own southern hemisphere "sweeps"? *Life and Letters of Adam Sedgwick*, II, J.W. Clark & T.M. Hughes (eds.) (C.U.P. 1890) 107.

120. Charles Darwin wrote to Herschel, 11 November 1859, about his *Discourse*, that "Scarcely anything in my life made so deep an impression on me" on the occasion of his presenting Sir John with a copy of The *Origin of Species*. Original letter in the Harry Ransom Humanities Research Center, Univ. of Texas. Reproduced in Schaaf, *Out of the Shadows* (Ref. 23) 16.

121. John Herschel, *Familiar Lectures* (Ref. 86) and Herschel's *Essays from the Edinburgh and Quarterly Reviews* (Longman, London, 1857).

122. Helmut Gernsheim, *Julia Margaret Cameron, Her Life and Photographic work*(Gordon Fraser, London, 1975) Plates 121, 153.

123. For the text of the published Hoax, which appeared in the New York *Sun* in 1835 and where Herschel was accredited with the discovery of intelligent beings on the Moon, see Richard Adams Locke "The Great Astronomical Discoveries Lately made by Sir John Herschel at the Cape of Good Hope", in *The Man in the Moon*, Faith K. Pizor & T. Allan Camp (eds.) (Sidgwick & Jackson, London, 1971) 190-216.

124. The *O.E.D.* gives William Whewell's *Philosophy of the Inductive Sciences*, I (1840) Intro. 113, as the source for "scientist": "We need very much a name to describe a cultivator of science in general. I should incline to call him a scientist."

125. Herschel's "Diary" (Ref. 15) indicates that he returned to the Camera Lucida by the mid-1840s. His skill with the Camera Lucida is examined and illustrated by Larry J.Schaaf, *Tracings of Light, Sir John Herschel and the Camera Lucida* (Friends of Photography, San Francisco, 1989). See also, Schaaf, *Out of the Shadows* (Ref. 23). 16.

126. *D.N.B.* states that William Huggins received the Olivera Bequest from the Royal Society in 1870, which provided him with a 5-inch Grubb refractor and an 18-inch Cassegrain reflector. Because "his private means were not large", Huggins also received a £150 p.a. Civil List pension in 1890. I am also indebted to Barbara Becker of Johns Hopkins University, Baltimore, U.S.A.,for further information about Huggins, on whom she is completing a Ph.D. thesis.

127. Thomas Hardy, *Two on a Tower* (1882), Wessex Edition, introduced by F.B. Pinion (Macmillan, London, 1975). Swithun St. Cleeve went to South Africa to study those skies "but partially treated by the younger Herschel", 261. Hardy's novel is perhaps the only major work of English literature in which astronomy both provides a plot motive and is competently dealt with on a technical level.

Originally published in an uncorrected form in *Vistas in Astronomy* 36, no. 1 (1993) 71–116, and its present form and pagination, as a corrected supplement, 1994.

XIV

PRIVATE RESEARCH AND PUBLIC DUTY: GEORGE BIDDELL AIRY AND THE SEARCH FOR NEPTUNE

The failure of the British scientific community to seize the initiative in the discovery of Neptune in 1845–46 has generated much heat and recrimination for well over a century. In the initial flurry, which blew up in the Royal Astronomical Society during the closing months of 1846, criticisms were voiced concerning the failure to pursue the calculations of John Couch Adams, and posterity has largely upheld the charges. The generally accepted scapegoats were James Challis, Director of the Cambridge University Observatory, and George Biddell Airy, Astronomer Royal. Airy in particular took the brunt of the fire at the time, through his pre-eminent status in British astronomy, and subsequent historians of the Neptune affair have continued to imply varying degrees of guilt.

Though the basic chronology of the events that first led Adams to investigate the perturbations of Uranus, to the actual discovery of Neptune based on Le Verrier's independent calculations, is well known, I believe that Airy's role in the affair is in need of re-appraisal. Though it is unfortunate that the Royal Greenwich Observatory file containing the original documentation dealing with the discovery of Neptune has been missing for at least thirty years, there are other materials elsewhere in the Airy papers that lend a new perspective.[1] I refer in particular to the "Astronomer Royal's journal", and his official correspondence with the First Assistant, the Reverend Robert Main.[2] Though neither of these sources specifically mentions Adams's role, they enable us to understand how the Royal Observatory worked as a public institution in the 1840s, and how the Astronomer Royal's time was spent and his official duties defined. The general corpus of documents upon which historians of the Neptune incident have relied over the past few decades has been the letters published by Airy in an attempt to explain his conduct, in the *Memoirs of the Royal Astronomical Society* in November 1846.[3] In addition, further extracts were published in the collected *Scientific papers* of Adams in 1896,[4] while Challis had already made his defence.

The most recent modern historical account of the incident is to be found in Morton Grosser's *The discovery of Neptune* (1962), although in essence it still re-uses the above-mentioned printed sources and attempts no interpretations based on original manuscript materials. Grosser's book contains some important errors of fact which derive from his failure to notice certain Airy manuscripts, and he proceeds to present the actions of the Astronomer Royal as those of a hidebound bureaucrat without first attempting to examine the official responsibilities of this early scientific civil servant.[6]

Airy entertained an exalted conception of his post as Astronomer Royal, though this was the product of circumstances that placed him in contrast to the six men who had preceeded him in that office. For one thing, he was the first Astronomer Royal to be entirely dependent upon his salary for a livelihood.[7] With the exception of Halley, all previous incumbents of the office had been in Holy Orders, and in receipt of ecclesiastical revenues. Halley, the only layman, enjoyed an independent Navy pension, as well as private means. Airy came into the office as a 'working scientist' in every sense, having risen from a relatively poor background, wholly without private means and also – as a layman – ineligible for Church preferment.

In the world of 1835, when he became Astronomer Royal, this gave him an awkward social status, he being neither a member of an acknowledged 'ancient' profession, nor a gentleman of independent means.[8] Science still hardly ranked as a profession in its own right, and as he was no longer a Cambridge don, he saw himself as a civilian, secular servant of the Crown.

Another significant factor in Airy's overall social status was the growing technical complexity of the government process, and the State's need for experts on matters of science, technology and engineering. Before becoming Astronomer Royal, Airy had demanded a substantial salary increase above that of Pond, his immediate predecessor, which would make up part of the absent ecclesiastical portion of his income. His total monies from government sources amounted to £1,100 per annum, and once in receipt of this, he was willing to place the whole of his working time at the disposal of the state.[9] Not only did he reform the working of the Royal Observatory root and branch, from the lax administration of Pond, but he made himself available to serve upon whatever Commissions or Inquiries Parliament deemed necessary. By the mid-1840s, Airy had created for himself the role of chief scientific consultant to the government, as well as "British Astronomer".[10] In this latter role, he saw the Royal Observatory as the national and international provider of all manner of useful scientific constants. Foremost amongst them were the astronomical constants required by the Admiralty, the department of state that employed the Astronomer Royal, and stipulated in the Royal Warrant by which he was appointed. To provide these constants and still find time to do Parliament's extra-astronomical bidding, Airy organized the Royal Observatory with a new and greatly improved instrumentation, operated by a diligent, disciplined and thoroughly efficient staff.

Airy was always aware of the statutory conditions of his office, and of its utilitarian, essentially service role to the exigencies of public life.[11] Most definitely, he did *not* see the Royal Observatory as a research institution, or a place where the frontiers of knowledge should be pushed back. This was the duty of the academic and private observatories, operating on independent endowments designed to finance 'philosophical' enterprises. Greenwich, on the other hand, was paid for out of taxation, to provide essential services to the Navy and Empire, and absolute standards to the world in general, and not to squander the citizen's money on abstract projects arising from private investigations.[12] As he reminded the Board of Visitors many years later,

> ...the Observatory is not the place for new physical investigations. It is well adapted for following out any which, originating with private

XIV

Airy and Neptune   123

investigators, have been reduced to laws susceptible of verification by daily observation.[13]

While a man who always maintained a lively private interest in pure research and kept up an extensive correspondence with scientific colleagues across Europe, Airy never lost sight of the conditions which his Royal Warrant placed upon the Greenwich Observatory. Had Adams, therefore, applied to a man who was less aware of his public duty, his planet might well have been looked for.

On the other hand, one might argue that while the statutory duties of the Astronomer Royal were essentially utilitarian, Airy interpreted them far more narrowly than any of his predecessors had done. What is more, there is nothing in his Warrant that necessitated his becoming Parliament's technological factotum, and his multifarious non-astronomical activities could have been largely avoided had he chosen to do so. Airy undoubtably moulded the character of his office and largely defined his own priorities, within the broad Warrant conditions, though he did this in the midst of unique circumstances. No previous Astronomer Royal had received so many official solicitations as he did, nor felt that his time was so well paid for. None of his predecessors, moreover, had been quite so committed to the overall propagation of useful knowledge in all its branches, or shared his passion for organizing facts. Though an astronomer by training and official designation, Airy was, by instinct, an engineer and administrator rather than a research scientist. What interested him most, was not the elucidation of new knowledge, but the systematic refinement and application of what was known already. It might be argued that such a man was not the best choice to head a major scientific institution, though when he was appointed in 1835, it was these very qualities of thoroughness, usefulness and willingness that appealed to his superiors. Though Airy was criticized by many persons within the scientific community for not backing Adams in 1846, his own conscience remained clear because he had maintained unswervingly a policy that he continued to define each year to the Board of Visitors, and which was approved by all his official masters.

It is possible to trace Airy's daily commitments and professional priorities over the year 1845–46 from the "Astronomer Royal's journal" currently in the Airy archives in the Royal Greenwich Observatory. The "Journal" informs us how he spent his time on a daily basis, and where he was on those crucial days when Adams was hoping to call on him. It is important to note, moreover, that the "Journal" contains no references to the new planet or any of the characters involved in its discovery, until October 1846, when its location had been announced in Berlin, and the storm was brewing in London. Even at this stage, one looks in vain for any references to Adams in the "Journal", though there are sections dealing with the planet itself, and the controversy. One might suggest that Airy's personal estimation of the strength of Adams's claim is signified by the total omission of his name from "Journal" entries for the eighteen months from the summer of 1845 to Christmas 1846. Though the "Journal" does contain several references to a "new planet" just before Christmas 1845, these relate to the discovery of the asteroid Astrea, and not to the body disturbing Uranus.[14]

124

The "Astronomer Royal's journal" relates the activities of an extremely energetic and wide ranging scientist, involved in a swirl of business, only part of which was astronomical. It is, moreover, one of the ironies in the history of science, that if there was one year in Airy's forty-six-year official career when he did not require extra demands being placed on his time, then this was 1845–46. After ten years in office at Greenwich, Airy had revolutionized the operations of the Royal Observatory, though his time was becoming increasingly burdened with non-astronomical official commitments placed upon him by the government or its agents. During this hiatus year of non-astronomical commitments, he was to spend between ninety and one hundred days in Railway Gauge business, along with some thirty further days devoted to other official consultancies. These included the Harbours Commission (nine days), marine engines (eleven days), his design for "my saw mills" at Chatham Docks (four days), ship launches (two days), Colonial and Foreign Office business (three days), and the design of the future "Big Ben" clock (one day). Bearing in mind the six-day working week recorded in the "Journal", this means that 130 out of a possible 310 working days of his time had been devoted to non-astronomical duties, most of them away from the Observatory, in London, Portsmouth and elsewhere.[15] The so-called "inflexible routine" with which he ruled the Observatory was much more than a personal love of bureaucracy; it was an essential element in ensuring that the Admiralty still received its statutory constants from an Astronomer Royal whose office depended on trustworthy delegation.

Without doubt, though, the commitment that placed the greatest demands upon his time was the Railway Gauge Commission, which set out to advise Parliament on the optimum gauge for new railways. What is more, most of the pressures of this Commission fell in the autumn of 1845, at precisely the time when Adams was endeavouring to engage him on the problem of the pertubations of Uranus. Airy was appointed to the Gauge Commission, as Second Commissioner, on 15 July 1845, and from that date to his leaving England for a journey to Europe on 25 August 1845, he wrote off almost every working day to "Gauge Business".

John Couch Adams first attempted to make contact with Airy during the last week of September 1845, when, armed with a letter of introduction from Challis in Cambridge, he made an unannounced call at Greenwich. That the call was unannounced and unexpected is evident from the wording of Challis's letter of 22 September 1845, recommending "My friend Mr Adams (who will deliver this note to you) [who] has completed his calculations respecting the perturbations of the orbit of Uranus".[16] Though Airy and Challis had already corresponded about Adams's work, and Royal Observatory data had been forwarded for his use,[17] Adams can hardly have felt too much disappointment if his chance call failed to find the Astronomer Royal at home. Airy returned from France on 26 September, and can only have missed Adams by a couple of days since his letter of introduction was dated 22 September. Prompt in all things, Airy wrote to Challis in Cambridge on 29 September to say that he and Adams had not met.[18]

One presumes that the Astronomer Royal spent the next fortnight clearing away the accumulated backlog of the month's business, for most days are designated to "Ordinary business", which generally meant routine Observatory affairs. From 12 October to the third week in December some thirty days in the

"Journal" are written off to "Gauge business", while others deal with such things as "I examined the airpumps on the Croydon Atmospheric Railway", and visits to Didcot and other places to observe railway trials.[19] Though the great bulk of his railway commitments fell into the latter half of 1845, he continued to take evidence and interview experts, such as Daniel Gooch, until April 1846. Indeed, Railway Gauges and other consultancies not only made serious inroads into the Astronomer Royal's time, but also took their toll on his general stamina, which was usually excellent. 13 November saw him engaged in "Ordinary business. I not quite well", while next day "I in bed weak all day". I suggest that the physical and mental strain which the additional work placed upon him over the course of that fateful year was also instrumental in his failure to pursue Neptune, as he was to record in the late summer of 1846, and as will be discussed presently.

The last dozen days of October 1845 were fateful ones for Airy. Since the 12th of that month, he had been engaged almost solidly on railway matters, and 21 October, the day later said to be that on which Adams sprang his second and third visits, was one upon which the "Journal" specifically mentioned his being on gauge business in London. One assumes that when Adams, on his way back to Cambridge from Cornwall, turned up to make his second unannounced call, the Astronomer Royal was still absent in town. Adams was later to recall that he left his card, went away for not much more than one hour, and came back. It was on this second occasion – his second visit on the same day – that he found Airy returned and at dinner (the Airy's usually dined in the late afternoon), and the servant refusing him entry.[20] Adams later wrote that he felt rebuffed and disappointed on this occasion, though one can fully understand the butler's possibly protective attitude towards his eminent and overworked master. On this occasion, before leaving, Adams wrote a three-page letter to Airy, specifying the elements of his hypothetical planet and giving its position. The letter contains no address and no date, beyond "1845 October".[21] Though Adams later seems to have recalled that this incident took place on 21 October, Airy could only ascribe it to the last days of October.[22] Airy was a meticulous man with a legendarily photographic memory, and if he could recall no date for the incident, one assumes that it must have made very little impression at the time.

If the date of Adams's visit was indeed 21 October, a glance at the "Journal" for the next ten days clearly focuses where Airy's thoughts must have been. Seven of those days were consumed by railway business, including the 23rd when a deputation from the Greenwich and Blackwall Railway visited the Observatory to discuss the problem of train vibrations on the instruments. Airy was naturally an interested party in the effects of a railway running through Greenwich Park, not only because of his official interest concerning Observatory instruments, but also because of his earlier experiments on the motion of trains in tunnels.[23] The following day, 24 October, he was testing the Croydon atmospheric line.

Yet there were two other events taking place during the week following Adams's visit which must have occasioned Airy grave concern, neither of which seems to have received proper historical notice. For one thing, at the age of forty-two, Mrs Richarda Airy was about to be confined with her ninth child, Osmund, who was born on 29 October. In view of Victorian attitudes towards childbirth, this could well have been the reason why a discreet butler was reluctant to admit a stranger to the family dining room when Adams made his second call on 21 October. To a man so deeply attached to his wife as was Airy – they used to correspond daily when business took him away from home – this must have been an anxious time, and may have been a contributory factor in Airy's two-week delay in replying by letter to Adams's call.[24]

In addition to the press of business and a confined wife in the Astronomer Royal's residence, a shameful criminal incident was brought to light in the Observatory on 27 October, just as Mrs Airy was about to go into labour. The "Journal" records: "Investigated a very serious charge of incest against Mr Richardson, and suspended him from his office." William Richardson was an Assistant at the Observatory, and a man whom Airy had described as a "valuable assistant" in 1841.[25] One can imagine, however, how such a scandal could threaten the prestige of Airy's reformed Observatory in the public eye. The loathesomeness of the affair was also reflected in the fact that the word 'incest' was inserted as an addition above the original line after the full "Journal" entry for the day had been completed. The "Journal" faithfully reports the progress of the sordid business from the police court hearing to the Old Bailey trial. Airy's embarrassment must have been especially acute, as the principal witness seems to have been the First Assistant, Robert Main, who was an ordained clergyman. The case possessed all the elements of a penny dreadful, for on 24 February 1846, the "Journal" recorded: "Mr Main absent today at the Old Bailey before the Grand Jury, on the trial of W. Richardson for the wilful murder of his incest child. True Bills found." Main's absences from the Observatory at various hearings of the trial were dutifully recorded, culminating in Richardson's acquittal – fortunate for all concerned – on 13 May 1846. Irrespective of Richardson's guilt or otherwise, such a trial was very bad publicity, and one can appreciate the repercussions that Airy must have foreseen when it was first brought to his notice towards the end of October. With the Gauge Commission, a parturient wife and the prospect of a nasty criminal case all coming within days of each other at the end of October 1846, the last thing the Astronomer Royal could have wanted was to be unexpectedly solicited by a junior Cambridge scholar with a mathematical speculation about the perturbations of Uranus.

What was impressive, under the circumstances, was the speed with which he did respond to Adams's undated letter left at the Observatory. In the midst of a block of days devoted to the Gauge Commission, in early November, he dispatched a letter to Adams on the 5th, requesting more information about the radius vector of the hypothetical planet, and the source of the value he ascribed to it.[26]

The radius vector, which constitutes the imaginary line connecting the centre of the Sun to a planet, is of course a fundamental element in gravitation theory. It is essential for establishing the physical and orbital characteristics of any planet, and without a knowledge of its *length*, it is impossible to calculate the mass, velocity or period of a planet. When searching for an unknown planet it was naturally impossible to know the value of the radius vector beforehand, so it was necessary for Adams and Le Verrier to ascribe it a theoretical one, congruent with Bode's Law. Though lacking any analytical basis, Bode's Law provided an empirical sequence of numbers which fitted the distances, orbits and radius vectors of the *known* planets. Adams, therefore, began his investigation for the new planet with a theoretical radius vector value based on Bode's Law, from which he could extract the theoretical mass and other elements, by Newton's Laws. The validity of his investigation, however, demanded a leap of faith in the assumption that Bode's Law would also apply to the unknown planet.

In Airy's opinion, this was not good enough, and before he, the Astronomer Royal, was willing to devote more time and energy to Adams's ideas, he required physical, not analogical substantiation for this key element of the new planet. Airy's stress on the value of the radius vector, which he described as an "experimentum crucis" for the credibility of the investigation, stemmed from two sources: his attitude towards mathematics as an investigative discipline, and his duty as Astronomer Royal not to become involved in theoretical projects. Though a former First Wrangler at Cambridge and a mathematician of brilliance himself, Airy had a strongly physical conception of the sciences, in which mathematics, as a body of intellectual constructs, only assumed meaning when rooted in observed physical phenomena. His apparent obsession with "daily observation" to substantiate the laws of nature lay at the very heart of his way of thinking, and in Adams's calculations, as presented in October 1845, there was too much that seemed to be taken for granted. The radius vector, after all, was central to Kepler's Laws and Newton's dynamics, and without a rock-solid value for this element, the rest of the work was little more than a brilliant exercise in abstraction. That Airy was genuinely interested in Adams's work cannot be denied, for as he later claimed:

> I waited with much anxiety for Mr Adams's answer to my query. Had it been in the affirmative, I should have exerted all the influence which I might possess ... to procure the publication of Mr Adams's theory.[27]

Yet no reply came, and Adams was not to contact the Royal Observatory for another eleven months, when he wrote to Airy on 2 September 1846, asking for further data to help refine his calculations of the planet's node. By this time, Airy was abroad again, and the business was dealt with by Robert Main, who deputized in the Astronomer Royal's absence.[28]

Airy was to return to the radius vector question many times in the future, when the Neptune affair had erupted and he was under fire. While it has been general to ascribe Airy's *post facto* comments about this quantity to an attempt to extricate himself from a mess, it was indeed the "experimentum crucis" and the source of his reluctance to act further on Adams's behalf, which in turn was rooted in the Astronomer Royal's conception of physical science. As he pointed out to Challis in December 1846, if no adequate radius vector value had been available "*then the theory would have been false, not* from any error of Adams's *but* from a failure in the law of gravitation. On this question therefore turned the continuance or fall of the law of gravitation".[29] As a ruthlessly open-minded scientist, Airy was willing to admit no sacred theories into astronomy, even in the form of Newtonian gravitation. So far, gravitation had worked because it had been in accordance with the "daily observations" of nature, but to predicate a radius vector derived from the convenient empiricism of Bode's Law was intellectually sloppy. "The progress of science has always depended on questions of this kind", he reminded Challis, and cited the phlogiston theory of oxidation and Laplace's theory of light as conspicuous examples. Chemists in the eighteenth century had used the phlogiston theory as a generally accepted 'true' explanation of combustion, yet if phlogiston was given off when a thing burned, why did the calx, or oxide, of the burned metal weigh more than the metal in its unoxidised state? This had been the Achilles's heel which had wrecked the eventual plausibility of phlogiston. Likewise, if Newton's laws were universal, then the solar system must be held together by precise, measurable phenomenon susceptible to mathematical *description*, and not to a mere chance sequence of numbers, as seemed to be the basis of Bode's law. It was not that Airy in any way doubted gravitation, but that one might well come to doubt it if new planets could be discovered by what he saw as a species of unsubstantiated numerology.[30]

Because of this doubt over a crucial element of Adams's work, Airy was not willing to proceed further, especially in his public capacity as Astronomer Royal. As the "British Astronomer", it was his duty to foster and pioneer the science in all of its useful departments, though the kind of service he would render must depend on the merits of individual solicitations. He would, for example, gladly supply Adams with as many planetary data as he needed, although this did not imply that the Astronomer Royal was willing to pursue the results of such calculations. Had Airy been wholly confident of the value of Adams's work, he would have done everything in his power to "procure the *publication*" (my italics) of the computations. One notes, however, that even under fire in November 1846, Airy did not say that he would ever have *looked* for the planet at Greenwich. When, in July 1846, after receiving Le Verrier's congruent values, he became convinced that an undiscovered planet really was present and would soon be found by someone, it was to Challis that he wrote to urge on the search; he did not involve his own official foundation. It is true that the Northumberland Telescope at Cambridge had almost twice the aperture of the largest instrument at Greenwich, but that was not all.[31] The Cambridge Observatory was a privately endowed academic institution, which existed for the furtherance of astronomical research and inquiry. Greenwich, conversely, was a functional public institution with a clear warrant to quantify nature once its phenomenon had been discovered elsewhere. With his finely tuned sense of duty, one might hazard the suggestion that he foresaw angry admirals and questions being asked in the House about taxpayers' money being spent on non-utilitarian projects, if such a 'philosophical' discovery was made in the Royal Observatory.

This enables us to explain why, by July and early August 1846, he was talking quite openly of *Le Verrier's* work on the planet; its impending discovery was being discussed at Board of Visitors' meetings, and by John Herschel at the British Association, and yet Airy was not willing to initiate any kind of search at Greenwich.[32] "Planet X" had become a possibility only in the wake of Le Verrier's work, and his two published studies on the subject. Adams, however, was little more than a memory from the previous autumn: a man who could not get his materials together, reveal his constants, or even reply to letters. In Airy's book, there was no doubt whose discovery it was going to be, and no reason whatever why he should feel guilty about not trying to hurry Adams along. Le Verrier was not only a mathematician of brilliance, but a forceful, businesslike man, who first took trouble to get his work published; and these were characteristics to which the Astronomer Royal could relate.

On 10 August 1846, Airy and his family travelled to Ramsgate, and from there to the Continent, as was his usual summer practice. This annual summer visit generally took the form of a working holiday, in which the Astronomer Royal would visit foreign scientific institutions, stay with fellow scientist friends and attend meetings of the Continental learned societies. It says much, moreover, about Airy's lack of sense of responsibility regarding the future of "planet X" that he could leave Greenwich at this time.

On the 1846 European trip, which was spent mostly in Germany, Airy stayed with friends in Göttingen and Altona. Airy had particular interest in the excellently equipped observatory at Altona, for only two years previously he had gone to great trouble to establish its precise longitude east of Greenwich. He stayed with his friend Professor Peter Andreas Hansen in Gotha (reciprocating a visit that Hansen had made to Greenwich earlier that summer), and made a brief visit to the aged Caroline Herschel in Hanover.[33] Yet this trip was intended to be more than just a scientific holiday; it was to provide a convalescence for a man who had just come through an exhausting year. As Airy stated: "We stayed for some time at Wiesbaden, as my nerves were shaken by work on the Railway Gauge Commission, and I wanted the Wiesbaden waters."[34] In an age that laid such store on the taking of mineral waters for vaguely defined illnesses, one suspects that Airy may well have been on the verge of what a later age would style a nervous breakdown. Whatever strength he derived from the Wiesbaden waters would soon become subject to a new set of demands, as soon as the Uranus-disturbing planet was located.

On 29 September, his "Journal" recorded: "While staying with Professor Hansen we received news of the discovery of the new planet beyond Uranus." The planet has been discovered on 23 September, at the Berlin Observatory, on the basis of Le Verrier's coordinates. According to Grosser's account in *The discovery of Neptune*, Airy, at Gotha, learned of the discovery of the new planet several days before the news reached England. Grosser states, moreover, that Airy "did not write to anyone about it for more than two weeks",[35] thereby implying that he was caught off his guard and pondering plausible excuses. This interpretation is contradicted, however, by two letters which Airy promptly dispatched to Main at Greenwich and Challis at Cambridge respectively. The day after receiving the news, 30 September 1846, he wrote to Main:

> While I was sitting yesterday at dinner with Professor Hansen we received from Encke the news of the discovery of the planet supposed to disturb the motion of Uranus. In all probability some notice of it has already reached Greenwich; but to provide as far as possible against failure I send the following places....[36]

A similar letter was dispatched the same day to Cambridge.[37]

Considering Airy's general expectation of the imminent discovery of a new planet, with the investigation of which he had been wholy conversant in July, along with his confirmed belief that such a search was not the duty of the Royal

XIV

Observatory, there are no grounds on which one should automatically assume
that he was suddenly gripped with anxiety concerning possible public reproach.
His two letters to Main and Challis are factual and straightforward in tone and
quite lacking in any sense of fear before a rising storm. The Airy family were
already homeward bound when the news was received in Gotha, though their
pace was still sufficiently leisurely for them to stop off at Altona and Hanover en
route for Hamburg. The North Sea crossing, unfortunately, was horrendous,
and through the combined efforts of fierce gales and engine trouble on board,
the ship took 107 hours, so that the party did not arrive back at Greenwich until
the afternoon of 11 October.[38]

Once home, however, the new planet became an almost daily preoccupation,
according to the "Journal" entries, though it would be incorrect to construe this
as reflecting an awareness of danger ahead. Airy was now interested in the new
planet in an official capacity, as he would have been with a newly discovered
comet or asteroid; for once discovered, it had ceased to be a mathematical
conjecture, and had become an observable natural phenomenon, like anything
else in the solar system. He now applied himself to the preparation of an
"Account" of the events leading up to the discovery, but if this appears like a
subtle way of writing a defence, there are two *private* statements which confirm
his own clarity of conscience.

The first of these is a pair of entries in his "Journal". On 22 October 1846 he
recorded, "In the evening I collected papers relating to the new planet beyond
Uranus", while on the 23rd, he wrote again, "In the evening began a paper on
the history of the discovery of the new planet". Not only are these entries
couched in plain, factual language, hinting at neither concern nor guilt, but they
were compiled "in the evening" after the real business of the day was done. One
might suggest that no man fearing for his future professional standing would be
likely to write his defence as a evening recreation.

Secondly, a week earlier, on 14 October, he had already informed Challis that
even "before leaving Germany", and before any hint of the embarrassing
Adams affair could have reached him, he had decided to place the facts leading
to Neptune's discovery before the public. He went on to give four reasons why
he took this task upon himself, the fourth of which says much: "Because I knew
of the history and have had no concern in the operations theoretical or
observing."[39] In short, Airy saw it as his duty, as the head of the British
astronomical establishment and immediately retired President of the Royal
Astronomical Society, to act as the impartial recorder of a story in which he felt
himself to be in no way involved. No matter how the scientific world was
subsequently to interpret Airy's actions, it is clear that the Astronomer Royal
himself felt no personal guilt, and was aware of no sins of omission which he
was obliged to hide.

The news of the new planet's discovery reached England during the last days
of September 1846. In his morning post of 30 September, J. R Hind in London
received word from Dr F. F. E. Brünnow in Germany, and promptly wrote to
Challis in Cambridge informing him that "Le Verrier's planet is discovered".[40]
On 1 October, Robert Main at Greenwich received a letter from Encke, written
on 25 September, announcing the same, and also passed the word on to Challis.
A postscript to Main's letter went on to add that he had received word that

Hind had *observed* the new planet the previous day, 30 September.[41] It would appear that the news travelled from Berlin to London almost as quickly as from Berlin to Airy in Gotha. Perhaps the international mails were faster than those within the German states, which would account for Airy's remark to Main, that in all probability news had already reached Greenwich, when he wrote on 30 September.[42]

Yet none of this correspondence made any reference to Adams, whose work so far had been confined to a very small circle of astronomers. On 1 October, however, both Challis in Cambridge and Herschel in London sent communications to the press, thereby entering Adams's name on the lists. Challis wrote to the *Cambridge chronicle* (following this sixteen days later with another letter),[43] while Herschel sent a piece on "Le Verrier's Planet" to the *Athenaeum*, which was published on 3 October. It was Herschel's announcement to the wider world that threatened controversy.

Though it is not my purpose in this article to trace the well-rehearsed international débâcle which ensued, and in which Airy and Challis received so much public blame, one must not lose sight of the fact that Airy never doubted Le Verrier's real primacy in the discovery. Though Adams had got his results first, credit went to Le Verrier because he published and acted on his. In Airy's mind, scientific knowledge was, by definition, public, open, international and broadly useful. The prize, therefore, went not to the man who hoarded his discoveries, but to him who followed them through to public announcement and the advancement of learning. In this respect, Le Verrier had undisputedly earned his laurels.

Adams himself does not seem to have done much further work on his computations until the late summer of 1846, when, on 2 September, he again wrote to Airy requesting orbital data. The Astronomer Royal being abroad, the data were promptly dispatched by Main.[44] It is important to notice, however, that Main did not seem to consider Adams's resurrected interest in the planetary problem to be of sufficient import to inform Airy, in his weekly letters of 9 and 16 September 1846.[45] Whenever Airy was abroad, it was the First Assistant's duty to keep him regularly informed about work in progress, though to accuse Main of being "intellectually emasculated" by Airy for not taking further initiatives, as Grosser does, is an absurd claim, made in ignorance of prevailing priorities.[46]

Though it is true that Airy felt no personal responsibility for England's failure to discover Neptune, this derived not from any implicit arrogance, but from a genuine conviction that he had done his duty. While it is true, as Grosser points out, that Airy faced major criticisms from such figures as Sir David Brewster and Sir James South, those who worked closely with him and understood his motivation were much more sympathetic. One important person in this respect is E. W. Maunder, who was to write in 1900 with the insight of a former Assistant of Airy's (though dating from the latter years of his career) and historian of the Royal Observatory. It is Maunder, with his access to intimate Observatory gossip, who seems to have been the source for many of the more 'pedantic' Airy legends. It is Maunder who recorded Airy's love of petty detail and his staffing policies. Yet in spite of these aspects of Airy's career which Maunder recorded, he reveals a profound respect and admiration for the

Astronomer Royal. When discussing the attacks that Airy suffered during his career, concerning the Neptune affair along with other battles within the scientific community, Maunder records how criticisms which would have agonised Flamsteed or Maskelyne, merely bounced off Airy. The context shows that this was not intended to indicate Airy's insensitivity to criticism, but rather his self-assurance once he had done his duty.[47] Grosser, however, not only quotes this section from Maunder somewhat out of context so as to reflect adversely on Airy, but even omits fourteen crucial words of the original so as quite to alter the thrust of the passage. Discussing Airy's attitude to the backlash of the Neptune affair, Grosser cites from Maunder's *The Royal Observatory Greenwich*: "He had done his duty, and in his own estimation ... had done it well."[48] What Maunder actually said in this contracted passage, relating to Airy's critics in general and not just those who attacked him because of Neptune, was: "He had done his duty, and in his own estimation – and, it should be added, in the estimation of those best qualified to judge – had done it well."[49] Maunder, who had worked at Greenwich during the last seven years of Airy's reign, and who wrote his history eight years after his superior's death, apportioned no blame and certainly imputed no negligence regarding Neptune.

Another of Airy's late contemporaries who dealt with the Neptune affair was James Whitbread Lee Glaisher (1848–1928), who wrote the biographical "Notice" prefixing the *Scientific papers* of his Cambridge colleague, J. C. Adams. J. W. L. Glaisher was admirably placed to be familiar with any possible grievances concerning the Neptune affair. He was the son of James Glaisher, the astronomer and meteorologist who had been on the Greenwich staff from the 1830s to the 1870s. James Glaisher senior, moreover, had come to enjoy an independent scientific reputation which occasionally provoked conflicts with the Astronomer Royal over his forty years on the Observatory staff. Not only was J. W. L. Glaisher the son of one of Airy's Assistants, but he himself rose to an early eminence in astronomy and mathematics, coming to know both Airy and Adams through the Royal Society and Royal Astronomical Societies, to which he was elected during the 1870s. J. W. L. Glaisher's renown was such, moreover, that when Airy retired in 1881, he was offered (but declined), at 33, the post of Astronomer Royal. He also pointed out, when writing his Obituary of Adams in *Monthly notices* in 1893, that Adams had been satisfied with Airy's conduct in 1846, and in no way held him responsible.

Like Maunder, Glaisher had no illusions about Airy's shortcomings, while at the same time admiring him for his great strengths. Glaisher considered that Adams himself was as much to blame as anybody, for not publishing his own work and taking the initiative to make sure that something was done.

> It is a most striking fact in the history of science that researches of such novelty and importance could have been known to two official astronomers [Airy and Challis] besides their author for nearly a year without any steps being taken to make them public.[50]

If any censure is implied in this passage, it derives from Airy's failure to take steps to make Adams's work "public" (that is, to have it published), rather than his failure to institute a search. J. W. L. Glaisher (like his Assistant father) was fully conversant with the Astronomer Royal's responsibilities when it came to observing the heavens.

In conclusion, Glaisher argued that three factors had cost Adams the priority of discovery: his youth, his faith in Newton's Laws (the radius vector), and his modesty.[51] Nowhere, however, does he apportion blame to the Astronomer Royal or criticise his official conduct. It must be remembered that Adams himself never seems to have lost the great respect and even awe with which he regarded Airy. This deference may well have lay behind Adams's refusal of a knighthood, made to him on the Royal visit to Cambridge in 1847; Airy had also just received his *second* offer of a knighthood on the same visit, and turned it down.[52]

Towards the end of 1846, Adam Sedgwick of Trinity College, an old friend of Airy's, entered into correspondence with the Astronomer Royal on the now awkward topic of Adams's recognition in the discovery. Several points emerge in that correspondence which indicate how un-enterprising Adams had been in the whole affair. The most significant feature to which Sedgwick drew attention was Adams's youth and modesty. To men burdened with massive commitments as were both Challis and Airy, Adams was ill-equipped in terms of personal force. Adams was not the type of man to demand attention from the mighty, even if possessing the greatest gifts of intellect. Sedgwick himself confessed to Airy that Adams had "acted like a bashful boy rather than like a man who had made a great discovery", while James Glaisher was also to state that Adams's youth had been one of his main shortcomings.[53]

If Adams was not an arresting or persistent man – as all evidence suggests that he was not – it explains many things, such as Airy's subsequently forgetting meeting him on St John's bridge in July 1846.[54] Indeed, one might well forgive Airy for having failed to register Adams, for he does not seem to have made a powerful impression on Challis himself. When Challis wrote to the *Cambridge chronicle* on 1 October 1846, pointing out Adams's claim in the discovery of the new planet, he stated that Adams and Le Verrier had come to their independent conclusions about four months earlier. This would have been around June 1846, or the month in which he would have read of Le Verrier's analysis of the Uranus problem in *Comptes rendus*. A fortnight later, he wrote a correction letter to the same newspaper saying that as his first letter had been sent "without consulting memoranda", he wished to state that Adams had first presented his results in September 1845, or some thirteen months earlier.[55] This in no way reflects adversely on the extremely busy and impeccably honest Challis, though it implies that the impression created by Adams on the Director of the University Observatory could have been neither strong nor sustained.The mistake in dating Adams's results to June 1846, moreover, makes it likely that Challis's own serious interests in the new planet were stimulated by Le Verrier's publications, rather than Adams's manuscripts.[56]

Not until after his return to England on 11 October 1846, when questions were coming to be asked in the Royal Astronomical Society, does Airy seem to have known much about Adams beyond his name. Most importantly, the Astronomer Royal had been quite incorrect concerning Adams's age and academic status. As he told Adam Sedgwick on 4 December 1846, he had not been aware that Adams was still *in statu pupillari* and only a B.A. Airy had assumed him to have been at least an M.A., and had forwarded a letter addressing him as the "Reverend W. J. Adams".[57] In short, Airy believed himself to be dealing with an older, senior Fellow of St John's, an ordained and possibly beneficed clergyman, and not what would now be called a junior research fellow. If a beneficed M.A. could not bother to reply to important correspondence, then he deserved no more of the Astronomer Royal's time. On the other hand, one suspects that Airy, who was a great encourager of promising scientists, would have taken further initiatives to help had he been more fully acquainted with the circumstances. The tenuous nature of Airy's relation with Adams is also indicated by the incorrect initials "W. J." instead of "J. C.".

In spite of what might appear to be an attitude of legalistic conservatism towards his duties, Airy was nonetheless the most innovative and dynamic man to hold the office of Astronomer Royal since its creation in 1675. In addition to having established the finest meridian observatory in the world, to fulfil the *prima facie* conditions of his Warrant, it was under his rule that the Royal Observatory established Magnetic and Meteorological Departments, and used submarine telegraphy to locate the longitudes of numerous international observatories and so pave the way to the worldwide acceptance of the Greenwich Meridian. He used electricity to re-determine the constant of gravitational attraction, and pioneered the transmission of telegraphic time-signals for railway and Post Office use around England, as well as bringing about numerous innovations in engineering and technology. All of these developments, though, still enshrined the Observatory's strictly utilitarian function to the navy, and the ever widening orbit of official business. Geomagnetic phenomena were seen as relating to the behaviour of compasses in iron steamships, while meteorology was thought to be connected with the incidence of epidemic mortalities recorded by the Registrar General. The Earth's density, which Airy tried to determine afresh by means of an elegant set of experiments using electrical pendulums at the top and bottom of a deep coalmine, related to the values ascribed to celestial bodies when computing tables.[59] Innovative as all these things were, they were all practical and capable of being studied by Airy's "daily observations". They belonged to a wholly different set of priorities to searching for a new planet, which was, after all, of no practical use to anyone.

Indeed, the only function of the Royal Observatory which seemed to lead away from the Warrant conditions during Airy's reign, was the establishment of an Astro-Photographic and Spectroscopic Department in 1874. This was the department that Maunder came to manage, and as the Astronomer Royal was careful to point out to the Board of Visitors, was organized "with a view to mapping out the [solar] prominences", thereby re-emphasising the cartographic, mensurational role of the Observatory and its daily monitoring of natural phenomena.[60]

Airy's attitude to the Neptune affair also brings out another feature in his approach to science which was by no means universally shared amongst his contemporaries: his lack of patriotic commitment. One of the reasons why Airy's conduct touched upon raw nerves, especially in the autumn of 1846, was his openhanded willingness to back a Frenchman's claim while seemingly doing nothing to back that of an Englishman. It illuminates an aspect of Airy's character which deserves more extensive historical examination in its own right than is possible in the present paper. Though proud of his position of Astronomer Royal and Director of the most efficient observatory of the age, Airy was wholly international in his approach to science as an enterprise. He travelled widely on the Continent and enjoyed the membership and recognition of many foreign universities and learned societies. He had been elected a Corresponding Member of the French Academy, was a Lalande Medalist, and was soon to be offered an Imperial Russian decoration.[61] In 1860, Airy was to lead a European scientific delegation to observe a solar eclipse in Spain. He viewed science as an international activity conducted by a corps of liberal, public-spirited gentlemen.[62] To a man of Airy's dispassionate and essentially logical cast of mind, patriotic issues were irrelevant in such matters, and I think he was astonished when his less impartial friends and colleagues, on both sides of the Channel, pointed accusing fingers in the wake of the discovery of Neptune. Nor could he ever consider that because he was a public servant, who fulfilled his legal responsibilities to the letter, it was his *ex officio* duty to champion the private researches of Englishmen.

Had Adams placed his work in the public domain, however, even by means of a letter to the *Times* conveying the planet's predicted place – as he had done in his investigation into the orbit of de Vico's comet in a letter to that newspaper of 15 October 1844 – then Airy might well have acted differently. Such a letter would have made it a prospective piece of human knowledge which its author had troubled to lift out of his private researches and place in the public realm. While it is futile to speculate what might have happened had certain persons acted differently, what cannot be denied is that the search for Neptune was the first major incident to highlight the distinction between a scientist's private research and what was seen to be the state's public duty.

Acknowledgements

Particular thanks are due to Miss Janet Dudley, former Librarian and Archivist of the RGO, and to Adam Perkins, the Observatory's current Archives Officer, for access to the Airy papers and for their unfailing help and support. I also wish to extend sincere thanks to the Librarians and staff of the Cambridge University Institute of Astronomy, the Royal Astronomical Society, and the Museum of the History of Science, Oxford.

REFERENCES

1. Private communication (1986) from Miss Janet Dudley, formerly Librarian and Archivist, Royal Greenwich Observatory, Herstmonceux. When the Airy papers were re-catalogued as part of the Philip Laurie project in the early 1980s, the Neptune file, being missing, was never given an RGO 6 reference number. Its old catalogue number had been RGO 694. It is hereafter cited as "RGO Neptune file".
2. George Biddell Airy, "Astronomer Royal's journal, Jan. 1836 – Dec. 1847", RGO 6; hereafter cited as "Journal". Also, "Astronomer Royal's correspondence with the First Assistant, 1835–1848", RGO 6, 28.
3. G. B. Airy, "An account of some circumstances historically connected with the discovery of the planet exterior to Uranus ...", *Memoirs of the Royal Astronomical Society*, xvi (1847), 385–

136

414 (dated 13 November 1846); hereafter cited as Airy, "Account". This article was also printed under the same title in the *Monthly notices of the Royal Astronomical Society* (hereafter cited as *MNRAS*), vii (1846), 121–44. The *Monthly notices* and *Memoirs* of the RAS were intended to represent a complete record of the Society's proceedings. From 1847 onwards, both journals were to be issued in different formats, of octavo and quarto, and published each year as a bound volume. The new policy, along with prices, is included in the "Explanatory notice" following the title-page in the 1847 volume of *MNRAS*. The significance of Airy's "Account" is indicated by its inclusion in both *MNRAS* and *Memoirs*.

4. James W. L. Glaisher, "Biographical notice", in *The scientific papers of John Couch Adams*, ed. by W. Grylls Adams, i (Cambridge, 1896), pp. xv–xlviii. Hereafter cited as Adams, *Scientific papers*.

5. James Challis, *Special report of proceedings at the Observatory relative to the new planet*, being part of the *Cambridge University Observatory Syndicate reports*, 12 December 1846 (hereafter Challis, *Special report*), and *Second report of proceedings in the Observatory relating to the new planet (Neptune)*, 22 March 1847, Cambridge Observatory Library, R. 2932 and 2933. Challis also presented his case before the meeting of the RAS on 13 November 1846, and published "Account of observations at the Cambridge Observatory for detecting the planet exterior to Uranus", *MNRAS*, vii (1846), 145–9.

6. Morton Grosser, *The discovery of Neptune* (Cambridge, Mass., 1962). W. M. Smart's "John Couch Adams and the discovery of Neptune", *Occasional notes of the Royal Astronomical Society*, no. 11 (August 1947), 33–88, is a much better work in many ways, drawing as it does on original documents. The article still perpetuates the 'heroes and villains' approach to the Neptune question, though its main value lies in the citation and publication at length of several letters (probably in the RGO archives) which were still available in 1947, but which went missing as part of RGO 694, the "RGO Neptune file". Smart's greatest weakness is his failure to cite the locations or references for his sources, which makes them difficult to trace. A more sympathetic appraisal of Airy's situation is found in H. Spencer Jones, *John Couch Adams and the discovery of Neptune* (Cambridge, 1947). In this little 43-page publication, Spencer Jones drew attention to Airy's extra-astronomical commitments, a point to which "P. J. M." added informed amplification in his review of the pamphlet in *The observatory*, lxvii (1947), 233–4. Spencer Jones further championed Airy in "G. B. Airy and the discovery of Neptune", *Popular astronomy*, lv (1947), 312–16, where he criticised Smart's article in *Nature*, clviii (1946), 648–52. Though citing several of Airy's letters, Spencer Jones also fails to give their archival location. Some of them, however, are in the Neptune file, Cambridge University Observatory Library, hereafter "Cambridge Neptune file".

7. The need to work for his living, and absence of private funding, is clear in the *Autobiography of Sir George Biddell Airy*, ed. by Wilfred Airy (Cambridge, 1896), chap. 3. Airy had been discussing salary matters in the event of his taking up a Greenwich appointment, since 1832: Lord Auckland to Airy, 8 October 1832, RGO 6, 1/151–2. Two years later, money figured prominently when he was negotiating the Office of Astronomer Royal: Airy to Lord Auckland, 10 October 1834, RGO 6, 1/153.

8. This could have played a part in Airy's refusal of a knighthood in 1835, *Autobiography*, 111–13.

9. This sum was made up from a salary of £800 per annum, Airy to Lord Auckland, 10 October 1835, *ibid.*, and a £300 p.a. Civil List Pension, offered by Peel and eventually settled on Mrs Airy. See *Autobiography*, 105, and correspondence with Peel, 18–19 February 1835, *ibid.*, 107, 108. John Pond's income from Office as Astronomer Royal had only been £600 p.a., Airy to Duke of Sussex, 19 May 1834, RGO 6, 1/145.

10. This self-adopted title was used by Airy in his *Report to the Board of Visitors*, (1842), 7.

11. It is true that, in many ways, Airy defined the *details* of his Office, though its functional character was implicit in the Royal Warrant under which he was appointed, William IV, 11 August 1835, and re-confirmed by Victoria in 1837, RGO 6, 1/193–4 and 195.

12. In the last months of his Office, before retirement in 1881, Airy wrote: "The Royal Observatory, as appears from the original inscriptions and the official warrants, was founded expressly for a definite utilitarian purpose.... And this utilitarian purpose has been steadily kept in view for two centuries, and is now followed with greater rigour and expansion than ever before", Airy to *The English mechanic*, part 831 (25 February 1881), 586–7 (letter dated 10 February 1881). In this same letter, which was published in connection with the discussion of the state endowment of scientific research, Airy also emphasised his belief that original research should be paid for out of *private* endowments, for only when pursued thus could it retain its freedom from state interference. This item is not included in the list of publications in Airy's *Autobiography*, printed on pp. 373–403.

13. Airy, *Report to the Board of Visitors*, (1875), 26. These *Reports*, which Airy delivered to the Visitors each summer, not only give an excellent record of the Observatory's achievements over the previous year, but are rich in policy statements by the Astronomer Royal on the nature of his Office.

14. Airy, "Journal", 22–23 December 1845. Airy did not search for Astraea, but true to his conception of his Office, circulated material about it and made "daily observations" after its discovery (in Germany on 13 December) had confirmed it as a fact of nature.

15. The "Journal" records two "general holidays" for the Assistants, 25 February 1846 and 26 June 1846. The precise durations of the Christmas and Easter holidays for the staff are not easy to ascertain, as Airy was absent for 4 and 6 days respectively. In addition to his late summer trip to Europe, which lasted around one month, some 24 further days were ascribed either to business trips outside the London area, or holidays, between 23 September 1845 and 10 August 1846.

16. Challis to Airy, 22 September 1845, reprinted in Airy, "Account", item no. 9, 394. I have been unable to trace the original letter in the RGO archives, and presume it to be part of the missing "RGO Neptune file".

17. Challis, *Special report* (ref. 5), 2–3.

18. Airy to Challis, 29 September 1845, in Airy, "Account", item no. 10, 395.

19. Airy, "Journal", 20 November 1845 (Croydon); 17 December 1845 (Didcot).

20. Airy usually dined at 3.30 pm, according to Wilfred Airy in "Personal sketch", in *Autobiography*, chap. 1, p. 8. For Adams's account of the visit see *Scientific papers*, p. xxviii.

21. Original untraceable at RGO, presumed in missing "RGO Neptune file". A facsimile was published in Spencer Jones's *John Couch Adams* (ref. 6), 15–17.

22. The 21 October date is cited without a manuscript source by Glaisher in Adams's *Scientific papers*, p. xvii. Airy's reference to the visit being at the *end* of October is in his "Account", 395. As neither Airy nor Adams left any documents specifying the exact date of the visit of which I am aware, one must accept Adams's subsequently remembered date of 21 October as *probably* correct, and the most precise available source.

23. On 18 January 1846, Airy's "Journal" records that he was conducting vibration experiments at Kensal Green tunnel, and on the 23rd mentions that these experiments related to the proposed Greenwich line. His Assistant Edwin Dunkin was also dispatched to study train vibrations, 28 February and 9 March 1846, etc. As early as 25 January 1836, Airy had been conducting experiments on train vibrations as a possible threat to the Observatory, see *Autobiography*, 126.

24. Airy's affection for his wife is given testimony by Wilfred Airy in the *Autobiography*, 57. Though not much of Airy's private correspondence with his wife is known to have survived there are some items amongst the RGO 6 papers at Herstmonceux.

25. William Richardson had been a Warrant Assistant at Greenwich since either 1822 or 1824 (sources differ), until his formal dismissal on 30 October 1845. His dismissal allowed promotion amongst the lower Assistants, Richardson being immediately succeeded by Thomas Ellis, and he by the newly Warranted Dunkin; see Edwin Dunkin, "Autobiographical notes by Edwin Dunkin FRS, FRAS" (1894) in the Royal Astronomical Society, Additional MSS 55, pp. 129, 163–4. Airy had tried, unsuccessfully, to have Richardson discharged for dishonest practices in 1835, see A. J. Meadows, *Greenwich Observatory*, ii: *Recent history, 1835–1975* (London, 1975), 2. Irrespective of whatever criminal traits may have lurked in Richardson, he possessed talent. Rising up from obscure Yorkshire origins, he had received the Gold Medal of the RAS in 1830, see *MNRAS*, i (1830), 165. Airy also spoke well of Richardson's abilities in a letter to Captain Beaufort, the Navy Hydrographer, on 5 July 1841: "Mr Richardson; a valuable assistant but who requires a good master to keep him in order." Richardson's good work earned him a salary of £190 in 1841, RGO 6, 72/34v–35. Orders for the payment of the outstanding salary for the incarcerated Richardson exist for February 1846, though so great was the family shame about the incest murder, that relatives refused to accept it, RGO 6, 73/88. Richardson's name does not appear on any of Airy's published staff lists presented in his annual *Reports* to the Visitors. Though Airy generally thanked and praised his staff in these *Reports*, he did not commence the practice of specifying each man by name, along with his duties, until the June 1846 *Report*, by which time Richardson had been discharged.

26. Adams, *Scientific papers*, p. xxii. Airy had also written with a similar question to Le Verrier, following the publication of his *Comptes rendus* article, Airy to Le Verrier, 26 June 1846. Le Verrier responded with a detailed reply, see Smart, "John Couch Adams" (ref. 6), 58–59.

27. Airy, "Account", 397.

XIV

138

28. Adams to Airy, 2 September 1846, see Airy, "Account", item no. 20, 405. Adams and Main corresponded on 5 and 7 September, see Airy, "Account", 408.

29. Airy to Challis, 21 December 1846, "Cambridge Neptune file", 7, reproduced in Adams, *Scientific papers*, p. xxxi (emphasis in original).

30. *Ibid.* Even after Neptune's discovery, Airy felt quite vindicated in his actions because the new planet's orbital elements were found (by the summer of 1847) only to have corresponded with the discovered position by a fortunate coincidence. Bode's Law, indeed, failed for Neptune, as Airy pointed out in *MNRAS*, vii (1847), 270. Airy's scientific reservations had been wholly correct, and only his luck had been at fault.

31. Airy was quite adamant that conditions in Cambridge were the most suitable for a search, the most important being the Northumberland Telescope, followed by "declination, latitude of place, feebleness of light and regularity of superintendence", Airy to Challis, 9 July 1846, "Cambridge Neptune file", 2. Airy did also appreciate that Challis's establishment at Cambridge was much smaller in contrast with the total staff of around a dozen at Greenwich, and offered the loan of an assistant. It seems, though, that one of the Greenwich junior Assistants was already being considered for an appointment at the Cambridge Observatory *before* the possible search for Neptune prompted his offer of 9 July 1846. On 30 June 1846, Airy had written to Challis offering either Breen or Ellis for consideration, Airy to Challis, 30 June 1846, "Cambridge Neptune file", 1. See also D. W. Dewhirst, "The Greenwich-Cambridge axis", *Vistas in astronomy*, xx (1976), 109–12.

32. Airy, "Account", 339.

33. Hansen had been visiting Airy between 10 June and 4 July 1846, according to the "Journal".

34. Airy, *Autobiography*, 183.

35. Grosser, *Discovery*, 119.

36. Airy to Main, headed "Gotha", 30 September 1846, RGO 6, 28/176.

37. Airy to Challis, "Gotha", 30 September 1846, "Cambridge Neptune file", 13.

38. In addition to details given in the "Journal", see also *Autobiography*, 183.

39. Airy to Challis, 14 October 1846, "Cambridge Neptune file", 18.

40. Hind to Challis, 30 September 1846, "Cambridge Neptune file", 12.

41. Main to Challis, 1 October 1846, "Cambridge Neptune file", 14.

42. Airy to Main, 30 September 1846, RGO 6, 28/176. The Neptune correspondence tells us much about the speed and efficiency of the postal services in 1846, from dates and replies to letters. Greenwich to Cambridge took a day or less, Paris 2 days, and Berlin about 5 days. English railways were also fast: Airy made a day return trip to Cambridge from Greenwich with Hansen, see "Journal", 2 July 1846.

43. Challis to *Cambridge chronicle*, 1 and 16 October 1846. Cuttings of these letters are in "Cambridge Neptune file", 15, 16.

44. Airy, "Account", 408.

45. Airy to Main, 9 and 16 September 1846, RGO 6, 28/173–4.

46. Grosser, *Discovery*, 114.

47. E. W. Maunder, *The Royal Observatory, Greenwich* (London, 1900), 116. Airy's invulnerability came not so much from arrogance, as from assurance. Even making allowances for a son's affection for his father, Wilfred Airy sums him up nicely: "He had a remarkably well-balanced mind, and a simplicity of nature that appeared invulnerable. No amount of hero worship seemed to have the least effect upon him." See "Personal sketch", in *Autobiography*, 5. One might suggest that not only hero worship, but also criticism, left him unaffected.

48. Grosser, *Discovery*, 137.

49. Maunder, *Royal Observatory* (ref. 47), 116.

50. Glaisher, "Biographical notice", in Adams, *Scientific papers*, p. xxiv.

51. Glaisher, *ibid.*, p. xxxi.

52. Adams, *Scientific papers*, p. xxxiii, and Airy, *Autobiography*, 187.

53. Though several contemporaries refer to Adams's "youth" and even "extreme youth" (Glaisher), one must not forget that in 1846 he was 27 years old, the age at which Airy, in 1828, had *resigned* the Lucasian Professorship to take up the Plumian Chair along with the Directorship of the Cambridge University Observatory. One suspects that these references may have more to do with Adams's manner and the way he appeared to people, than with his age in years.

54. Grosser, *Discovery*, 105, mentions that "His [Airy's] otherwise excellent memory invariably failed him in matters involving Adams". Grosser states that Airy "totally forgot" his first meeting with Adams in December 1845, and had only the vaguest recollections of having met him on St John's bridge eight months later. No source is given by Grosser for these interesting incidents, but as they are cited in Smart's "John Couch Adams" (ref. 6), I suspect that they were taken from him. As Smart gives none of *his* sources, one suspects that he may have got the information from the missing "RGO Neptune file". According to Airy's "Journal", he was in Cambridge 4–6 December 1845, in the midst of a very busy schedule, while the chance meeting on St John's bridge in July 1846 occurred on his 'day return' visit when he was entertaining Hansen on the 2nd. These meetings were only chance and fleeting, and in no way given to discussion. One must remember that in the year 1845–46, Airy met and interviewed so many people in the course of his official business that it would be unwise to give a perjorative meaning to the forgetting of two chance encounters to which, at the time, he had no need to attach any significance.

55. See ref. 43. Le Verrier's "Recherches sur les mouvements d'Uranus", *Comptes rendus*, xxii (1846), 907–18, was published on 1 June 1846. The paper reached Airy at Greenwich "about the 23rd or 24th of June", see Airy, "Account", 398. It is likely that internationally dispatched copies of *Comptes* reached Cambridge around the same time, thereby dating quite accurately the genesis of Challis's *serious* interest in the new planet to some four months prior to October 1846.

56. Challis virtually admitted the same. Between September 1845 and the midsummer of 1846, "...I had little communication with Mr Adams respecting the New Planet. Attention was again called to the subject by the publication of M. Leverrier's [*sic*] first Researches in the *Compte Rendu* [*sic*] for June 1 1846", Challis, *Special report* (ref. 5), 4.

57. Airy to Sedgwick, 4 December 1846. I have not been able to trace the original of this letter, cited in Smart's "John Couch Adams" (ref. 6), 71–72. This letter is not amongst the surviving Airy/Sedgwick correspondence in the Cambridge University Library (which is primarily concerned with geological matters for the years 1839–42 and some items from the 1850s), nor is it in the RGO archives. Airy made duplicate copies of all his official correspondence, and one is left to assume that Smart used Airy's duplicates, which in 1947 would still have been in the now-missing "RGO Neptune file".

   Adams took his B.A. at the late age of 24, in 1843, and his M.A. in 1846 (J. A. Venn, *Alumni Cantabrigiensis*, pt II: *1752–1900*, i (Cambridge, 1940)). Adams never took Holy Orders, and like Airy, determined upon a secular scientific career.

58. Reference to these multifarious activities is to be found in Airy's annual *Reports*, as well as in the over 500 published works listed in his *Autobiography*, 373–403.

59. Though the eventual values obtained were neither as consistent nor as accurate as expected, Airy's experiments formed a model example of how to apply the latest technology from one field to render advances in another: G. B. Airy, *Account of pendulum experiments undertaken at the Harton Colliery for the purpose of determining the mean density of the Earth* (London, 1855).

60. Airy, *Report*, (1874), 8.

61. This was the Order of St Stanislas, about which Airy was approached by Count Ouvaroff in September 1847. For diplomatic reasons, however, he was not permitted by the British authorities to accept the honour, correspondence in *Autobiography*, 190–3.

62. Airy's international standing in science is evident in the seven volumes of documents and correspondence covering the Spanish eclipse, RGO 6, 123–9.

XV

THE PIT AND THE PENDULUM: G. B. AIRY AND THE DETERMINATION OF GRAVITY

Summary

GEORGE BIDDELL AIRY'S attempts to 'weigh the Earth' not only rank as one of the most novel adaptations of a horological principle to be devised up to that time, but also constitute one of the most dramatic physical experiments of the nineteenth century. By monitoring the vibrations of a pair of identical pendulums placed at the top and bottom of a deep mine, Airy set out to measure the slight differences in the gravitational force acting upon them some 1,200 feet apart on the same Earth radius line, in an attempt to measure the mean density of the Earth. His two sets of experiments, performed originally in the 1820s and repeated in 1854, moreover, furnish an interesting yardstick alongside which one can trace the rapid development taking place in precision mechanics and chronometry over this important twenty-eight-year period. What is more, they indicate how available physical resources had changed between 1828 and 1854, as the advent of railways, electric telegraphy, and a rapid postal service had made the latter set of pendulum experiments seem almost routine in the technology which they employed.

ASTRONOMERS needed to know the mean density of the Earth (expressed in relation to a stable substance such as water) so that the globe's already well-known orbital characteristics with regard to the Sun, Moon and planets could be ascribed a precise gravitational weighting. As a consequence of Newton's *Principia* in 1687, it was clear that the Earth's gravitational attraction was greater than the Moon's, while both were dominated by that of the sun, though this could only be expressed as a series of proportions. If one could determine the actual physical density of any one of these bodies, however, then one could extract the density of each of the others (from their proportions), and establish thereby the gravitational masses of all the planets in the solar system.

The obvious planet to test was the Earth, and since the eighteenth century astronomers had been trying to derive an accurate figure from a variety of ingenious experiments. In 1774, for instance, Airy's predecessor as Astronomer Royal, Dr Nevil Maskelyne, had measured the deflection caused by the Scottish mountain, Schiehallion, upon a precision plumbline, and compared it with the position of a zenith star.[1] Maskelyne had extracted an extraordinarily good result, finding the Earth to be 4.5 times the density of water, which he later corrected to 4.713. The physicist Henry Cavendish obtained an even better value from a torsion balance experiment in 1798, with which he measured the mutual attraction between two metal balls, to derive the figure 5.48, which is extremely close to the value 5.517 which is generally accepted today.[2]

Pendulums had been swung all over the world since 1672, when Richer discovered that a pendulum regulated to beat seconds in the Paris Observatory lost two minutes per day when swung on the Caribbean island of Cayenne, though the same pendulum beat true seconds again when returned to Paris.[3] It appeared that the Earth's attractive force was slightly different in the two locations, and when Newton announced his Theory of Universal Gravitation to the Royal Society in 1686, and published his *Principia* the year after, an explanation was offered for the vibrational differences between pendulums of fixed length

1 Nevil Maskelyne, 'An Account of Observations made on the Mountain Schehallien, for finding its attraction', *Phil. Trans. R. Soc.*, 65 (1775) 500-542. See Derek Howse, *Nevil Maskelyne. The Seaman's Astronomer* (Cambridge, 1989) 129-141.

2 Allan Chapman, *Dividing the Circle, the development of critical angular measurement in astronomy 1500-1850* (Chichester, 1990) 105-106.

3 J. Richer, 'Observations Astronomiques et Physiques faites à l'Isle de Caienne', (1672) *Mémoires de l'Académie Royale des Sciences depuis 1666 jusqu'à 1699*, 7, 1 (1729) 233-326: 320.

swinging in different latitudes. This was, of course, because the Earth is rotating, and the combined effect of gravity and rotation is to give the Earth the shape of an oblate spheroid, flattened at the poles.[4]

The effective downward force on a pendulum bob is affected by differences of distance from the Earth's centre, by the non-spherical distribution of the Earth's mass, and by centrifugal force. When all of these are included, the attractive force on a pendulum rises towards the poles, so that it is slightly stronger in Spitzbergen than in Jamaica. At Spitzbergen, therefore, the pendulum will beat marginally faster, and the clock will gain. As pendulums are extremely sensitive to gravitational changes, there was no reason why such instruments should not be used to map the Earth gravitationally.

The biggest drawback to ultimate accuracy was seen as being the pendulum itself. No matter how fine the craftsmanship, slight variations in the effective length of the pendulum could always take place due to temperature change, barometric pressure, or density irregularities in the metal of the pendulum itself.

But in 1817, Captain Henry Kater designed a new type of pendulum that took advantage of a fundamental law of physics. He realised that in a pendulum, the point of suspension and the centre of oscillation are interchangeable in physical terms, and that there is no reason why the pendulum should not swing isochronally from either end. He therefore designed a brass rod just under four feet long with two sets of precision knife-edge suspension points set facing each other 39.4 inches apart (the appropriate length to swing seconds) with a pair of counter-weights and a centrally located adjusting screw. In this way, a weight-adjusted pendulum could be made to swing either way up without any difference so that errors could be cancelled out by reversing the suspension. It was, of course, a free

pendulum, attached to no clockwork, though conventional regulator clocks were used to monitor its swings.[5]

Kater's pendulums were swung in many places around the world from Australia to northern Russia after his description of the instrument to the Royal Society in 1818, and made possible considerable advances in geophysics. In 1826, however, Airy saw a new application for the Kater pendulum. Instead of swinging the device at points upon the Earth's surface to detect gravity changes across lateral continental masses, he realised that it could be used to detect changes along a **vertical** line, or Earth-radius, to see how much gravity changed as an observer approached the centre of the globe.

If the Earth was homogeneous in its composition, the pendulum should beat **slower** at the bottom of the mine than at the top, for while it was closer to the Earth's centre at the bottom, the outer 'shell' of the globe above the pendulum was no longer present beneath it, and so the total gravitational attraction should be less. One should be able, therefore, to 'weigh' the inner core of the Earth against the outer shell by using the lower pendulum as the fulcrum. By the 1820s, however, most geologists agreed that the inner core of the Earth was probably denser in composition than the outer shell, and if this was the case, the lower pendulum (being closer to the point mass centre) should swing **faster**. But the principle of weighing the shell against the core could still stand, provided one could assume that the composition of each subterranean zone was itself homogeneous.[6]

As the newly-appointed Lucasian Professor of Mathematics at Cambridge at the age of twenty-five (and FRS 1836), Airy began to look for support within the scientific community, as well as for a mine of appropriate depth. Various deep mines were considered until Sir Humphrey Davy and Davies Gilbert in the Royal Society put him on

---

4   I. Newton, *Principia Mathematica* (1687) translated by Andrew Motte (London, 1725) Book III, Prop. 20, Prob. iv.

5   Henry Kater, 'An Account of Experiments for Determining the Length of the Pendulum vibrating Seconds in the Latitude of London' *Phil. Trans. R. Soc.*, 108 (1818) 33-102.

6   *Account of Experiments Made at Dolcoath Mine in Cornwall in 1826 and 1828 For The Purpose of Determining the Density of The Earth*, anon., [ascribed to William Whewell and G. B. Airy] privately printed pamphlet not intended for general publication, (Cambridge, 1828). The 'fulcrum' principle is discussed on page 16. [Copy in the Airy Papers, Royal Observatory Collection, Cambridge University Library, RGO6 211/193].

to the Dolcoath copper mine near Camborne, Cornwall.[7] The owners and managers of the mine were eager to participate in the novel experiment, and the Royal Society secured the loan of the necessary Kater pendulums, regulators, chronometers, and other instruments.

The journey to Cornwall had about it the aspect of a holiday as Airy and his Cambridge mathematical friends, Ibbotson and William Whewell, went by coach to Devonport and sailed to Falmouth by steamer on a journey that took several days.[8]

But once on site, they got down to business in earnest, erecting **two** sets of Kater free pendulums and regulator clocks at the top and bottom of Dolcoath mine. This was a job in itself because the mine had no proper cage mechanism and only a rough steam hoist was used to lift up the ore. The scientists with their precision instruments had to climb down the 1,200 foot shaft using fifty tiers of sharply-raked ladders![9]

At the heart of Airy's experiment was the observing technique for using the Kater reversible pendulum. Being a free pendulum, it was set to swing freely on its own steel and agate knife-edge bearing until it came to rest. When in motion, the observer counted how many free swings it would make in a given period of time. To help in the monitoring of the swings, the free pendulum was set in front of an astronomical regulator which had a conventional pendulum, of approximately the same length as the free one, controlling the running of its escapement. A white disc of paper, about one inch in diameter, was placed upon a black background at the centre of the clock pendulum bob. Attached to each end of the Kater free pendulum was a thin sliver of wood about one inch across and a foot long. This sliver of wood was painted black and

served as an almost weightless pointer projecting below the free pendulum's bob.[10] When the entire apparatus was in adjustment, the clock stopped, and the Kater free pendulum hanging dead, the black wooden pointer on the free pendulum obscured the white disc on the clock pendulum. The observation of the black pointer against the white disc was made through a low-power telescope screwed firmly down to its stand some feet in front of the apparatus.

When both the Kater and the clock pendulums were set in motion, an observer looking through the viewing telescope would see the pointer and disc flashing past alternately. But moving slightly out of phase in the arcs which they described (the powered clock pendulum maintaining a constant arc while that of the free pendulum gradually diminished), the black pointer and the white disc would **coincide** in the viewing telescope's line of sight every two minutes or so. In consequence, the white dot would appear to miss a beat in its otherwise perfectly isochronous sequence. It was these coincidences which interested the astronomers, for they were caused by local gravity acting upon both pendulums.[11]

The free pendulum was allowed to run for abour 530 one-second swings, and if its arc had become too small after this time to properly monitor the coincidences, it was given a new push. As the Kater pendulum was not being used to measure a **continuous** period of time, but rather an accumulation of individual courses of swings against the regulator, its stopping and starting could have impaired the eventual quality of the results. These courses of swings were timed against a second regulator clock, as the free pendulum obscured the face of the main clock in front of which the Kater pendulum swung.

---

7   It is not clear why Airy did not become FRS until 1836, though almost immediately he was nominated for the Council; G. B. Airy, *Autobiography*, edited by Wilfred Airy (Cambridge 1896) 127. Airy discusses the mines being considered in *Autobiography*, page 66. In an unpublished autobiographical note written at the end of his life about 1889 and in the possession of the Airy family, Airy states that his friend Whewell was his main contact in the Royal Society; Enid Airy Papers, 'Family History of G. B. Airy'. The Royal Society loaned one of the Pendulums.

8   No exact route survives for the 1826 journey to Cornwall, at least up to Devonport, but in 1828, it seems to have been: Cambridge, London, Worthing, Portsmouth, Plymouth and Falmouth. As part of the journey was by steamer, one assumes that they coached to Worthing and sailed from there; Pierce Morton to G. B. Airy, 4 June 1828, RGO6 802/84.

9   *Life and Letters of Adam Sedgwick*, eds. J. W. Clark and T. M. Hughes (Cambridge, 1890) vol. I, 329-332, for Sedgwick's description of Dolcoath Pit. In his Ms. poem 'Dolcoath' Airy wrote of the shaft 'Behind a chain ladder from hooks was descending', 18 October, 1828; RGO6 211/291.

10   Kater, 'Account' (ref. 5) 43. *Account of Experiments* (ref. 6) 6-7.

11   For coincidence technique, see '1826 Coincidence Book', Airy Papers, RGO6 211/1.

Fig. 1. Kater's free pendulum swings in the alcove. The small telescope on the stand before it, observed the 'coincidences' between its swings and those of the regular pendulum. The second regulator denotes the time. (ref. 5, plate V).

Monitoring the coincidences must have been extremely boring, and counting sheep sounds exciting by comparison. None the less, Kater and various explorers from 1818 onwards had shown that the method was very sensitive as a geophysical measuring device, if one counted long enough courses of swings on different parts of the Earth's surface. To make his own experiment work, however, Airy needed two sets of everything – two Kater pendulums and four regulators. One set of pendulums and regulators was to be placed at the top of Dolcoath mine, and the other in a little whitewashed shed at the bottom directly beneath it. To make sure that the observations were absolute and not subject to the errors of a given instrument, the pendulums were not only reversed on their suspension points but brought up and taken down the shaft to be swung in both places.

But if one thinks that sitting for eight hours at the bottom of a damp mineshaft counting coincidences by lamplight was brain-deadening, one should remember that reversing the apparatus was back-breaking. The success of the whole enterprise hinged on the fraction-of-a-second discrepancies which gradually accumulated between the two sets of identical apparatus, one of which was 1,200 feet closer to the centre of the Earth than the other.

The time-keeping rates of each set of apparatus had to be regularly compared with those of the other, and the only way of doing this in the days before electric signalling was to compare the bottom clock with the top by means of portable chronometers. The Royal Observatory loaned four Admiralty chronometers to Airy and three more were obtained from other sources.[12] The chronometers were used to convey absolute time between the upper and lower stations. Multiple chronometers

12 Airy, *Autobiography* (ref. 7) 67-68.

were used to minimise errors, and these had to be carried in a bag up and down the mineshaft at regular intervals to monitor any discrepancies between the two sets of apparatus. To climb between the upper and lower stations was likened to ascending a set of ladders that extended nearly four times the height of the dome of St Paul's Cathedral!

On 26 June 1826, the experiment ended abruptly.[13] One of the pendulums (called 'Foster' from its owner) crashed to the bottom of Dolcoath as it was being hoisted up to be reversed with its companion 'Hall'. Nothing but burning straw packing remained, and whether the straw had been ignited by falling sparks, spontaneous combustion, or a superstitious miner flinging a candle-end into the rising tub, no one ever knew.[14] In consequence, no decisive results were obtained from the suddenly curtailed Dolcoath experiment of 1826, though Airy believed that as the lower pendulum had gained four seconds per day upon the upper one, it was possible to extract a provisional terrestrial density of 7.73. His caution about this figure was emphasised however, by his prior expectation of a lower pendulum acceleration of only **two** seconds per day, as based on Hutton's values for the density of the Scottish mountain, Schiehallion, as derived from Maskelyne's observations, and specific gravities of mineral specimens taken from the mountain.[15] None the less, Airy believed that the method had been shown to work and should be repeated as soon as fresh funds and instruments could be assembled.

He was delighted to return to Dolcoath in 1828 with a much fuller strategy in mind. One can only assume that the boring nature of the work was not too discouraging, for he now returned with a veritable galaxy of Cambridge mathematical talent, including the eminent geologist Adam Sedgwick. In 1828 the observation runs lasted longer with more observers working the shifts to monitor the pendulums constantly. From 127 hours of constant swinging, Airy and his team confirmed that the pendulum at the lower station gained upon that at the top. This showed that the gravitational force was stronger, and therefore had a greater **accelerating** effect, on the pendulum swinging closer to the centre of the Earth.

But as the 1826 observations were spoiled by fire, so those of 1828 were brought to a halt by water. A rock 'many times the size of Westminster Abbey' slipped and a thunderous shock reverberated through the mine. No one was hurt, but the pump could not properly cope with the rising water, especially as the lowest galleries of Dolcoath went down well below sea level. The astronomers soldiered on for a further eleven days, but when the water was only a few feet below their station, they were forced to strike camp and hoist their clocks and pendulums up the shaft.[16]

Yet one can hardly help wondering how the miners themselves regarded the influx of Cambridge pendulum-swingers into the piskie-realms where they dug the earth. Adam Sedgwick, the geologist, who had a sharp eye for local colour and could talk easily with quarry men and miners, recorded several incidents. 'I think they're up to no good', one workman told Sedgwick about the scientists, 'there must be something wicked about them – the little one [that was Airy] especially. I saw him stand with his back to the Church and make strange faces'. On another occasion, when Sedgwick was descending the shaft to work at the lower station, he jokingly asked a miner how quickly one could get to Hell. 'Let the ladder go, Sir, and you'll be there directly' was the reply![17] Ironically, Airy was the only member of the expedition not to be an Anglican Clergyman, while Sedwick later combined his geological Chair with a Canonry of Norwich Cathedral.

But Airy did not possess sufficient confidence in the two sets of Dolcoath results to publish them, especially after he discovered an oscillation error in one of the pendulums upon reducing the coincidence figures back in Cambridge. Instead, the results were set out in a pamphlet intended for private circulation that was produced in conjunction with his

13 *Account of Experiments* (ref. 6) 6.

14 *Account of Experiments* (ref. 6) 8. *Autobiography* (ref. 7) 68. Airy also includes an account of 'Foster' in his detailed Ms. 'Essay' account of the Dolcoath work dated 31 August 1828 in RGO6 211/270-290. See also his poem 'Dolcoath' ref. 9.

15 *Account of Experiments* (ref. 6) 8.

16 *Account of Experiments* (ref. 6) 12-13.

17 Sedgwick, *Life and Letters* (ref. 9) 332.

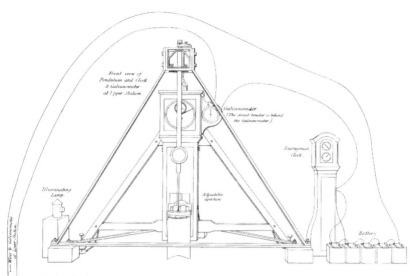

Fig. 2. Pendulum, clock and galvanometer at the upper station with wires leading to the galvanometer at the lower station.
(ref. 20)

friend and collaborator, William Whewell.[18] It was to be in that inexact state that the pendulum measurements were to remain for the next twenty-six years.

Airy's thoughts returned to pits and pendulums in the wake of the massive technological revolution which had taken place in the ensuing quarter of a century, by which time he had resigned his Cambridge professorship to take charge of the nation's foremost scientific institution, the Royal Observatory, Greenwich, with the title Astronomer Royal. By 1854 the country had been covered with a fast and efficient railway network making it possible to get to virtually anywhere in England within fifteen hours of leaving London. In 1826, Michael Faraday had not yet discovered the laws of electro-magnetic induction, whereas by 1854, an extensive commercial telegraph system covered not only Britain, but also Europe and half of the United States of America. Electro-magnetism had started to be used to collect and regularise scientific

data with the 'American method' of taking star transits, and the use of automatic pen-recorders. The Royal Observatory under Airy had pioneered the use of these electrical devices, and already possessed an Electrical Department supervised by Charles Todd who had charge of the electro-magnetic apparatus in the Establishment, including that which, after 1852, transmitted Greenwich Time into the commercial telegraph network.[19]

Airy knew that one of the flaws of the Dolcoath experiment had been the difficulty in making comparisons between the upper and lower observing stations in the mine, since it was necessary to man-handle box chronometers up and down the ladders of the 1,200 foot shaft to monitor the regulators against which the Kater free pendulums were checked. But now there was no reason why the upper clock should not transmit an electrical time signal to a galvanometer needle at the lower one, to facilitate constant, synchronous monitoring.

18 Sedgwick's editors in the *Life and Letters* (ref. 9) 331 claim that it had been Whewell who actually wrote the *Account of Experiments* (ref. 6) and cite their source as Isaac Todhunter's *Life of William Whewell* (Cambridge, 1876) vol. I, 37. Airy, moreover, does not list it amongst his published works for 1828 in *Autobiography* (ref. 7) 374, though to Airy's mind, the privately printed pamphlet was never formally published. It is clear, however, from both the *Autobiography* and from his letters in RGO6, that Airy was the initiator and manager of the Dolcoath project. As he was also intimately connected with the subsequent reduction work on the measured figures, it is likely that the *Account of Experiments* was a collaboration.

19 Derek Howse, *Greenwich Time and the discovery of the Longitude* (Oxford, 1980) 91-100, 109.

In the Autumn of 1854, Airy recommenced the pendulum experiments in the 1,260 foot deep Harton coal mine, at South Shields, on Tyneside. The observations ran for the first three weeks of October 1854, and when one reads the large body of correspondence, along with the results that were published in *Philosophical Transactions* in 1856, one becomes aware of how far the world had changed and speeded up since the 1820s.[20] Mr Anderson, the lessee of Harton Colliery, was delighted to be involved, and not only provided a specially built dust-proof chamber underground to house the horological equipment, but constructed a thermally-regulated upper room at the pit-head so that the clocks and Kater pendulums at the upper and lower stations were maintained at the same temperature. Mr Mather of the Electric Telegraph Company supervised the installation of the necessary wiring and galvanometric apparatus whereby both the upper and lower stations were monitored electrically, while the railway and daily Penny Post facilitated rapid contact with Greenwich.[21] A man, or a letter, could do the journey from London to South Shields comfortably in twelve hours, while a telegram could communicate urgent information instantaneously. Needless to say, Harton Colliery could set its clocks to GMT directly from the Clock Room of the Royal Observatory via the Electric Telegraph Company's wires.

As Astronomer Royal, and a pillar of the scientific establishment, Airy was no longer inclined to do the donkey work at Harton, while his old friend Whewell now reigned as Master of Trinity College, Cambridge. Instead, a picked team of assistants was appointed to undertake the horological work at South Shields and once Airy had overseen the installation of the apparatus, he returned to London and directed operations from behind his desk in the Royal Observatory. The six assistants whom Airy appointed belonged to the growing class of 'professional astronomical assistants' who were coming to do the routine work at Greenwich and many other English observatories by the 1850s. Bright young men without university degrees who were virtually apprenticed into the arts of practical observation and computation after leaving grammar school in their late teens. They were led by Edwin Dunkin (FRS 1875), a forty-three-year-old Senior Assistant at Greenwich, and William Ellis, a junior Greenwich man.[22] The other four were borrowed from the Oxford, Durham, Cambridge and Redhill Observatories, and were, respectively, Norman Pogson, George Rumker, G. S. Criswick and G. H. Simmonds.[23]

Though Airy is popularly reputed to have been a hard taskmaster, I have found little evidence for it in those documents pertaining to the administration of the Royal Observatory staff, while his handling of the assistants on the Harton experiment was positively generous. They were accommodated at the Golden Lion, which was the most comfortable hotel in South Shields, while explicit arrangements were made with the proprietors to ensure that they had good beds, substantial, meaty meals, and an abundance of hot water, soap, and clean clothes when they returned from the pit at any hour of the day or night. Norman Pogson was reimbursed first-class train fare for his journey from the Oxford Observatory, while George Rumker, whose clothes appear to have suffered damage, was compensated. The wealth of social detail in the documents, especially in Dunkin's 'Journal' and letters to Airy, is enormous, and provides an important insight not only into the daily running of the experiment, but how these young gentlemen found life in a community where the value of Mr Rumker's spoiled coat was equivalent to almost a month's wages for an ordinary Harton collier. It is a pity that Adam Sedgwick, who was still very much alive and well in 1854, did not visit South Shields from Cambridge or Norwich, to record the local gossip, as he had done at Dolcoath. The

20 G. B. Airy, 'Account of Pendulum Experiments undertaken at the Harton Colliery, for the purpose of Determining the Mean Density of the Earth', *Phil. Trans. R. Soc.*, 146 (1856) 297-342.

21 Airy-Mather letters, August-September 1854; see RGO6 212/290-296. Airy-Dunkin letters for Harton; see RGO6 212/226-274.

22 Mr Dunkin's Book of Expenses at South Shields and journeys to and from S. Shields and Hotel Bills & Co.' RGO6 213/258 ff. 'Journal Kept by Edwin Dunkin for S. Shields trip', RG0 6 213/275-305. Dunkin also provided an independent account of the expedition and his colleagues' duties in 'Auto-Biographical Notes. By Edwin Dunkin, F.R.S., F.R.A.S. . . . '; Royal Astronomical Society Additional Ms., 55, Library No. 22837.

23 Dunkin, 'Expenses' and 'Journal' accounts (ref. 22). Accommodation at the 'Golden Lion' cost 35 shillings [£1.75] per week per man. 'Expenses', 6 Oct. 1854.

Harton assistants received a generous fee of £10 for their trouble in addition to their salaries.

In spite of the omnipresent grime of which Dunkin complained, the experiment had proceeded better than anyone expected. After three weeks of unbroken running (apart from the pendulum reversals) the readings had already become so consistent that it was deemed unnecessary to carry them out over the entire month, although the assistants still received the full month's pay. The apparatus at the lower station had **gained** a regular $2\frac{1}{4}$ seconds of time per day against that at the top, which not only demonstrated that gravity increased as one descended into the Earth, but that one could 'weigh' the upper shell against the lower core by using the lower pendulum as a fulcrum.[24] On Saturday, 21 October, 1854, the Astronomer Royal, his wife, and entourage, travelled up from London to terminate the successful experiment, breaking their journey overnight at Durham, so that they could attend Service at the Cathedral on Sunday.[25]

The experiment also caught the public imagination. Not only did many dignitaries come to South Shields to see what was going on, but the main newspapers of the day, including the *Illustrated London News* and *Punch*, carried accounts.[26] Punch's satirical 'Airy and the Coal-Hole' by Percival Leigh greatly amused the Astronomer Royal! Before everything was dismantled, the dignitaries were taken below ground to see the pendulums, while the accompanying ladies were recommended to cover their voluminous crinolines with capacious mackintosh capes and caps, and protect their hands with sturdy housemaids' gloves. Harton Pit, one should add, possessed a steam-operated cage for easy descents and ascents.

When the experimental work was over, the festivities began. Various dinners were given in honour of the visiting scientists, and it should be noted that Airy always made sure that the six who had done the work were invited. Airy was also prevailed upon to give a lecture on the experiment in the South Shields Central Hall, where some five hundred people seem to have been packed into a room measuring 46 by 30 feet, and the assistants demonstrated the pendulum apparatus and batteries. As a man who was very involved with what we would now call 'the public understanding of science', Airy was concerned that tickets were distributed to working men.[27]

But it was back in Greenwich that the real work began for Airy when he set about reducing the thousands of coincidences observed at the upper and lower stations at Harton to extract a density figure. In his reduction, he attempted to hunt out all possible sources of error, collating the pendulum swings with barometric pressure and temperature, and obtaining extensive samples of the minerals to be found in the Newcastle measures along with their precise geographical distribution. Professor Miller at Cambridge determined the specific gravities of a box of mineral samples, so that, in conjunction with the geological survey maps of the Tyne Valley, Airy could locate the changes in density in the surrounding rocks and compute the effects that they may have had on the pendulums.[28] Upon their return to London, the Kater pendulums and clocks were re-erected in the basement of the Royal Observatory, so that further courses of swings could be observed, to provide control runs upon the apparatus.[29]

24 G. B. Airy to James Mather, S. Shields, 2 December 1854, published in *Lecture at South Shields on the Pendulum Experiment at Harton Pit* (South Shields & London, 1854) 25. A copy of this pamphlet is included in *Radcliffe Tract* 51.9, Museum of the History of Science, Oxford. Also, Airy to G. W. Arkley, S. Shields, 29 November 1854, RGO6 212/73-74.

25 G. B. Airy to Professor Mitchell, London, 17 October 1854, RGO6 213/77-78. Airy, 'Astronomer Royal's Journal', 21-28 October 1854, RGO6 24/1.

26 *Illustrated London News*, 4 November 1854, 446-447. *Punch*, xxvii, 11 November, 1854, 199. For a full account of Airy's lecture, see *South Shields Gazette*, 27 October, 1854. The lecture was delivered on 24 October.

27 E. Dunkin, 'Journal' (ref. 22) Monday, 16 October, 1854, for details of S. Shields Central Hall. In his 'Auto-Biography' (ref. 22) 257, Dunkin states that 'several hundred' working men attended Airy's lecture.

28 For Airy's correspondence with Professor Miller on the specific gravities of rocks, see RGO6 213/48, 75 etc. Airy's 'Account of Pendulums . . . at Harton', (ref. 20) includes published geological surveys for the Harton region. See also RGO6 214/307-311.

29 In a 'Supplement' to his 'Account of Experiments . . . at Harton' (ref. 20) 243-255, Airy describes the control runs made on the Kater and galvanic apparatus at Greenwich in November 1854.

After a month of meticulous desk work which gave due weighting to every conceivable aberration, Airy announced the Harton results. At a depth of 1,260 feet, the pull of gravity increased by $\frac{1}{19\,190}$, thereby indicating that the Earth must be 6.565 times more dense than water. Over the last days of November 1854 this result was communicated to the Royal Society and notifications sent to the world's principal scientific institutions. Individual letters of thanks containing the newly-computed density value were also sent to each of the participating assistants.[30]

Considering the scrupulous care and attention to detail that went into the work at Harton and the apparent congruity of the results at the time, it seems strange that Airy obtained a value that was around 20% in error from the correct one of 5.517 times the density of water. Though he was confident of the result in his own mind, the figure was met with some caution elsewhere, differing so widely as it did from previous determinations of the Earth's density. Colonel Henry James, for instance, suggested that Airy's surprisingly high value arose from having ascribed an excessive density to the surrounding rocks in the Tyne Valley, though Airy retorted by reminding the Colonel of the care with which he had determined the specific gravities of the local minerals.[31] Even before commencing the work at Harton, in August 1854, Airy had requested Faraday's advice as to whether Earth-currents could affect a pendulum swinging underground and received a reply in the negative if the pendulum was made of brass.[32] Professor Stokes sent a letter to Airy in which he argued that no proper compensation had been applied for the effects of centrifugal force upon the pendulum.[33]

But the real cause of error probably stemmed not so much from a fault in horological principle or technique as from a lack of adequate knowledge of the Earth's interior. The logic of the Earth weighing method hinged upon the assumption that the interior zones of the globe would at least be homogenous in their densities, whereas in reality, they are not. A local survey of the Tyne valley rocks was too narrow in its range to give an accurate density for the outer shell against which the inner core was being weighed. In consequence, it was not really possible to obtain a reliable figure for the total terrestrial density by this method unless pendulums were swung in dozens of pits around the world and a much fuller body of geological knowledge was available.[34] In a way, Colonel James's remarks had contained more than a grain of truth.

In spite of the anti-climactic character of the results, however, Airy's pendulum experiments represented a bold and even dramatic attempt to explore the physical properties of the Earth that was just as daring, in its own way, as the voyages of the *Beagle* or *Challenger*. They provided an elegant example of how the pursuit of physical discovery could borrow a familiar piece of physics from horology, backed up by the latest technological innovations, and apply it to the weighing of the Earth itself. But what most of all captured the imagination of the Victorian public was the presence of a group of astronomers taking their instruments into the depths of the Earth to quantify the fundamental force of the universe.

Acknowledgements

I wish to thank many people for assistance with the research which preceded the writing of this article. In particular, I wish to thank Adam Perkins of the Cambridge University Library, Peter Hingley of the Royal Astronomical Society Library, A. V. Simcock, of the Museum of the History of Science and Dr Geoffrey Brooker of Wadham College, Oxford. I also thank Sheila Edwards and her staff at the Royal Society Library. I am also indebted to the Airy family for access to material in the 'Enid Airy Papers'.

30 Airy informed the French Académie and *Compte Rendu* respectively on 7 and 11 of December 1854; RGO6 212/5, 6, 7.

31 G. B. Airy to Colonel Henry James, 22-23 February 1856, RGO6 212/343-345.

32 G. B. Airy to Michael Faraday, 19-21 August 1854, RGO6 212/327-329.

33 Professor G. C. Stokes to Airy, 4 February, 1856, RGO6 213/165-171. I am indebted to Dr Allan Mills of the University of Leicester for his suggestion that the greatly increased humidity below ground could also have affected the period of the lower pendulum.

34 Richard Glazebrook, *A Dictionary of Applied Physics* vol. III (London, 1923) 279-280.

XVI

Sir George Airy (1801-1892) and the Concept of International Standards in Science, Timekeeping and Navigation

The need for exact, and uniformly accepted, standards in the interpretation of natural phenomena, has been a fundamental requirement of scientific research since the seventeenth century. The early Royal Society emphasised the point, while the rise of the mechanical philosophy had been based upon the axiom that mechanical analogies found to operate in one part of nature should be applicable to others. Flamsteed, Hooke and Newton investigated the universal constants of nature, while the scientific men of the eighteenth century had established those of gravitation, aberration, nutation, and planetary, lunar and stellar ratios, as well as seeing the application of related concepts even to biology and chemistry.

Though this approach to natural enquiry obviously transcended national barriers, England nonetheless came to take a major lead by 1800. Our highly developed scientific instrument making trade emphasised the close relationship between pushing back the frontiers of knowledge and the creation of precision research tools,[1] but it was also our revolution in industrial manufacture which set so much of the tone for the sciences of the nineteenth century. Through the exploitation of steam, wind and water power, working in conjunction with a set of relatively simple mechanical devices, it seemed that Bacon's vision of 'relieving man's estate' by the application of technology had at last become possible.

It also suggested new possibilities for man's investigation of nature herself, for if the application of the same principles of reason, order and regularity could produce such bountiful results in 'the philosophy of manufacture', then why could they not do likewise when applied to the reaping of nature's harvest in scientific research? It is in this context, of the application of 'industrial' methods to the gaining, distributing and utilising of natural knowledge, that the career of George Biddell Airy assumes its historical significance.

Born in the first year of the nineteenth century, Airy represented, in many ways, the intellectual entrepreneur; energetic, self-possessed and astonishingly single-minded in the career which he mapped out both for himself and for the institution which he was to head for forty-six years. As a youth, he had seized the opportunity to leave his obscure parental home, to secure the patronage of his socially elevated uncle Biddell. He systematically worked his way through Cambridge as a sizar, scholar and Fellow of Trinity, eventually holding the University to ransom for a substantial increase in salary before he

XVI

would accept — at the age of twenty-seven — the Directorship of the Observatory and Plumian Professorship. By the time that he was thirty-five, he was Astronomer Royal, married into a well-connected clerical family, and on sufficiently good terms with Peel and Russell to be able to secure a Civil List pension for his wife and calmly refuse the offer of a Knighthood on the grounds of expense.[2]

Such ruthlessly well-planned conduct may seem more in keeping with the career of a rising politician than a scientist, but Airy applied the same principles of painstaking care to his [322] scientific work. Though a First Wrangler and theoretical mathematician by training, he was an engineer by instinct, and only placed complete trust in theory after it had been shown to work in practice. This led him, first in Cambridge and then at Greenwich, to set about the wholesale measurement and quantification of physical nature with an unprecedented thoroughness.

Airy must also rank as perhaps the best organised of all scientists. In private life, he kept exact records from his eighteenth to ninety-first year, while professionally, he recognised that merely to accumulate data is only part of the battle. It had to be reduced, interpolated and published to the world, although in hindsight, Airy often seemed more concerned with the physical management of data than with its application or use. At its highest level, Airy's approach made the Royal Observatory an internationally recognised standard for astronomical and physical data of all types, whilst at its worst, it engendered a punctilious love of bureaucracy, the excesses of which are still part of the folklore of the institution.[3]

From the outset of his career as Astronomer Royal, Airy took it upon himself not only to reform the lax administration of John Pond, but to introduce a new rigour and public accountability of operation, which had been absent in all the previous Astronomers Royal. Unlike his predecessors, Airy regarded himself as a public servant, not as an essentially private gentleman who occasionally supplied astronomical data to the government. If Britain was, after all, the dominant nation in the world after 1815, it was absurd that its scientific services should be run on an irregular basis, especially when Britain's power was so rooted in maritime affairs. Airy was impressed by the power of contemporary industrialists and engineers to transform both society and the manufacturing arts by the imposition of an efficient, rational discipline based upon the utilisation of mechanical devices, and all that separated him from them was the fact that his profit was measured in terms of public utility and scientific prestige, rather than Pounds Sterling.

In the introduction of a 'factory mentality' into the Royal Observatory, Airy hoped to eliminate some of the major encumbrances to scientific data collection. Firstly, he established a rigid hierarchy of staff, in which all personnel knew their places in a strict chain of command. Beneath the Astronomer Royal himself came a small band of Cambridge graduates, the chief of whom supervised the daily running of the Observatory, yet who himself received a salary less than half of that of Airy, and was not responsible for policy decisions. Airy's relations with his 'gentleman' assistants were often strained, for Victorian university graduates did not easily accommodate themselves to a factory mentality, and sometimes complained officially. It was only in the wake of such a complaint that Airy formally designated these gentlemen with the prefix 'Mr' in the official staff list of the Observatory.[4]

Beneath the graduate assistants came the observers, computers and 'obedient drudges', who were generally appointed as very young men, and trained exclusively in the performance of one or other specialised task, such as the calculation of lunar places or the operation of a particular instrument. These men were usually admitted at fifteen or sixteen years of age, fresh from school, and selected by competitive examination. They were lowly paid, kept intellectually blinkered, and encouraged to leave to take up clerkships in city offices by their mid-twenties.[5] Like many contemporary factory owners, Airy recognised the advantages of 'juvenile labour', in recruiting a permanently changing workforce, where a simple training could utilise sharp eyes and minds, before they came to demand more from life. By reducing all the principal tasks of observation and calculation to a series of specialised rote exercises, Airy helped to eliminate error, and speed up the output of data.

The way in which this method bore fruit is indicated by the fact that both Adams and Le Verrier applied to Airy for essential planetary data when they undertook their independent [323] quests for Neptune. It also indicated how freely such information was to be obtained for the asking, for by 1846, Airy was willing to supply astronomical constants to all comers, irrespective of his personal belief in the value of their investigations. Furthermore, Airy was a ceaseless international correspondent, and it is not without significance that he was in Germany when the discovery of the new planet was announced.[6]

Secondly, Airy hoped to improve the standard of accuracy of all types of observations by placing crucial stress upon the instrument, rather than the man using it. His concern with the personal errors of his observers, and their hoped-for elimination by such devices as the galvanic micrometer, is symptomatic of this mode of investigation. It is also to be seen in his lifelong interest in instrument design and the application of mechanical, electrical and photographic self-recording devices to astronomy, meteorology and time-signalling at this time.

Thirdly, Airy recognised that the irregular manner in which all previous Astronomers Royal from Flamsteed to Pond had published their work was

XVI

another impediment to the Observatory's efficient function. Like a factory, Airy's Greenwich produced a massive output of annual reports, observations, monographs and tables, generally reduced and ready for use by scientists across the world.

Airy possessed a strongly mechanical turn of mind. He had a genius for seizing upon the significance of key contemporary inventions, and adapting them to the work of the Observatory. In the early 1840s, he shared John Herschel's hope that the newly-invented photographic process would simplify celestial cartography, and while the emulsions of the period were too slow to be of much immediate use in this respect, Airy did develop an early photographic magnetic recording instrument, as well as foreseeing the application of the process to other tasks where it would eventually supersede the human eye.[7]

Electricity, electromagnetism and telegraphy likewise stimulated his imagination, for they held out the possibility of reducing, to a large degree, the traditional dependence upon the nervous system of the observer in many types of recording work. By 1860, he had applied electricity to horology and transit astronomy, and had developed the self-recording 'drum' micrometer.[8] It was also mentioned by James Stuart, the engineering professor, who visited Greenwich during the mid-Victorian period, that Airy filled the Observatory with a range of mechanical devices to facilitate the smooth running of the institution.[9] The Astronomer Royal always found it easy to get on with engineers, for both he and they shared a common pragmatic approach to problems, and in addition to Stuart, Airy had been familiar with engineers of the older generation, such as George Stephenson and I.K. Brunel.

The first fifty years of the nineteenth century had witnessed a revolution in precision engineering, as fast-acting machines for the turning, cutting and dividing of mechanical artefacts bcame available on an industrial scale. Airy saw the application of these techniques to the production of the Observatory's instruments after 1835, so that bearings, scales and lenses were not merely made to the highest standards attainable by the painstaking efforts of an individual craftsman, as had been the case with Graham, Bird and Ramsden, but in accordance with the infinitely reproducible standards of a machine. It had taken Bird weeks to graduate an 8-foot quadrant by hand, whereas the 6-foot circle for the great transit of 1850 was machine-divided down to $1/1000^{th}$ of a minute in a few hours, under the supervision of an operative.[10] Indeed, no finer example of Airy's appreciation of the application of industrial techniques to instrument making can be found than in the production of the 1850 transit. The heavy parts of this instrument were tendered out to a firm of agricultural engineers — Ransome [324] and May's of Ipswich — while the delicate ones were made by William Simms of London, which was one of the world's first mass-production precision scientific instrument making firms.

Airy's fundamental task at Greenwich was to observe and quantify the comstants of nature, for the benefit of mankind in general, and for the British Empire in particular. In addition to the continuous celestial cartographic work, which he saw as forming the 'standard and staple' concern of his career from his first appointment at Cambridge in 1828 down to his retirement in 1881, he recognised the importance of monitoring related phenomena. Upon assuming office at Greenwich in 1835, he began to make systematic magnetic and meteorological observations, for while regular pressure and temperature measurements had been made at the Royal Observatory since 1750, Airy greatly extended the scope of the work. This was also to be an area where self-recording instruments soon came to play a major part. Clockwork barographs were already available, but Airy came to introduce photographic recorders in 1847, as well as other magnetic and meteorological instruments which placed only minimal demands upon the attention of the assistants.

Airy also attempted to give a precise weighting to the gravitation constant, and on several occasions tried to re-determine the terrestrial density in the wake of Maskelyne's original work in 1775. Between 1826 and 1828, he made laborious experiments with free pendulums at the top and bottom of the deep Dolcoath mine in Cornwall. Though the values which he obtained in these researches were less close to the modern value than were those of his predecessors, they show a concern with quantifying the major constants long before he became Astronomer Royal. One possible reason contributing to the error of these measurements was the loss of his first set of instruments when they were severely damaged in an accident in the pit shaft in 1826.

In 1854, recognising the improved signalling and recording possibilities of electrical devices, Airy performed further pendulum experiments at Harton pit, near South Shields, close to his native Northumberland. Winning the cooperation of the local coal owners in this work, he set up identical pendulums at the top and bottom of this deep mine, the swings of each being monitored by electromagnetic impulses, to see if the stronger gravitation at the bottom of the shaft produced measurable effects on the subterranean pendulum. While the arrangements and details surrounding these new experiments clearly show Airy to have been a splendid organiser of research, the final results still left much to be desired.[12]

The role into which Airy had cast the Royal Observatory by 1840 made it necessary for him to re-determine the relative longitudes of all the other principal observatories in the world. In 1844, he used forty-two chronometers and repeated crossings of the North Sea to establish the co-ordinates of Struve's Pulkova-Altona base line, while after 1850 the electric telegraph seemed full of possibilities.[13] Instead of the laborious chronometer method of fixing longitudes for foreign stations, telegraphy seemed to be faster, simpler and much more accurate. Following the laying of the first Channel submarine cable in 1851, he determined the longitudes of the Paris and Brussels Observatories, although his Paris values were shown to be as much as a second of time in error by John Herschel. Likewise, the first Atlantic telegraph in 1858, followed by a more reliable one in 1866, enabled him to establish the longitude of the Washington Observatory in addition to other American stations. By the 1860s, Airy had used the telegraph to re-define not only the longitudes of many British Isles observatories, but the major ones of Europe, Canada and America. The concern which Airy devoted to this painstaking work not only emphasised his role as 'The British Astronomer', but went a long way towards establishing the primacy of the Greenwich Meridian in 1884.[14]

Airy was also the first Astronomer Royal to see himself as a fully professional government scientist, rather than a private gentleman whose time was largely his own. In his stalwart [325] refusal to take Holy Orders, Airy broke the tradition of clerical directors of the Royal Observatory, and because he knew that no ecclesiastical sinecures could come his way, had to be more hard-headed than his precedessors in demanding a substantial increase in salary following his appointment. But in return, he was willing to devote himself wholly to the Observatory and the nation as a full-time scientific civil servant. During his long term of office, he applied himself with extraordinary energy to science and technology in the national interest. Having set up the Observatory to more or less run itself by discipline and bureaucracy, he was free to sit on divers commissions and enquiries to give advice to a succession of administrations of state.

Though the main problems of navigation had long been solved by 1835, Airy's role inevitably involved him in many aspects of maritime affairs. Having relieved the Observatory of much of the routine work of chronometer testing for the admiralty, he saw his greatest service to the navy in the production of the observational constants upon which the *Nautical Almanac* was based. But he was also involved in hydrography and a host of problems relating to navigational instruments and ship design that came in the wake of the iron and steam navy after 1840. He was the first scientist to experiment upon, and eventually solve, the problems of magnetic compass deviations in iron ships, made especially urgent when the-newly launched S.S. *Great Britain* ran aground in familiar home waters in 1846.[15] To a navy beginning to contemplate the introduction of steam and iron for major warships, upon which the nation's maritime supremacy would depend, the compass problem was a formidable one. In spite of the dire forebodings of Dr. Dionysius Lardner concerning the impossibility of ever being able to make a magnetic compass function in an iron vessel, Airy produced a simple solution to the problem, which is still, in its essentials, the one in use today.

As a dutiful government scientist, in office when industry and science were transforming national life, Airy had little opportunity to be idle. His capacity for work, and precise organisation, saw his involvement in such diverse problems as the optimum efficiency of railway gauges, the improvement of London's sewer system, the standardisation of weights and measures, the timekeeping accuracy of Big Ben and the discussion of a possible decimal currency system in the 1850s.[16]

Upon seeing Shepherd's electric master and slave clocks at the Great Exhibition in 1851, Airy determined to set up such a system at the Observatory. This was to lead to the installation of the electric 'motor' and corresponding slave clock on the Observatory gate in 1852, so that the main regulator could transmit a public time signal. This concept of electric time distribution was soon extended to the time balls at Greenwich and the main admiralty ports, and from thence to the South-Eastern Railway Company telegraph system. By the late 1850s, many towns which possessed a commercial telegraph line could receive daily signals from the Observatory, in the first public distribution of Greenwich Mean Time. Railway stations and telegraph offices now replaced church clocks as the foremost time sources in the community, and brought the role of the Observatory, and its Director, before the nation.[17] In addition to time signals, one must also recall that the increasingly frequent publication of astronomical and meteorological information to the newspapers also served to keep the British public informed about the Royal Observatory.

It has been said that while Airy was not himself a great scientist, he nonetheless made great science possible.[18] Yet this came about not only as a result of his 'factory' methods at Greenwich, but also in the exploitation of crucial processes and inventions the character and utility of which demonstrate the singlemindedness of Airy's vision. Precision demanded instruments, which in turn were the products of new techniques of engineering. These subsequently led to a reduction in dependence on the physical skills of the observer, and the [326] replacement of that skill by the ceaseless automatic vigilance of a photographic plate or an electric pulse. Electricity and photography in their own right were next to lead to the regular dissemination of signals and images, so that the *concept* of precision would produce an *invention*, which in turn would lead to a *public utility*. Though a mathematician and astronomer by training, Airy was an engineer by inclination, and an organiser and public disseminator of useful information by compulsion.

Four-square in the Baconian tradition, he sought to discover knowledge and use it 'for the glory of God and the relief of man's estate', and one

XVI

wonders what earlier devotees of that tradition, such as Robert Hooke, would have done, had they had the electric telegraph and other resources of the industrial revolution at their disposal. During his uncommonly long productive life, Airy saw science and technology transform the world, and one should not be too hard on him if his genius for accumulating and distributing information outpaced his imagination in how best to use it.

It is unfortunate that this man, who occupied a place of such eminence in his own time, moved familiarly amongst the scientists, statesmen and engineers of the world, and invented the modern concept of data processing, should still await a major contemporary evaluation. Perhaps the sheer bulk of his output, consisting of over five hundred publications and some seven hundred volumes of manuscripts in the Royal Observatory archive alone, scare researchers away. On the other hand, it is probably true to say that the generally held view of Airy's very Victorian personality — as a slave driver and bureaucrat — does not endear him quite so much to our less authoritarian age. Even so, he was a scientist no less well known in his own time than Darwin, Huxley and Herschel, while his popular books and Mechanic's Institute lectures carried him before audiences which many of his better known contemporaries failed to reach. Stiff and unapproachable as he may appear in many ways today, one must remember that he held some remarkably advanced opinions on a wide variety of subjects, extending from Roman archaeology to social reform and the higher education of women. In his private capacity, he was said to possess a sense of humour and a fondness for singing traditional English songs. He lived plainly and preferred English beer to any other drink in the world. Airy was certainly an efficient and straight-sighted man, but on the other hand, he was no puritan. It seems incongruous, once again, that this man who spoke of his assistants as 'obedient drudges' could also discuss fossils with coal miners and travel around the country delivering packed public lectures to working men. He was a figure of surprising contrasts.

By making the Royal Observatory the world's principal repository for physical data of all kinds, Airy not only created an institution which foreign nations had no choice other than to emulate, but went a long way towards creating the professional scientist, unfettered by other callings, and willing to place the fruits of his research at the disposal of the state and mankind. Having placed so much valuable material before the world, Airy should neither be blamed nor neglected if it demanded men of wider scope and imagination to use it creatively and make it bear fruit. Though he was in retirement in 1884, George Biddell Airy, First Technician of the British Empire, had done more than any other single individual to bring about the prime Meridian of the World, and make our 'Longitude Zero' possible.

[327] **Notes and References**

1. For further discussion of the concept of instruments providing a 'frontier' to research, see my 'The accuracy of angular measuring instruments used in astronomy between 1500 and 1850', *Journal for the History of Astronomy* XIV (1983) 133.

2. The details of Airy's life are recorded in his *Autobiography* (Cambridge, 1896), edited by his son, Wilfred. This is a very unsatisfactory work insofar as it restricts itself to the cataloguing of events, and says nothing about motives or personal matters. See also the article on Airy in *Dictionary of National Biography*. A good modern account is to be found in *Dictionary of Scientific Biography*, by Olin J. Eggen. J.G. Crowther presents a perceptive, if somewhat stilted, portrait of Airy in *Scientific Types* (London, 1968) 359-92.

3. A variety of stories are still handed down in the Observatory relating to Airy's obsessive concern with often ludicrously petty administrative matters. One such story relates how Airy once spent an entire afternoon labelling disused packing cases 'empty': E.W. Maunder, *The Royal Observatory, Greenwich* (1900) 116.

4. Airy discusses the matter of giving recognition to his assistants in his *Report* for 1872. E.W. Maunder, who had worked at Greenwich in his earlier days, does not paint a flattering picture of Airy, who kept staff wages 'discreditably low', was 'despotic to an extent which would scarcely be tolerated at the present day' and instituted a régime of 'remorseless sweating', in *The Royal Observatory, Greenwich* (1900) 117. See also A.J. Meadows, *Greenwich Observatory II, Recent History (1836-1975)* (London, 1975) 1-12. Airy really 'industrialised' the Royal Observatory when he instituted clocking-on for the assistants, to make sure that they did not cheat on their timings.

5. Meadows discusses the low pay rates, and Christie's wish to improve them upon his becoming Astronomer Royal in 1881, in *Greenwich Observatory II* [n. 4] 9-10.

6. In the Cambridge Observatory Library is a letter dated 'Gothar, 1846 Sept. 30', which Airy sent to Challis. He says that he received news of 'the Uranus-disturbing planet' the day before, as he was dining with Professor Hansen. Camb. Univ. Observatory, Boxfile '1846-1847. Discovery of Neptune'; letter 13.

7. For Airy's magnetic instruments and their photographic registration, see Derek Howse, *Greenwich Observatory III, The Buildings and Instruments* (London, 1975) 123 and Fig. 113.

8. Howse, *Greenwich Observatory III* [n. 7] 45-6.

9. James Stuart, *Reminiscences* (1911) 159.

10. For an account of Airy's transit circle, see G.B. Airy, *Astronomical Observations at the R.G.O. 1852* (1854), Appendix II, p. 17. See also E.W. Taylor and E.S. Wilson, *At the Sign of the Orrery* (undated, c. 1950). Discussion of the increasing skill of scale dividers from Bird to Simms is to be found in A. Chapman, *Dividing the Circle. The Development of Critical Angular Measurement in Astronomy 1500-1850* (2nd edn, Wiley-Praxis, Chichester and New York, 1995). [328]

11. Bradley had first realised that temperature and pressure could affect refraction in 1750, and from that date recorded them as part of an astronomical observation: see James Bradley, *Astronomical Observations Made at the Royal Observatory,*

Greenwich, edited by T. Hornsby (Oxford, 1797) p. iii.

12. G.B. Airy, *Account of pendulum-experiments undertaken at the Harton Colliery, for the purpose of determining the mean density of the Earth* (London, 1855), being the published version of a lecture delivered in South Shields following the conclusion of the experiments.

13. Meadows, *Greenwich Observatory II* [n. 4] 59.

14. There are many documents relating to the use of electricity for time distribution and longitude finding in Airy's papers at Herstmonceux. These papers are currently being catalogued, along with those of the previous Astronomers Royal, in the 'Phillip Laurie Project'. See No. 1204, on Galvanic Longitudes; No. 1182 on Galvanic time balls, and Nos. 1205-7 on the determination of the Paris, Brussels and Edinburgh meridians. In 1862, Airy determined the Valentia meridian electrically (No. 1208), and others followed later. In the *Mechanic's Magazine* No. 1816, in 1858, Airy wrote an article on the laying of the first Atlantic telegraph. Airy seems to coin the title 'The British Astronomer' for himself in his *Report* to the Board of Visitors in 1842.

15. Airy had first addressed himself to the problem of compass deviations in iron ships as early as 1837, when experiments were conducted in the iron vessel *Rainbow*: see Airy papers, No. 1257. Airy's solution to the problem is outlined by Ewan Corlett in *The Iron Ship* (1983) 40-1.

16. In his *Autobiography*, Airy catalogues the numerous commissions on which he sat. He wrote several articles and advised government committees on metrication, particularly in 1868-9, while he produced articles 'On decimal coinage' for the *Athenaeum* magazine in May and September 1853. He advised Gladstone on the possible division of the Pound Sterling on a metric basis on 26 March 1853, as well as giving advice on Exchequer Standards in 1868.

17. For details of these clocks, see Howse, *Greenwich Observatory III* [n. 7] 137-44. See also J.A. Bennett, 'G.B. Airy and Horology', *Annals of Science* 37 (1980) 269-85.

18. Cited by O.J. Eggen, *Dictionary of Scientific Biography*, under 'Airy'.

Reprinted from Vistas in Astronomy 28 (1985), pp. 321–8. Numbers in square brackets indicate original pagination.

XVII

William Lassell (1799-1880): Practitioner, Patron and 'Grand Amateur' of Victorian Astronomy

The fundamental concerns of most astronomers in the year of Lassell's birth were similar, in many ways, to those which had occupied Hipparchus; the refinement of celestial cartography and the positional measurement of the stars. These concerns had been enshrined in the appointing Warrants of all the Astronomers Royal since Flamsteed, and in addition to dominating the subject intellectually, also dominated it technically, for the instrument-making trade was far more concerned with improving ways of 'dividing the circle' than with producing better optics.

While it is true that physical astronomy and the study of astronomical bodies in their own right, as opposed to the positional measurement of light sources in the sky, had received great attention in the seventeenth century in the wake of Galileo's discoveries, this impetus had largely expended itself by 1700. It might be argued that physical astronomy languished because, by 1700, the development potential of the simple refracting telescope had reached its limit, and could provide no new data. One might further suggest that the limit was not merely instrumental, for the reflecting telescope, which had been devised by Newton in 1670 and improved by Hadley and Short, remained largely neglected until the peculiar genius of William Herschel demonstrated that it could be developed to send astronomy down a new avenue of inquiry. Herschel's work, which depended on the greatly enhanced 'space penetrating power' of reflecting telescopes, created a new set of relationships between the questions an astronomer could ask and the tools with which he might answer them, to produce thereby a dynamic science of physical astronomy.[1]

I believe that William Lassell's significance derives from the fact that he was the man who, more than any other, carried on the physical astronomy tradition from Herschel, and recognised the role of the large reflecting telescope within it. Yet Lassell is important for a variety of reasons, and not all of them concerned with 'internalist' issues in the history of astronomy. He is also grist to the social historian's mill, and what is truly remarkable is that he has received so little historical attention beyond the fulsome obituary tributes which followed his death in 1880.[2]

By the early nineteenth century, astronomy had become a hard subject for an outsider to break into on a serious academic level. Being intellectually dominated by Newtonian dynamics, one needed a systematic education if one aspired to rise beyond the ranks of the backyard stargazer. Though only a handful of men actually earned their livings by astronomy at this period, the creative forefront of the discipline was composed of 'gentleman amateurs' who had still received a sufficiently learned education at one of the ancient universities in their early years to easily master the higher branches of mathematics as they needed to.

Because he was a dissenter, and perhaps for other family reasons as well, Lassell could not have conformed with the Stuart *Test Acts*, which still restricted access to the ancient [342] universities to communing Anglicans. He came, indeed, from the same social background as many of the young men who became Computers and Assistant Observers at the Greenwich, Oxford and Cambridge Observatories; modest middle-class dissenters whose parents could afford a good education up to the mid-teens, but whose religious scruples, and perhaps sense of social identity, precluded them from the English degree-granting universities in the days before the founding of University College, London.[3]   Where Lassell differed from the various Methodists and Independents who worked for Airy and Challis, however, was in his talent for commerce. As a wise young man of conscience, he relegated astronomy to the status of a very serious hobby until he had made his fortune by his late thirties. Then rich, secure, and a pillar of Liverpool society, he proceeded to spend thousands of pounds of his own money on his 'hobby', defining his own research and devising the great reflecting telescopes which came, at last, to exceed the creations of Sir William Herschel of fifty years before. Able as he now was to give full rein to his interests, he made a major mark not only as a scientist of international standing, but as a private patron of 'big' science.[4]

Lassell was born of moderately prosperous parents, both of whom descended from trading families which, like their religious denominations, could be characterised as Congregationalists and Independents. He was born in Bolton, Lancashire, on 8 June 1799,[5] where his father Nathaniel traded in partnership with his brother-in-law, James Gregson, as timber merchants and builders.[6] William Lassell's mother, Hannah Gregson, came from Liverpool where her family were in business, and it is also likely that his father came from a family with Liverpool branches, as one sees the regular occurrence of the name Lassell in the local registers of the early nineteenth century. This could account for the astronomer's long-standing practice of styling himself William Lassell 'junior' in written documents, which he continued to do down to his late forties, and makes him easy to identify. There was, for instance, a William Lassell trading in the City as a wine merchant in 1821,[7] and it is possible — but by no means proven — that it was in his 'merchant's office' that the astronomer was apprenticed in either 1814 or 1815.[8] No record survives, to my knowledge, of the trade into which Lassell was apprenticed, but as he was being described in Gore's *Liverpool Directory* as a Brewer by

1825, it could well have been to a relative in the alcoholic drinks trade.

The brewing trade was a rapidly growing one in early nineteenth-century England, especially in the teeming cities of the industrial north,[9] and it made Lassell a substantial fortune. In spite of the many thousands of pounds which he spent on the building of three major astronomical telescopes, he was able to bequeath an estate valued at between £70-80,000 at his death in 1880,[10] and by his late thirties, had accumulated sufficient capital to leave his house at 18, Norton Street, in Liverpool's business district, where he had resided for fifteen years, and build 'Starfield', the first of his three suburban mansions.

There is no evidence that Lassell was ever poor in the real sense of the word. As early as 1822, when only just out of his apprenticeship, he was recording observations made with a 2¾-inch Dollond refractor, while his scientific friends referred to in his notebooks had similar pieces of equipment at their disposal.[11] In 1828, for instance, Lassell recorded an 'observing party' in which his friends Dawes and Nixon stayed up until nearly 4 a.m. comparing the images of their respective telescopes, which included 9- and 7-foot Newtonians along with 3¾- and 2¾-inch-aperture achromatic refractors.[12]

But whatever financial assistance he may have received at the outset from the Lassell and Gregson families, he was soon making his own way in the world and trading under his own name by the time he was twenty-six. His principal commercial venture, the brewery in Milton Street, seems to have been trading as Bagnold and Lassell in 1829, but under his sole name or senior [343] partnership after 1832.[13] Though the trade directories are not always clear about dates, there seem to have been several junior partners culminating in the firm trading as Lassell and Sharman in 1872.[14] As Lassell's daughter, Maria, married Nathaniel Sharman (described on the Marriage Certificate as a 'Shoe Manufacturer') in 1860, one suspects that a shoe-making fortune subsequently mingled with a brewing one.[15] Clothing and drinking, after all, are fundamental human requirements, and men who earn their livings by providing for them are less exposed financially than purveyors of luxuries.

Much of the success which provided the ample economic base from which he could prosecute original research stemmed, no doubt, from the traditions of thrift and diligence within which he had been brought up. Characteristic, as one might think, of the nonconformist middle class of the day, he had been taught not to throw things away which still might be useful, and this approach to life is still visible in his surviving papers. He was never to get out of the habit of making up notebooks from folded sheets of foolscap paper stitched within covers fashioned from discarded wrapping paper. It is ironic to think that observations made with the most advanced reflecting telescopes of their day were recorded in such home-made exercise books, with original postage and stamp marks still visible on their covers.[16]

In his early teens, however, Lassell had been given a handsomely bound ledger in which he had started to make sermon notes, before re-using the volume — by starting from verso end — as an astronomical notebook several years later. 'Texts preached from by different Ministers at Duke's Alley Chapel, Bolton', between January 1814 and June 1815, is the earliest surviving Lassell document in his own hand.[17] The elegantly written notes in this volume, which he had signed and dated on the original endboard, were clearly intended as exercises in accurate memory training and fine penmanship, which he had most likely received at the Rochdale Dissenting Academy where he had previously been sent to complete his education.[18] From the dates on the sermon notes recorded in the volume, however, it appears that Lassell was resident in Bolton over most of 1814 and early 1815, and could not yet have begun his Liverpool mercantile apprenticeship which some obituary writers dated to 1814.

Lassell's early notebooks, preserved by the Royal Astronomical Society, provide invaluable information about his life and influences. By 29 November 1817, however, he was residing in Liverpool, as indicated from the cover note in R.A.S. Lassell Ms. 6. This invaluable volume records a wide range of literary and scientific interests by the time that Lassell was in his early twenties, including items as diverse as a transcription of the life of the poet Andrew Marvell, attendance at a lecture in November 1819, chemical experiments and the making and using of a 7½-inch-aperture Gregorian reflector in 1821. Throughout his life Lassell enjoyed literary and religious pursuits as well as those of a scientific character, and as a young man was willing to devote hours to transcribing works such as the 'Life of Sir Samuel Romilly', the reformer.[19]

Because the brewery was the source of Lassell's fortune, I have tried to find if any correlation existed between its profitability and his investment in large telescopes, though unfortunately, I have been unable to find any detailed accounts for the business. There were partnership names mentioned in 1829, 1835, 1857 and 1872, but whether these were actual changes and brought in the substantial sums of money which were necessary to build the Starfield 9- and 24-inch telescopes, and then the 4-foot Malta equatorial, is impossible to prove. It would have been interesting to see, had figures been available, whether 1838, 1844 and 1858-60, when he built his great telescopes, had been bumper years for beer sales.

The only notes relating to the brewery in the surviving Lassell papers are some twelve pages of what might be called 'vat records' from the mid 1820s, when he had just gone into [344] business.[20] They survive because Lassell used up the presumably superseded volume as an observing notebook, in the same way as he had used his 1814 sermon book, by re-starting it at the verso end. These pages do contain some interesting material about the running of his business in 1824-6, however, not least being the information that he was already making beer by November 1824, and not 1825, which was the date given for the founding of his business by several obituarists. Rather interestingly, one of his customers was a 'William Lassell', who could well have been the vintner mentioned above and to whom he *may* have been related. At this early stage, his plant contained three vats labelled 'B, H, Q' respectively for 6^d, 8^d, and 10^d beers, while the difference between the 'when brewed' and 'when sent out' columns indicates an interval of about three weeks.

In 1827, William Lassell married Maria King in Toxteth, Liverpool. The marriage certificate is interesting firstly because the marriage was solemnised in Toxteth, and secondly because it was witnessed by Alfred and Jos.[eph] King.[21] Toxteth seems to have been a place well stocked with Lassells, several of whom were confusingly called William. One of them, describing himself as a 'gentleman', had his will proved for that parish in August 1816, while in 1834, another local William Lassell, a clockmaker by trade, was a signatory to the purchase of a piece of burial ground for the use of Toxteth Chapel.[22] In Gore's *Directory* for 1845, moreover, three Lassells are mentioned as operating in the wine and beer trades, two of them called William, and the astronomer identifiable by his usual designation of 'junior'. His private address is given as 'Starfield' — his observatory residence — and his brewery as '20, Milton Street, Liverpool'. As there was also a firm trading as 'Lassell & Co. wine and brandy merchants', it is not unlikely that this could have been part of a family consortium. Whether these Lassells were related to William the astronomer cannot be proved at present, though a family association with the area seems certain.

Alfred and Joseph King, who were most likely brothers of Lassell's wife Maria as they shared her maiden name, are referred to several times in his early astronomical notebooks, and seem to have been amateur astronomers in their own right.[23] Indeed, it is not unlikely that it was through a prior friendship with Alfred and Joseph that Lassell met his wife. On 11 August 1828, for instance, Lassell was observing sunspots with Alfred, and applying tests to the latter's Tulley refractor, while he and the two Kings observed the 'Great solar eclipse' of 15 May 1835 together.[24]

Lassell does not seem to have lacked congenial scientific company in Liverpool in the 1820s and 1830s, for his notebooks are full of occasional references to local craftsmen and businessmen with whom he shared his interests. Where he was unique lay in the degree of commitment which he lavished upon his 'hobby', along with his own rather spectacular commercial success whereby he was lifted out of the ranks of comfortable tradesmen to become very wealthy indeed.   The Starfield observatory and the Malta reflector indicate not only engineering ability and scientific vision, but a sense that astronomy was a subject he was most conspicuously endowing with his ample fortune.   In this respect he was unique in Liverpool and perhaps England, for as the writer of Lassell's *Daily Post* obituary observed, it is a 'pity that wealthy merchants with taste and capacity for science are so few ... [and that generally] ... wealth it feels is its own sufficient reward'.   But 'Mr. Lassell was fortunate in combining, in his own case, the taste for scientific research and the ability to prosecute it, and the pecuniary means needful for its prosecution.'[25]

His wealth, the exemplary cause to which he devoted it, and the academic accolades [345] which his work earned produced another visible sign which indicated success and respectability to a Victorian tradesman who had 'made it'; the transition from 'Brewer' to 'Gentleman' as his style in legal documents.[26]

Yet for all the intellectual stimulus and honour which Lassell got out of science he never permitted it to cloud his judgement where matters of money were concerned.   When he came to draw up his Will in 1875, and was creating trust funds for his family, he recommended that the principal should be invested in such regular earners as government stock, incorporated English and Indian railways.[27]   There was no mention of investing in the new science-based industries that were coming into being such as synthetic dyestuffs and electricity, for while as an eminent F.R.S. he must have been familiar with them and may have sensed their massive future investment potential, they were not yet well enough established to attract his money.

Liverpool was well equipped with intellectual resources during the fifty-odd years that he lived there, and one can understand why his notebooks bristle with the names of scientifically minded friends.   It is true that few were 'wealthy merchants', though the directories of the period indicate that many of them were enlightened shopkeepers and craftsmen with comfortable businesses, not to mention clerical and medical men.[28]

There was Mr. Roskell the watch manufacturer, whose premises in Church Street were 6.92 seconds west of Lassell's residence, while the shop of his friend Mr. Davies was 5.52 seconds west.[29]   In Bootle, there were Mr. Findlow and Miss Harrison, the exact latitudes of whose houses are recorded in his notes.[30]   Indeed, the latitudes and longitudes of the homes of many local people are recorded in his notebooks, and one suspects that they were fellow devotees of astronomy whose observations may have been collated with his own. When he travelled to Sweden to observe a total eclipse of the sun in

1851, he mentioned being accompanied by his astronomical friend Mr. Stannistreet, though as he fails to provide a Christian name for this gentleman, it is impossible to say which Stannistreet he was among the several mentioned in the directories.[31]

Yet two of his most important and long-standing friends were not Liverpool residents. William Rutter Dawes was a medical man who had taken on the ministry of a dissenting chapel in Ormskirk a few miles outside Liverpool.[32] I do not know when the two men first met, but they were to become firm, lifelong friends sharing religious and possibly political beliefs as well as their mutual devotion to astronomy. Dawes clearly possessed a good library at Ormskirk in the 1830s, and Lassell sometimes mentioned copying out articles from its books. In April 1829 he was transcribing a list of stellar objects '... extracted from Mr Dawes's Book', while he also used the latter's set of Rees's *Cyclopaedia*.[33] He further related that around 1830, Dawes was working with a 3¾-inch-aperture refractor of 58 inches focus.[34]

When Lassell completed his first major telescope, the 9-inch Newtonian of 112 inches focal length, before mounting it in the new observatory at Starfield, he left it on loan to Dawes at Ormskirk, bringing it back to Liverpool on 7 August 1837.[35] Perhaps he had left it with Dawes, with whom it had 'been for a considerable time' so that the skilled observer of double stars could give it a critical testing. Dawes and Lassell certainly saw a great deal of each other, according to the notebooks, and when the house at Starfield was completed, by 1837, he frequently stayed as a house-guest as they observed together.

By the late 1840s, when the growing urban sprawl of Liverpool was making seeing progressively worse at Starfield (so that Lassell would be forced to move out in the mid 1850s to his new observatory residence of 'Bradstones'), Dawes left an amusing note in the observing log. In September 1847, Dawes was spending a lot of time at Starfield where both men were trying to [346] obtain confirming observations for Lassell's discovery of one of the satellites of Neptune, and finding the weather exasperating. Dawes wrote in the observing log for 26 September 1847 'Disgusted and despairing we closed the observatory and came in at 11 ... to end my unfortunate attempts at observing with the truly excellent 20-foot equatoreal at Starfield (qu. *Cloudfield*) in September 1847. W.R.D.'[36]

Lassell's other great friend was the engineer James Nasmyth, who was not only a skilled astronomical observer, but whose manufactuary at Patricroft, Manchester, produced the precision polishing machines which made the 24- and 48-inch telescopes possible. The two men met around 1840, and their warm friendship was to last forty years.[37] Lassell had been so impressed by the 10-inch mirror which Nasmyth made that 'it made his mouth water' upon seeing it,[38] and he found that he could learn from the professional engineer. Though Lassell was already a highly skilled caster and figurer of hand-ground specula, working (as his early notebooks testify) by patient empiricism, it was Nasmyth who opened up the prospect of devising a geared machine, which could reproduce the delicate hand movements which Lassell had perfected for his specula after he had been disappointed with an earlier machine based on the prototype by Lord Rosse.[39]   Much of the secret of Lassell's success in producing mirrors of perfect figure had been the use of short curved strokes with the polishing tool. Nasmyth's mechanical ingenuity made it possible to reproduce these strokes to their finest delicacy when polishing a 4-foot mirror with a steam-driven tool.[40]

Lassell's sense of humour is never far below the surface when writing to his intimate friends, and on one occasion when writing to Nasmyth, he lamented the folly of intelligent men who enjoyed 'killing something' on the sports field, saying that 'For my part I would rather take to the bicycle and do my 17 miles within the hour.' As Lassell was seventy-six years old at the time of writing, however, one presumes that he was speaking metaphorically.[41]

By the 1840s Lassell was corresponding with the leading British, European and American astronomers in terms of familiarity. He was already an F.R.A.S. by this time, and clearly respected as both a patron and practitioner of high quality astronomical research.   In one surviving copy letter book, he was writing to Le Verrier, Professor Struve, Professor Nichol and Lord Rosse.[42]   Other notebooks indicate that he was on good terms with Sheepshanks, Baily and G.B. Airy, the Astronomer Royal, while Airy's own 'Journal', in turn, contains several references to Lassell.[43]   One letter of 1850 provides an illuminating aside on the informality with which he moved in the upper echelons of the astronomical community by this time. It seems that Professor Struve, when visiting England, had travelled to Starfield and stayed with Lassell. Upon leaving, after a very happy and successful visit, he forgot to pack one of his shirts which was later given to Mrs. Lassell by a servant.[44] At this time, however, G.B. Airy happened to stop by for a one-hour visit to Starfield en route for Greenwich after a visit to Glasgow. The Astronomer Royal was pressed into service by Mrs. Lassell to personally transport Struve's forgotten shirt to its owner in London prior to his return home to Russia.[45] Lassell's intimate 'My dear Struve' mode of address combined with concerned inquiries about a return of the latter's 'spasms' says much about his relationship with the leading men of European astronomy, not to mention his use of the Astronomer Royal as a laundry boy. By this time, Lassell had already been a guest of Lord Rosse at Birr Castle, in 1843, where he had been greatly impressed by the mirror-making machinery in use there, though when two years later Rosse sent him an offprint of a recent article, he replied in a friendly tone which still deferred to the Irish astronomer's aristocratic standing.[46] It is clear that by 1850, however, Lassell occupied an established and respected place in the republic of learning. [347]

His entry into that world and its learned societies had taken place in 1839, when he was forty. In that year, he had become a member of both the Liverpool Literary and Philosophical Society and the Royal Astronomical Society.[47] He was to serve these institutions well over the next forty-one years of his life, being made a life member of the former and a President of the latter. The esteem in which the Liverpool Lit. and Phil. held him is apparent from the notes concerning his subsequent academic and foreign honours which were dutifully entered after his name in the membership book, while the R.A.S. awarded its Gold Medal to him in 1849.[48] In 1849, he was also elected F.R.S., and became in turn one of its Royal Medallists in 1858.[49]

One might argue that Lassell's formal entry into the Courts of Learning in 1839 represented the great transition not only in his scientific status — from mercantile amateur to gentleman of science — but in his life in general. Though it is true that he was still to complain, as he did to Struve, in 1848, that 'owing to my necessary occupations in business my leisure rarely affords opportunity for me to do more than make observations ...',[50] astronomy was now to be the real occupation for the rest of his life. The year 1839 was also important in other respects as well. He had recently moved his residence out of central Liverpool to Starfield in what was still then the countryside, and in its grounds set up one of the finest telescopes for physical astronomy in England: his equatorially mounted 9-inch Newtonian.[51] The possession of this new and remarkable instrument automatically brought Lassell before the eyes of the astronomical establishment, especially as he had designed and built it himself. It was this peculiar combination of self-made private wealth, optical and mechanical ability and skill as an observer and interpreter of the heavens which made Lassell so remarkable. Sir John Herschel put his finger on it when he styled Lassell as belonging 'to that class of observers who have created their own instrumental means, who have felt their own wants and supplied them in their own way'.[52]

Perhaps the nearest parallel had been that of John Herschel's own father, Sir William, who started off his own revolution in telescope technology and cosmological research on the proceeds of a comfortable musical income, though even he had needed ample Royal grants to provide the necessary money and leisure to fully develop his work.[53] William Lassell on the other hand remained wholly self-sufficient, providing all his own resources, and in this respect was probably unique.

Yet both Herschel and Lassell were 'grand amateurs' in the noblest tradition, migrating into astronomy in early middle age with established reputations in non-scientific careers already behind them.[54] Both were highly independent men, having largely educated themselves outside the formal academic structures of the day and being essentially freelance in their approach to life. In addition to their conspicuous intellectual talents and capacities for hard work, moreover, they shared a natural charm and ability to make lasting friendships with a diversity of people. The ease with which both men were accepted into the scientific establishments of their day says much in this respect, for apart from their purely intellectual gifts, both men started as social outsiders: Herschel rose from the ranks of 'forreign fiddlers'[55] while Lassell was a nonconformist provincial tradesman. Both represented types about whom stock jokes were made in that polite and still predominantly Anglican Oxbridge society out of which the metropolitan learned societies drew their Fellowships. Yet both won the approbation of their respective Sovereigns; Herschel was taken under the personal wing of George III, and when Victoria visited the Croxteth district of Liverpool [348] 'Mr Lassell was the only person who was invited to meet her by her Majesty's own request, and it was said that she rose and advanced to meet the astronomer as he entered the room'.[56]

Quite apart from parallels of character and social circumstance, Herschel and Lassell shared a common approach to astronomy. Unlike the great majority of their colleagues, they were not primarily concerned with positional astronomy, and though his notebooks show that Lassell did a little transit astronomy at Starfield, it was only a minor part of his work.[57] Herschel and Lassell were concerned with the physical examination of heavenly bodies, and in consequence, required not the finely divided circles of Ramsden and Troughton, but 'space penetrating power' and light grasp. At very early stages in their respective careers — 1774 for Herschel and 1821 for Lassell — they had come to recognise that nothing was available on the commercial telescope markets that fulfilled their needs, and turned to the fabrication of Newtonian reflectors as a solution.

Both men were fundamental innovators in big telescope technology. In the strictest parenthesis, it might be said that Herschel had shown that big reflectors *were* possible and could be used to draw startling conclusions about the universe, while Lassell showed that the technology of the Industrial Revolution could take them yet further, especially when mounted in the equatorial plane.

That the young Lassell was inspired by Herschel is clear from his early notebooks, and the way in which he borrowed and transcribed some of Herschel's published papers.[58] On the other hand, there were two significant differences between the two men in their predominant interests. There is no evidence, for instance, that Lassell shared Herschel's fascination with cosmological issues, and one looks in vain for those brilliant interpretative leaps from observational data to fathom the 'Construction of the Heavens' as Herschel had done throughout his career. Lassell undertook no 'reviews' of the sky, and showed no systematic interest in gauging space or trying to unravel the Milky Way.[59]

Lassell's approach was guided by no broad philosophical assumptions and was quite pragmatic in character. His main concern was to inspect nature with telescopes which gave increasingly fine definition, rather than ponder upon cosmological causes and effects. Although he did a little 'deep sky' work in Malta in the 1850s and 1860s, this consisted of descriptions of specific objects, such as the Orion Nebula, from which he was at pains *not* to draw wider conclusions. Although he published a catalogue of 600 new nebulous objects in 1866, which had been seen from Malta, the secondary status of the work was made clear in his opening paragraph, for the nebulae had been looked for only when 'the objects I especially wished to observe were not to be seen at all'. The catalogue, moreover, was 'entirely the work of Mr Marth on those occasions when the telescope was not otherwise engaged'.[60] Perhaps Lassell realised that, in an age when photography and spectroscopy were in their earliest infancies, there were no new tools with which accurate measurements could be made or new data recorded, so that one could only, in effect, duplicate the existing work of Herschel.

Lassell's real interest lay in the astronomy of the solar system; an interest which is immediately obvious from his observing books and regular communications to *Monthly Notices*. From the late 1840s onwards, Saturn and the outer planets became the real foci of his interest. On the other hand, I do not believe one can say that Lassell was specifically engaged in a deliberate planetary programme, in the way that William Herschel had been with his stellar reviews, for the interest seemed to grow out of a series of almost serendipitous associations. Given that remarkable alchemy of superlative equipment, patience, insight and good luck, he [349] could scarcely help making visual discoveries of new satellites and planetary features, and once they had been discovered, he naturally wished to follow them up.

But fundamental to all these discoveries was his unrivalled expertise in the making and using of metal mirrored reflecting telescopes. If his notebooks might be considered as an authentic guide to his predominant interests, then it is obvious that Lassell devoted at least as much time to making telescopes as he did to using them. While one senses in the case of Herschel that a concern with telescope making was a means to the wider end of cosmological research, one sometimes gets the sense that the skies were, to Lassell, a collection of test objects upon which to try optics and mountings of increasing

XVII

sophistication. It was a priority which he shares with many present-day amateur astronomers, and we can understand how he and Nasmyth had so much in common, as they pored over plans for polishing machines and compared the quality of each other's specula.

It was probably in 1821 that Lassell cast his first mirror, for in his notebook for 18 September of that year, he completed copying out the Revd John Edwards's article on the casting and figuring of specula, from the 1787 volume of the *Nautical Almanac*.[61]   Herschel, one must remember, had published nothing on the manufacture of his optics, so that it was necessary to go to a source printed thirty-four years previously for the state of the art on this still neglected technology. Edwards's article, on the other hand, is superb, and provides what was probably the only complete published set of step-by-step instructions for making a reflecting telescope, from setting up the casting furnace for the speculum, to applying tests to the finished instrument. Mirror making was a quasi-alchemical art, and devotees, even at the level of Herschel, tended to hoard their secrets of manufacture even when publishing their results, so it must have been hard for a raw beginner to find advice. Though Lassell, when describing his great instruments in the 1840s and 1860s, ventures far more constructional information than Herschel ever did (especially on the exact proportions of his speculum alloys), one must remember that neither Lassell, nor his equally open colleague Lord Rosse, actually sold telescopes for profit, and so had no need for trade secrets.

It is in Lassell's notebooks, however, where one finds him gradually perfecting his art, and sees in him the marks of a born experimentalist. The notebook 'On the construction of telescopes' records his developments in this technology; the first eighty-six pages of it were written before 1833, and the latter part between 1833 and 1848.[62]   At the very outset of the volume, he stated that while there was a general prejudice in favour of refractors on the grounds of portability and convenience, nonetheless, given a 'proper degree of care', good specula could be made which would not tarnish.[63]   Reflectors, furthermore, became proportionately cheaper as soon as one required an optical aperture bigger than 3¾ inches. In this preliminary essay — composed sometime before 1833 — he said that while a 5-inch mirror was too small for serious use and a 9-inch too large for a beginner, a 7-inch speculum of 7 feet focal length was ideal.[64]   Generally speaking, he thought that the mirror diameter in inches should equal the focal length in feet to establish a good proportion.

In his section 'Of casting and grinding the metals', he recommends an alloy consisting of copper 32 parts, tin 15 or 16, and arsenic 1½ by weight. 'Grain' tin was better than bar, while the best copper came from the 'old copper bolts of ships [which] are very good and convenient for melting'.[65]

One should first melt the copper in the crucible and only when molten add the tin. When both were fused together, a small quantity should be ladled out and cooled with water. In a good mix, the mass would reduce to form brilliant white fissures on cooling. Only at this stage should the arsenic be added to the pot, wrapped up in a little paper packet, and stirred vigorously in, for the arsenic was not easily soluble. Lassell advised against adding too much [350] tin, which made the mirror dull and hard to work.[66] He preferred to smelt the alloy over a fire of oak chips.

This remained the basic metallurgy for his mirrors, although he experimented incessantly to get the best grinding surface and capacity to take a good figure without tarnishing. He published his 'perfected' technique in a 'Description of a Machine for Polishing Specula, & Co.' in 1848,[67] though when he came to cast the mirror for his 4-foot reflector a decade later, he omitted the arsenic, preferring to depend on the purity of the copper and tin alone to obtain the requisite degree of whiteness.[68]

Though he appears to have used sand in the casting of his earlier mirrors, by the time of the 24-inch, in the mid 1840s, he was experimenting with the so-called *chill* method employed by Lord Rosse for his 36-inch. This required the metal to be poured onto a warm iron plate which had the effect of cooling it in more regular strata and producing a harder surface for the grinder.[69]

For polishing and grinding, Lassell used emery, rouge and pitch.[70] When working the 24-inch specula, he found it advantageous to cover the polishing tool with a thin sheet of lead which, when under pressure, absorbed the emery grains and made a much more homogeneous tool.[71] From his notebooks, it seems that the optics of his 9-inch Newtonian reflector which came to be mounted at Starfield in 1839 were first commenced in early September 1833, for it was at this time that he started making important notes on the polishing of a 9-inch speculum.[72] This is the description of his renowned hand-polishing technique which Nasmyth's machine was devised to reproduce and enlarge for 24-inch blanks a decade later. His intention was to impart not *brilliance* but an even lustre to the surface of the speculum. This was best obtained by giving short cross strokes and advancing in a circular motion every six or eight strokes. Commercially made reflectors, he warned, were often erroneously judged by the brilliance of their mirrors, of which he complained, for 'The London artists send out their instruments with an exquisite polish and without a scratch on their surfaces but in attaining this high lustre they frequently *impair* the figure previously given.' Lassell often complained of how over-polishing often 'hurt' a good figure, and concluded by observing 'Indeed it is surprising with how low a polish they may be used if the figure be exquisite.'[73]

A variety of optical tests were applied by Lassell to his mirrors as he brought them to perfection, and he was familiar with those described by Mudge.[74] His approach to mirror testing, however, was that of a craftsman rather than a scientist, for in the long run, he preferred to rely on 'feel' and sheer technique. Bringing a 9-inch mirror close to perfection after polishing it from 3 p.m. to 6.30 p.m. on 4 September 1833, he tried it out on the star α Lyrae, but not being wholly satisfied, he removed it from the telescope and polished it for some hours more and obtained a better image later that same evening. Even so, his notebook contains further finishing touches and star-image tests which he applied over the next eight days until he was wholly happy with it.[75]

Like all accomplished craftsmen, Lassell knew when it felt right, and in the absence of modern optical bench tests, knew how to look critically at both celestial and terrestrial images. Some of these tests might sound irregular to modern readers, but in the hands of a master, they could be made to yield critical results: [355] 'Sept. 21 1837, 3 p.m. Atmosphere clear and favble. tried 9 in Met[al] B on chimney top ... Vision now very sharp and good apparently as good as A.'[76] and 'July 26 1841 7½ a.m. Examined on [watch?] Dial at Bath room window penumbra decidedly more distinct when focus shortened.'[77]

Indeed, chimney pots seem to have constituted favourite test objects for Lassell, for he used one to test the evenness of the 'zones' of his mirror in August 1837,[78] while twenty-two years later, on 13 May 1859, we find him trying out the newly polished 48-inch mirror on 'Mr King's chimney top in brilliant and powerful sunshine'.[79] Whether this was the chimney pot of Alfred or Joseph King is not stated.

One must not underestimate the rigour of his tests however, for they were generally conducted by placing diaphragms over different parts of a mirror surface, first at the centre, then working outwards and inspecting an object containing plenty of detail such as the dial of a watch or fine letterpress set up some distance away.[80] If the object produced the same image quality at the same focal point on all parts of the speculum, then the mirror was a good one with an 'even' surface. But the test which Lassell styled his *Experimentum Crucis* was on the image of a star, which he would put slightly out of focus, to see if the resulting penumbra of light was exactly the same shape and intensity in all parts of the mirror.[81]

Lassell was not the first astronomer since Herschel to turn his attention to big reflecting telescopes, but I believe he was the most significant. John Ramage had built a 15-inch instrument of 25 feet focal length which had been loaned to the Royal Observatory between 1826 and 1836, but it was only mounted on a wooden altazimuth like Herschel's telescopes of forty years previous, and was never used for serious work.[82] But it had been Lord Rosse who made the first significant optical innovations in the late 1830s, for when Lassell was still producing excellent 9-inch specula by hand, Rosse was machine-polishing mirrors of 36 inches.[83] Rosse was a pioneer in other ways as well, including the 'industrial' scale of his Parsonstown operation. On the other hand, Rosse mounted his first 3-foot speculum in an altazimuth mount almost identical to Herschel's, and giving no better tracking control.

Though much was learned from Rosse by Lassell in the early 1840s, the most significant lesson was the realisation that if he was going to make specula beyond 9 inches diameter, he must use power machinery. Having inspected Rosse's machine in 1843, he built one for himself and began to polish a mirror with it. Yet he was not pleased with the results, and found that no matter how the machine was adjusted, it was incapable of producing 'the evolutions of the hand by which I had been accustomed to produce very satisfactory surfaces on smaller specula'.[84] He considered that the revolving spindle which carried the speculum was too weak to give proper control in the Rosse pattern machine, and then took the problem up with Nasmyth. The result was the machine which formed the subject of his 1848 paper.

In addition to being much more heavily engineered than the Rosse machine, the Lassell-Nasmyth polisher was capable of various adjustments so that the respective motions of speculum and polishing tool could be made to figure parabolic mirrors of various diameters and foci with the greatest delicacy. It could be left on automatic once adjusted, and using best quality rouge as an abrasive, produce a 'lustre of polish [which] transcends even my best efforts by hand, and is the easiest quality to obtain'. He suggested that the machine might also be used for polishing large lenses.[85]

It is interesting to bear in mind that in the same decade as Lassell devised this machine to [356] automatically produce high-quality mirrors, so William Simms in London developed the first self-acting power-driven dividing engine for graduating circular scale mathematical instruments, from earlier hand-operated engines by Ramsden and Troughton.[86] The rapidly advancing engineering technology of the Industrial Revolution had suddenly rendered obsolete the two arcane trades of circle divider and optical polisher, upon which the positional and physical branches of astronomy depended. By improving and standardising the error parameters of precision instruments by machine manufacture, the 'ceiling of accuracy' which formed a barrier to practical research was suddenly lifted.[87] Not only were better instruments made available, but more reliable ones, once individual researchers were freed from the personal variations of different specimens produced by different craftsmen. This standardised manufacture has still to be properly assessed by historians of scientific instruments, for the part which it played in the massive advances of science and technology from the 1850s onwards was very great indeed.   Not only did it affect astronomy, but the new science-based technologies of photography and electricity.   While it is in no way my argument that Lassell was especially conscious of this wider historical process when he devised his steam-driven polishing machine, I believe that it was symptomatic of the direction in which scientific instrumentation was moving in the 1840s.

Important as William Lassell's work on the optics of large reflecting telescopes was, I believe that he was at least as, if not more, significant in another area: the mechanics of telescope design.   Herschel, Ramage and Rosse had been, to modern eyes, remarkably unconcerned with how their mirrors were mounted, and seem to have been happy to place the arduous fruits of their optical labour in contraptions of wood, rope and stone with little appreciation of how such mounts impaired mirror performance.   None of these big mirror users had found it necessary to devise equatorial mounts, which to a modern observer is tantamount to depriving their specula of any opportunity for sustained critical use.

This was a problem which Lassell recognised and tackled at source, and considering his lack of formal training in mechanics, his results display an astonishing perspicacity.   He recognised what the problems of mounting big equatorial reflectors were, how they could be overcome given the resources of the age, and went on to produce at Starfield the ancestor of all subsequent big reflector mounts.   It is hardly surprising that the instrument aroused the interest that it did, and immediately earmarked its builder and owner as a man of importance and promise in professional astronomical circles.[88]

In his description of the 9-inch instrument, which was published in 1842, Lassell drew particular attention to two traditional problems of reflecting telescope design which the Starfield instrument overcame.   The first was the problem of mounting relatively large specula so that when the tube was directed towards the zenith, the unsupported central part of the mirror did not sag and destroy the evenness of the field.   Lassell tackled this problem by devising a lever and plate compensator which backed the mirror, applying no pressure when the tube was in the horizontal, but giving progressive central support as it approached the zenith.   He informed the R.A.S. 'I have seen no mirror of this size which, when supported by antagonist screws placed behind those at its face, does not bend down in its middle at great altitudes, and produce distorted (generally three-cornered) images of stars.'[89]   The new mirror support provided a 'perfect cure', and gave even images of all elevations of the tube. [357]

The equatorial mount upon which the telescope was placed formed his second concern, and aimed to achieve rigidity of support combined with maximum delicacy of adjustment, neither of which would be attainable without

Fig. 1. William Lassell. Hitherto unpublished photograph in Bolton Public Library, Local History section, Bl, 405. Lassell was probably in his seventies when it was taken, and judging from details of hair and dress, may have been taken at the same photographer's studio session as the 'Presidential Portrait' in the R.A.S. Lassell was President 1870–1872.

Mem. Roy. Astron. Soc. Vol. XII Plate V.

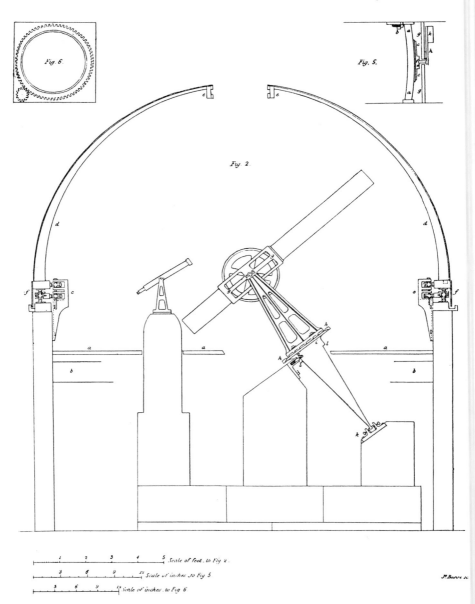

Fig. 2. Section of Lassell's 9-inch telescope and its dome at Starfield. Note the conical polar axis, carrying a fork mount. Figure 5 (top right) shows the anti-sag mirror compensating levers. Lassell left no pictures of his 24-inch, simply saying that its design was identical to that of the 9 inch. *Memoirs R.A.S.* xii (1842) Plate V.

Mem. Roy. Astron. Soc. Vol. XVIII. Pl. II.

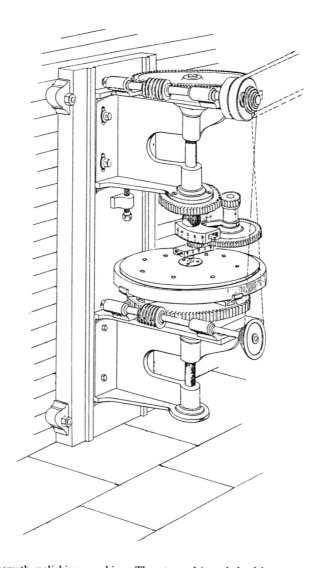

Fig. 3. Lasell–Nasmyth polishing machine. The steam-driven belt drive was capable of turning both blank speculum and tool through a series of adjustable convolutions, to reliably produce 24-inch diameter mirrors (or less) with precise optical surfaces. The machine was rebuilt to a scaled-up specification to produce the 48-inch mirrors for the Malta Equatorial. *Memoirs R.A.S.* xviii (1849) Plate II.

MR. LASSELL'S NEWTONIAN REFLECTING TELESCOPE

EQUATORIALLY MOUNTED IN SANDFIELD PARK, NEAR LIVERPOOL.

APERTURE 4 FEET. ———— FOCUS 36 FEET 7 INCHES.

JANUARY, 1860.

Fig. 4. The 48-inch Malta Equatorial. The circular platform and adjustable observing tower were movable around the fork mount, to cover any elevation or part of the sky. The telescope mount was so well balanced that it could be easily moved manually, and instead of a clock drive, a manually operated rachet and lever operated by an assistant enabled easy tracking. Several versions of this picture exist, such as in *Phil. Trans.* (1867). The present print is reproduced from a single unbound sheet in Lassell Mss. 13. It claims to depict the telescope as assembled at Sandfield Park, Liverpool, in January 1860 prior to export to Malta.

perfect balance.[90] The Starfield telescope was secured to its polar axis by a large, accurately machined declination bearing, while the axis itself terminated in a five-foot-long cast-iron cone resting in a ball joint. Motion in Right Ascension was governed by a set of friction wheels which supported the polar axis and made the whole responsive to the slightest deliberate pressure, while at the same time rock-solid when not touched.[91] As Lassell had not yet made the acquaintance of James Nasmyth when the Starfield observatory was set up in 1839, and his paper renders no acknowledgements for technical assistance (as he later did to Nasmyth), we may feel confident that we are seeing Lassell's own creative talents in action in the 1842 paper. Though he was much indebted to Nasmyth for the later development of his polishing machine, his telescopes, castings and mounts were in most cases the products of his own ingenuity, and not infrequently his own workshops.[92]

Though a 9-inch mirror would seem paltry for a major instrument today, and was only modest in 1839, it was Lassell's radical and all-embracing approach to mirror making, supporting and mounting which made his telescope so significant at the time. The basic precepts which he had established at Starfield by 1839 were to be carried over, after being scaled up, for his 24-inch in 1844-5 and 48-inch a dozen years later. Lassell used the same basic fork-type mount for the 48-inch mirror of 37 feet focal length at Malta in the 1860s that he had for the Starfield 9-inch, and found that it performed as well as its smaller ancestor.[93]

If Lassell does appear to have been a man who was, in many ways, more interested in the engineering of telescope design than he was in prosecuting a programme of research fitting to such superlative tools, part of the answer may well have derived from the rapidly deteriorating climatic conditions of south-west Lancashire. With the expansion of Liverpool, complemented by the rise of the soda and chemical industries in nearby St Helens and Widnes, there was often nothing to see. Reference has already been made to Dawes's joke about 'Cloudfield', while Lassell reminded Struve in 1847 that 'the climate of this district is singularly unfavourable to astronomical observation'.[94]

On the other hand, we must remember that Lassell was a sufficiently dedicated astronomical observer to make four changes of location after 1850, in search of clearer skies than those available at Starfield. Two of these were permanent changes of home and observatory within England, and two required extended sojourns to the Mediterranean, while all involved extremely costly moves of major capital equipment.

The first of these was in 1852-3, when he took the 24-inch Newtonian to Malta.[95] By January 1854, he was back at Starfield, though soon afterwards, in 1855, he moved to 'Bradstones', his second observatory-residence, which was a few miles further out from central Liverpool than Starfield. But any hope of making serious use of the 48-inch Newtonian, upon which he started work in 1857, in industrial Lancashire was impossible, and made it necessary to select a better climate in which to mount the finished instrument.[96] Because Malta was easily accessible by sea and had a strong British presence as a naval base, he shipped out the 48-inch in 1861.[97] He had also formed a most favourable opinion of its climate on his previous visit and was to stay until 1865, doing important planetary work.

The 48-inch was a truly impressive piece of engineering, like all of Lassell's telescopes. Its tracking machinery was manually operated by an assistant using a geared mechanism which dispensed with the need for a heavy clock drive, while the observer worked from an adjustable sentry box over thirty feet high to give access to the eyepiece at all elevations and protect him [358] from the weather.[98] The instrument stood in the open with no dome or covering, save for the optics. Lassell pioneered the use of a skeleton iron tube as a way of preventing hot air currents building up above the speculum[99] and had travelled to Malta taking his polishing machine and workshop, so that he could experiment with different polishing techniques on location, to obtain the optimum optical surface suitable to local seeing conditions.[100]

When he came back from Malta, however, he did not return permanently to Lancashire but set up the 24-inch on his newly purchased estate 'Ray Lodge' near Maidenhead, Berkshire.[101] It is not indeed without irony that the one place in the world where all the technical skills existed to build those giant telescopes in the mid-nineteenth century was perhaps the worst place in which to try to use them. The great 48-inch was never re-erected properly after its return to England, and it is tragic to think that this superb instrument — the finest and most versatile telescope of its day — could not even be given away, and was consigned to the breaker's hammer during its maker's lifetime.[102]

Though Lassell showed no signs of being interested in the speculative aspects of cosmology, he did share with Sir William Herschel a major observing interest in the physical astronomy of the planets and their satellites. Throughout the eighteen years of notebooks which preceded the building of Starfield, William Lassell had been a regular planetary observer. Yet unlike his friend Nasmyth, he showed no particular interest in the moon, and unlike Dawes, exhibited no special concern with double star work.[103]

But Lassell was a regular cometary observer, and used his excellent instruments in attempts to resolve their head and tail structures. On 20 September 1835, while still living at 18, Norton Street, in central Liverpool, he secured his first glimpse of Halley's Comet at 11.15 in the evening, after keeping vigil for it in the region of 48 Aurigae, where its appearance was expected. Because of its awkward location at the first appearance, however, he was able to bring no major instrument to bear on it, contenting himself with a view through his '2 ft. pocket Dollond telescope with neg.^{e} eyetube magnifying 21 times ... and that too out of the skylight of my workshop'. The comet had no tail and was still invisible to the naked eye.[104]

In April 1848, he was sending observations of Mauvais' Comet to Le Verrier in Paris, though his attempts to regularly observe this object, like most other things at Starfield by the late 1840s, were frustrated by poor seeing conditions.[105] But in the autumn of 1858, working with the 24-inch from his new Bradstones observatory, outside Liverpool, he secured some spectacular views of Donati's Comet. On 3, 4, 5 and 8 October, he traced evolving structures in the Comet's nucleus, including a clearly defined stratification of luminous contours, and the appearance of a curious black spot. True to form, as an observer and not a theoriser, he advanced no notions as to what these clearly delineated phenomena might be in his published communication to the Royal Astronomical Society.[106]

Lassell claimed that his first visit to Malta in 1852 had been for the purpose of 'observing the larger and more distant planets' with the 24-inch telescope.[107] Between October 1852 and February 1853 he made a series of studies of Saturn (his favourite planet), examining the rings, surface structures and satellites. As this must have been one of the earliest occasions on which anyone had observed the planet under ideal atmospheric conditions at such a southerly latitude — 36° N — and with such an excellent instrument, it is hardly surprising that new discoveries were made. Although Professor Bond (who made the co-discovery of Saturn's satellite Hyperion with Lassell in 1848) was working with an excellent 15-inch Merz refractor at this time, the latitude of his Harvard Observatory was 42°22′ N, and the air less tranquil than that of Malta. [359]

Working with powers around 565 and 650, Lassell made several prolonged surveys of the planet's belts and other surface details, although it was the rings which really interested him. Saturn's rings had given Lassell concern since December 1850 when observing the planet with Dawes's 6⅓-inch refractor (now in his new observatory in Kent), he noticed a thin 'crape veil' which seemed to extend from the inner ring towards the body of the planet. At the same time, Bond at Harvard was observing the well-placed Saturn, and had announced the existence of a third ring.[108] In Malta, however, on 16 October 1852, Lassell was able to provide firm substantiation for the 'crape ring', and observe that the body of the planet could be discerned through it.[109]

Lassell's Malta observations of Saturn, with their meticulous presentation of what he saw, combined with a strict reluctance to pay cognizance to what he was not certain that he had seen, provide clear substantiation for his status as a significant observer of nature. Always aware of the failings as well as the strengths of his instruments, and how purely local atmospheric fluctuations could influence the minute changes of colour, shading or location of objects on the planet's surface or rings, he comes over to us as careful, critical and sceptical. He was a highly skilled observer, with a great deal of practical experience in using the instruments which he had created. One also senses that he greatly enjoyed observation, and revelled in the clear air of Malta as soon as he was able to escape the polluted skies of Liverpool, for he wrote to De la Rue in November 1852 'The nights are as remarkable for their tranquility as transparency; and I have not encountered one of those nights so frequent at Starfield, on which I opened the Observatory to close it in disgust ... *Saturn* was exhibited in a style that beggars description: power 650.'[110]

One species of investigation for which the high resolving power and definition of the 24-inch Newtonian made it ideal was the identification of very dim objects. The crape ring of Saturn provided a good example of this power, but even better was his search for planetary satellites. Even working from Starfield with the 24-inch, he had been able to discover Saturn's satellite Hyperion on 18 September 1848,[111] along with two new moons of Uranus on 24 October 1851. He obtained confirming observations for the Uranian satellites on subsequent nights, and reported their discovery to the R.A.S. and on 17 November to the Liverpool Lit. and Phil.[112] These were the first Uranian satellites to be discovered since those observed by Herschel in 1787.

He also devoted considerable energy to the newly discovered Neptune in 1846-8. It is interesting to wonder, in hindsight, what the outcome would have been had J.C. Adams gone to Lassell with his co-ordinates for 'Planet X' in 1845, rather than to the Astronomer Royal. As I have discussed at length elsewhere, one of the reasons why Airy gave such low priority to the search for the new planet came from his conception of the public role of the Royal Observatory.[113] As a state-funded institution, it was the job of Greenwich to make 'standard' observations of known objects, and not waste its time on speculative investigations. This had been the reason why Airy had re-directed the search for Neptune back to the Cambridge Observatory which, as a private academic foundation, was more suited to investigative astronomy.

Starfield, on the other hand, possessed every resource which would have made it an ideal place for the search. It contained one of the biggest telescopes of the day, in private hands, unfettered by exhausting research programmes or official regulations, and owned by a leisured private gentleman who was an enthusiastic planetary observer. The new 24-inch, moreover, would have been powerful enough to reveal an immediate disk, making it unnecessary to have wasted weeks [360] of valuable time obtaining painstaking micrometric measures to see which star in the field moved, as Challis was persuaded to do at Cambridge.

The irony of the undisputed capacity to have found the planet had he been approached, was amply demonstrated only eleven days after Neptune's first sighting in England, when Lassell found it from its now publicly announced position on 10 October, and immediately discovered that it possessed a satellite.[114] For much of the eighteen or so months after the announcement of the discovery of Neptune, Lassell observed it whenever possible. By 26 November 1847, he was able to send to Struve a list of observations of the newly discovered inner satellite — later named Triton — made between its discovery and 9 November 1847, though complaining as always that the weather 'foiled' all attempts at regular observation.[115] The existence of the satellite had received confirmation by the autumn of 1847, when Lassell received a series of observations sent to him by Bond from Harvard.[116]

As soon as he began trying to obtain accurate measurements of the Neptunian satellite, so that its elements could be computed, he realised that he did not have a micrometer of sufficient delicacy to observe it. By February 1848, however, he had purchased an excellent prismatic micrometer from Merz of Munich, and as he wrote to Augustus de Morgan, it was invaluable for satellite work, for while 'no advantage in definition is gained by the prism, ... there is an obvious saving of *light*, which becomes especially valuable in searching for satellites'.[117]

In October 1846, only a few days after the new planet's discovery, William Lassell recorded seeing what he thought may have been a ring round Neptune.[118] Challis, Dawes and Hind, working with refractors, also reported glimpses of it.[119] At the same time, however, Neptune was very low in the sky, and Lassell, always wary of what he was looking at, seems to have entertained suspicions regarding its real existence. On 29 September 1847, for instance, he compared the image of the planet with 'Uranus at a very low altitude in order to verify the suspicion of a ring of Neptune — I fancy the planet a little oval in the same direction as Neptune appears'.[120]

It is possible that this 'ring' had led Lassell to entertain doubts about the perfection of the image produced by the 24-inch, for in *Monthly Notices* XI (1851) pp. 165-6, he described the way in which he had re-designed the mirror cell and support system of the telescope by elaborating the mechanical principles originally used in the 9-inch. In the new mirror cell, some twenty-eight supporting levers, each carrying fifteen pounds of mirror area, acted upon eighteen pressure disks to give a more balanced support. Yet even this did not reliably succeed in ridding Neptune of his ring, for Lassell mentions seeing it again in the southerly latitude of Malta. This ring, which has no existence in nature, was almost certainly the product of optical distortions in the mirror figure combined with the atmospheric problems of viewing dim objects low down in the sky. No reference to it occurs after 1852.[121]

Reference has already been made to Lassell's relative lack of interest in the study of deep-sky objects outside the solar system. Yet he was to produce a major R.A.S. *Memoir* on the Orion Nebula in 1854, based on observations made at Malta with the 24-inch. Lassell stated that his intention was to see if he could resolve the Nebula using powers sometimes over 1,000, and when he failed to do so, went on to describe the structure of the object as resembling several layers of pea-green tinged cotton wool.[122] Nowhere in these carefully documented observations, though, does one find any attempt to *infer* or *interpret* the structure of the Nebula, as Lassell confined himself to reporting, step by step, what he had seen on particular nights. One never ceases to [361] admire the controls which Lassell placed upon his work, and how he saw his task as faithfully describing astronomical objects and nothing more.

Much of Lassell's historical significance stems from the type of astronomer that he was, how his interests developed, and how the resources to pursue them were obtained. He forms a marked contrast with such contemporaries as G.B. Airy and Challis, one of whom operated from a tightly budgeted government institution, the other from a well-endowed academic one, yet both forced to work within the constraints of management committees and official decrees. Both of these astronomers were required, moreover, to perform officially designated meridian astronomy, aimed at providing useful information to sailors and mathematicians. Even such a figure as J.R. Hind, who had been the first man in England to see Neptune on 30 September 1846, was an employed astronomer in the South Villa Observatory of Mr Bishop, to whose establishment he had graduated from a Royal Observatory Assistantship.[123]

Of course there were astronomers whose scientific time was entirely their own, though these men generally enjoyed inherited family wealth like Admiral Smyth, or William Huggins, whose astrophysical researches after 1855 were made possible by the private observatory settled upon him by his brewery-owning father.[124] Sir James South, on the other hand, abandoned a surgical practice upon marrying an heiress in 1826, and proceeded to use her fortune to equip his major private observatory.[125] Others, like Dawes, combined modest family means with the pursuit of a profession. All of them, however, preferred to purchase the best instruments that Merz or Tulley could supply, and made no attempt to design or manufacture their own.

This, I believe, is how Lassell forms such a contrast, for the private commercial fortune which he started to acquire in the first forty years of his life gave him the leisure and ample resources to flourish as an instrument designer and front-rank researcher during the second forty. Though one might think of him as forming a parallel to his big-telescope contemporary, Lord Rosse, the parallel only really becomes valid for the second half of Lassell's life, when he had already turned himself into a rich man. A closer parallel could perhaps be made between Lassell and Hevelius in the seventeenth century, whose distinguished astronomical career and observatory, built up during the second half of his long life, were made possible by a fortune made in the Dantzig brewing trade.[126] Even the excellent instruments with which the forty-four-year-old John Flamsteed privately equipped the Royal Observatory after 1690, after struggling for fifteen years on a paltry salary and no official grants, were partly paid for out of the estate of his recently deceased father, a brewer and 'rich merchant' of Derby.[127]

Yet where Lassell differed most conspicuously from both Hevelius and Flamsteed was in his ability not only to *design* specific instruments, as they did, but to *fabricate* them as well. We must bear in mind that while Nasmyth built Lassell's polishing machine, most of the construction work on Lassell's telescopes took place in his own workshops, under his immediate direction and often with his own hands, as he made clear on several occasions.[128] I am aware of no other major scientific figure for whom the mechanical, conceptual and financial aspects of research were so much their own personal creation.

When J.L.E. Dreyer catalogued the Lassell manuscripts in the R.A.S. Library in 1921, he prefaced his index to the papers by stating 'There can be little doubt that anything of any value [in them] has been printed.'[129] While it is true that as soon as William Lassell had made his entry into academic [362] astronomy in 1839, he became a most prolific publisher of his results, his manuscripts still contain far more than Dreyer gave them credit for. It is true that over thirty years of publications, from the 1840s to 1870s, enable us to chart quite accurately his contributions as a front-rank astronomer, but they tell us little about the formative years before 1839, his personal associates, or the commercial wherewithal which made the astronomy possible.

In an age when no official, and very little academic, money was made available for non-utilitarian scientific research, Lassell formed a rare example of intellectual *laissez faire* by combining the unique talents necessary to command, direct and employ his own time and resources to produce a crop of remarkable instruments and discoveries. Perhaps because such commercial and scientific abilities are, as his obituarist emphasised, so rare in the same person, one might validly describe Lassell as practitioner, patron and 'Grand Amateur' of Victorian astronomy.

Acknowledgements

I have been assisted by many persons during the research and writing of this article, and wish to thank Peter Hingley and Mary Chibnall of the Royal Astronomical Society Library for their unfailing help with the Lassell documents. I also thank the other archivists and librarians who have provided material and granted access to sources. My greatest debt of gratitude, however, goes to my friends in the Liverpool Astronomical Society, who invited me to be the first 'William Lassell Memorial Lecturer' in 1988. Without the ready assistance of Gerard Gilligan of that Society, his invaluable knowledge of Lassell sources in local archives, and his unfailing willingness to look up and check things for me, my job would have been vastly more difficult, and I wish to give him my special thanks.

Note on Sources

The surviving Lassell astronomical manuscripts are deposited in the Royal Astronomical Society Library. Lassell left his papers in an ordered condition, though they were properly catalogued by J.L.E. Dreyer in 1921 into nineteen main divisions. Some of these divisions consist of single volume items, while others consist of over a dozen separate notebooks in one division. These multiple division items are cited with the main division number, followed by the item number, e.g. Lassell Mss. 11:14. Lassell sometimes numbered the pages of his notebooks, and sometimes did not. As they follow a clear chronological sequence, however, unpaginated items can easily be identified from their date sequence.

Notes and References

1. Several studies have been made of Herschel's instrumentation: see Angus Armitage, *William Herschel* (Nelson, London, 1962). Also James Bennett, '"On the power of penetrating into space", the telescopes of William Herschel', *Journal for the History of Astronomy* [*JHA*] vii (1976) 75-108; Allan Chapman, 'William Herschel and the measurement of space', *Quarterly Journal of the Royal Astronomical Society* [*QJRAS*] 34 (Nov. 1989).
2. Lassell's main obituaries appeared in *The Observatory* 43, 1 Nov. (1880) 587-90 by Margaret Huggins; and in *Monthly Notices of the Royal Astronomical Society* [*MNRAS*] xl-xli (1880) 188-91 by William Huggins. Detailed obituaries appeared in several newspapers, especially in the Liverpool area. See *The Daily News*, 7 Oct. 1880; *Daily Post*, 9 Oct. 1880; *Bolton Chronicle*, 8 Oct. 1880, etc. The newspaper articles seem to have borrowed much of their technical content from Sir John Herschel's Presidential 'Address' to Lassell on the occasion of his being awarded the Gold Medal of the R.A.S. in 1849, and which reviewed his career to date; see *Memoirs of the Royal Astronomical Society* [*MRAS*] xvii (1849) 192-200.
3. The *Test Acts* in relation to the Universities were not wholly repealed until 1871. From my work on the Assistants employed by Airy at Greenwich, several were specified as non-Anglicans. These included the extensive Breen family who were Roman Catholics, and who between them worked at Greenwich and Cambridge Observatories. The Greenwich Assistant William Rogerson was specified as a Methodist by Airy to Charles Wood, 15 July 1837, RGO 6, 2/213. Several dissenters also worked as Assistants at the Cambridge Observatory. [363]

4.   This appreciation comes across clearly in Herschel's Presidential 'Address': *MRAS* 1849, loc. cit.

5.   Bolton Public Libraries, Reference Section, see 'Nonconformist Registers. Duke's Alley Chapel 1785-1818', No. 1482.

6.   No contemporary reference; information included as part of a Lassell-Gregson family history published by J.J. Slater-Gregson in the *Bolton Journal*, 13 March 1914: Bolton Public Library file B1 426.

7.   William Lassell, wine merchant, 8, Hope St., Liverpool. See Gore's (later Kelly's) *Directory of Liverpool* for 1821 and 1823.

8.   The 1814 mercantile apprenticeship is stated in William Huggins's obituary article on Lassell, *MNRAS* 1880, 188. It was repeated in some later secondary sources, such as F.W. Peaples' article 'The great Bolton astronomer', *Bolton Journal*, 6 March 1914: Bolton Public Library file B1 424.

9.   Even before the dawn of the railway-building age, in the early Industrial Revolution, around 1806, 'Lancashire itself was one of the most rapidly expanding areas of consumption [of beer] in the land': Peter Mathias, *The Brewing Industry in England 1700-1830* (C.U.P. 1959) 182.

10.   Lassell's Will was signed on 30 July 1875. On 24 August 1878, however, he added a codicil which reduced his bequests as a result of '… Having suffered various heavy losses in my property since this my last Will was written'. These were, no doubt, the reasons why his estate which had originally been assessed at £80-90,000 was finally settled at £70-80,000. The Will was proved in Oxford (as he was then residing in Maidenhead) on 2 November 1880. I have not been able to discover the precise cause of Lassell's losses, but they were most likely incurred as part of the wider financial depression of the late 1870s, which economic historians have styled the 'end of Victorian prosperity', which fell to its lowest point in 1877-8. This crash is dealt with in most standard economic histories of the period: see Pauline Gregg, *A Social and Economic History of Britain, 1760-1965* (Harrap, London, 1965) 367-80; graph 369.

11.   Royal Astronomical Society Mss. Lassell 1:4 34r. On 1 October 1829, when his business was beginning to flourish, he recorded the purchase of a Dollond pocket sextant for £12, and using it a few days later with Dawes at Ormskirk: Lassell Mss. 10:1.

12.   Lassell Mss. 9:1 5v., 15 Sept. 1828.

13.   Gore's *Directory of Liverpool* gives some information on subsequent trading partnerships. Particularly useful is John Barge, *A Gazeteer of Liverpool Breweries* (Neil Richardson, Manchester, 1987) 16, 42.

14.   *Slater's Directory of North and Mid Wales* (Manchester 1895) 218, under 'Caergwrle', and 220. In the 1870s, Lassell and Sharman were trading from the Welsh town of Caergwrle. The business was being managed by Septimus Sharman.

15.   Marriage Certificate of Nathaniel Pearce Sharman to Maria Lassell at West Derby and Toxteth Park, Trinity Chapel 'according to the Rites and Ceremonies of the Independents', 9 August 1860. It was witnessed by William, Jane and Caroline Lassell, and Francis Sharman.

16.   Lassell's observations were recorded in several dozen home-made books, for almost all of the commercially-produced ones date from his early years, when they may

have been given as presents. The making of such notebooks seems to have been standard office practice in the Victorian era and later, for when my own father became a junior clerk in a Manchester mill office in 1921, he was taught to make them by his supervisor.

17. Lassell Mss. 1:4. The volume is signed on the verso (originally recto) end-board 'Wm. Lassell, Bolton, March 18th 1814'. The sermon notes cover a total of 36 pages from 2 January 1814 to 26 April 1815. One presumes that he had started to keep the notes on scrap paper until presented with the handsomely bound volume in March. The volume consists of a flyleaf watermarked 1814, though most pages are watermarked 'Allee 1810'.

18. I have no manuscript source for the Rochdale Academy. The reference comes from William Huggins's obituary in *The Observatory*, op. cit. [n. 2] 188. Both William and Margaret Huggins knew Lassell well during his latter years, and presumably had learned this fact by word of mouth.

19. Items in this paragraph from Lassell Mss. 6, 'Astronomical Observations 1821-1847'. Begin at verso end, marked 'C' on cover for early notes. The reference to Romilly's Life comes from Lassell Mss. 1:1.

20. Lassell Mss. 5: twelve pages at start of volume. [364]

21. Marriage Certificate of William Lassell to Maria King at St. Michael's Church, Upper Pitt St., Toxteth, Liverpool, 8 May 1827. Since he was a dissenter, the ceremony was performed 'by Licence'. It was witnessed by Irene Pearce, George Dover, Alfred, Jos and Jos (junior) King, and another King of unclear initial. Curiously, no Lassells acted as witnesses. One suspects that Irene Pearce may have been a relative, as indicated by Lassell's future son-in-law's name, Nathaniel Pearce Sharman.

22. *An Index to the Wills and Administrations ... now preserved in the Probate Registry at Chester, 1811-1820*, ed. William Asheton Tonge, Lancs. and Cheshire Record Society (1928); see p. 300 for Wm. Lassell 'gentleman' of Toxteth Park, 1816. For Wm. Lassell the Toxteth clockmaker, see Lawrence Hall, 'The ancient chapel of Toxteth Park and Toxteth School', *Trans. Lancs. and Cheshire Historic Society* for 1935 (Liverpool 1936) 52-3.

23. Lassell Mss. 9:2; date order.

24. Alfred (1797-1867) and Joseph King came from a commercially and mathematically inclined family. Alfred became an engineer of some eminence in the new gas industry, and was said to have been the inventor of the first commercially viable gas stove: see *The Liverpool Group Gas News* 1, 2 (1948) 38-9. Also 'Funeral Records of Liverpool Celebrities' for Alfred King: Liverpool Record Office.

25. Obituary, *The Daily Post*, Sat. 9 Oct. 1880. Cutting in Liverpool Record Office, V.1. 1879-1880.

26. He had come to use the style 'gentleman' by the time of his daughter's marriage certificate in 1860, and in his Will, op. cit., refs. 10 and 15. He nonetheless still appears as 'Brewer' in the 1851 and 1861 Census Returns for the West Darby district of Liverpool where he resided.

27. Lassell certainly possessed railway investments long before 1875. In Lassell Mss. 11:9, he wrote optical notes for his 9-inch reflector onto the verso side of an old half-yearly Directors' *Report* for the Leeds and Bradford Railway of 15 August

1851. There are no detailed financial documents amongst his papers in the R.A.S. Library, except those written on paper or in books later re-used for astronomical purposes.

28. By 1840, Liverpool had a thriving Lit. and Phil. Society, an Athenaeum and a Royal Institution to cater for its broarder intellectual needs. In 1837, the Liverpool Medical Institution was founded to provide specialised library and academic facilities for the medical profession: see John A. Shepherd, *A History of the Liverpool Medical Institution* (Liv. Med. Inst., Liverpool, 1979) 43-6. The Liverpool Astronomical Society was founded in 1881.

29. Lassell Mss. 1:1, written on end-board. These positions were probably for Starfield, as he mentions his observatory in the same set of jottings.

30. Lassell Mss. 2, 41-2. He was also making sextant observations at Miss Harrison's cottage in Bootle on 22 March 1838: Lassell Mss. 5, date order.

31. In Gore's *Directory* for 1841 there are four Stannistreets listed, including a watch engraver, an attorney, a cooper and a tin-plate worker.

32. *D.N.B.*, Dawes. Lassell stated the coordinates of Dawes's observatory at Ormskirk as 53°33'59" North and 11 mins 29.4 seconds West: Lassell Mss. 5, 7 June 1837.

33. Lassell Mss. 1:4, 39v, 40, etc.

34. Ibid. 76r. In this undated reference for c. 1829, he mentions that Smyth (with whom he seems to have been in correspondence) had a 6-inch Tulley refractor.

35. Lassell Mss. 9:2, 9 August 1837.

36. Lassell Mss. 9:8, fol. 20v.

37. *James Nasmyth Engineer, an Autobiography*, ed. Samuel Smiles (John Murray, London, 1883) 236-7.

38. Ibid., 312-3. When testing Nasmyth's mirror on 27 July 1849, Lassell 'concluded that it showed stars quite as well as any metal I ever saw made — disks exquisitely round': Lassell Mss. 9:11, date order.

39. Lassell, 'Description of a machine for polishing specula, & Co.', *MRAS* xviii (1849) 5. The paper was read on 8 December 1848.

40. Lassell, 'Description' [n. 39] 4, 5.

41. Smiles, *Nasmyth* [n. 37] 396. [365]

42. If Lassell kept copies of his letters on a regular basis, they have not survived, except those recorded in Lassell Mss. 8:2, which contains a dozen or so letters written between 1844 and 1855. Some of his letters from Malta were published in *MNRAS*, while originals are to be found in the papers of Dawes and other astronomers with whom he corresponded. Jottings and draft letters occur in several of his observing and technical journal books.

43. Lassell's name occurs sporadically in G.B. Airy's 'Astronomer Royal's Journal', in RGO 6 24-25, mainly in the 1850s and 1860s when he was observing in Malta.

44. Lassell to Prof. Struve, 7 Sept. 1850, copy in Lassell Mss. 8:2, fol. 39.

45. Airy mentions this visit to Lassell in his *Autobiography*, ed. Wilfred Airy (C.U.P. 1896) 196, and writes of inspecting the equatorial and clock drive along with the grinding apparatus. He fails, however, to mention the shirt. Lassell's planetary work is briefly mentioned with relation to Struve in Alan H. Batten's *Resolute and Undertaking Characters: the Lives of Wilhelm and Otto Struve* (Reidel, Dordrecht, 1988) 139.

46. Mentioned in an undated letter to Rosse, but written between December 1844 and April 1845, in Lassell Mss. 8:2, fol. 24. Lassell mentions visiting Rosse and examining his telescopes and apparatus in his 'Description of a machine' [n. 39] 1.

47. He was admitted to the Liverpool Lit. and Phil. Society as Member 254 on 4 January 1839, as William Lassell junior, Brewer of Starfield, West Derby Road: 'Roll of the Lit. and Phil. Society of Liverpool', Liverpool Record Office, 050 Lit. 3/1.

48. The above Lit. and Phil. Society 'Roll' also states that he gave his first paper on 'The construction and accuracy of astronomical instruments for time measuring' on 27 January 1840. Lassell's son, also called William Lassell junior, was admitted to the Society as Member 481 on 29 November 1852. He was also described as a Brewer residing at Starfield. He contributed no papers.

49. Lassell's obituaries catalogue his various awards and honours. See also the Herschel 'Address', *MNRAS* 1849 [n. 2]. Lassell was also admitted to the Honorary Fellowships of the Royal Societies of Edinburgh and Uppsala, and received an Honorary LL.D. degree from Cambridge.

50. Lassell to Struve, 25 Feb. 1848: Lassell Mss. 8:2, fols. 34-5.

51. The exact date of the building of the Starfield Observatory, as described in a 'Description of an Observatory erected at Starfield near Liverpool', *MRAS* xii (1842) 265-72, is not easy to ascertain. He read the paper on 7 April 1841, though he had composed it on 5 December 1840. He further mentions that it had been erected 'During the summer of last year' (265), which could be construed as indicating the summer of 1839. Herschel in his 'Address' [n. 2] gives it as 1840. Yet Starfield was his residence and was being used to make observations with a transit telescope as early as 4 October 1838, for on that day he recorded observations headed 'At Starfield' in his log: Lassell Mss. 9:4. As the 9-inch telescope was probably begun in 1833 (Lassell Mss. 3, fols. 87-88) and was being tested by Dawes at Ormskirk in the summer of 1837 (Lassell Mss. 9:2, 7 Aug. 1837) and used by Lassell in Liverpool in September 1837 (Lassell Mss. 3, fol. 130), it is likely that it was the mount, dome and fittings, rather than the optics, which had made the instrument so noteworthy by 1840.

52. This passage is cited in most of the obituaries and subsequent biographical sketches of Lassell — invariably un-referenced. It comes from the opening of Herschel's 1849 Presidential 'Address' [n. 2] 192.

53. By 1770, Herschel was making £400 p.a. from his musical 'receipts': *The Scientific Papers of Sir William Herschel* 1 (R.S. & R.A.S., London, 1912) xxi. His income also continued to rise. His pension from George III was only £200 p.a. with £50 for Caroline, while the Astronomer Royal's salary, with all its duties, was only £300, ibid., xxxvii.

54. John Herschel refers to Lassell as an 'Amateur astronomer' in the noblest terms; as a man who works purely for the 'enjoyment he receives in its pursuit', and whose love of intellectual recreation represents a 'higher phase of civilisation' than time spent in more worldly pastimes: 'Address' [n. 2] 198-9. The term 'amateur', one must remember, conferred a mark of respect and gentlemanly independence in 1849, and there would have been nothing at all patronising in Herschel's remark.

55. Professional musicians, even when outstandingly successful, did not always enjoy

positions of particular respect in eighteenth-century England, and this term of abuse was coined by the Oxford antiquary Thomas Hearne against no less a figure than Handel: V.H.H. Green, *A History of Oxford University* (Batsford, London, 1974) 114.   Lord Chesterfield and many others concurred upon the low status of musicians.

56. Obituary, *The Daily Post*, 9 Oct. 1880: cutting in Liverpool Record Office V. 1 1879-80.   Queen Victoria visited Croxteth over 8-9 October 1851, as part of a tour of Lancashire. [366]

57. The Starfield Observatory, described in *MRAS* 1842 [n. 51] contained a small transit as shown in Figure 2, ibid.   Several pages of transit observations occur in his observing books, e.g. Lassell Mss. 9:4, 4 Oct. 1838.

58. Lassell Mss. 1:4, fol. 53v (1829); Mss. 6, fol. 10v (1822), etc.

59. Herschel, in his 1849 'Address' [n. 2] 200, does commend, as future research objects for the 24-inch, the measurement of the closest double stars, and the examination of nebulae.

60. Lassell, 'A catalogue of new nebulae discovered with the four-foot Equatoreal in 1863-65', *MRAS* xxxvi (1867) 53.   Read 9 November 1866.

61. This was John Edwards's 'Direction for making the best composition for metals of reflecting telescopes ...', *Nautical Almanac* (1787) 1-48 and 49-60 respectively. The articles are copied out in Lassell Mss. 2, see fols. 21-22 for date.   In all probability, however, he was experimenting with mirror compositions prior to copying out this article, as indicated in a note of 23 May 1821 in Lassell Mss. 6: see beginning, headed 'Minute Book W.L. 1821', date order.

62. Lassell Mss. 3.   It states that the first 86 pages of this notebook were written before 1833, and the remainder from 3 Sept. 1833 to 8 April 1848.   It is his most detailed notebook on mirror making.

63. Ibid. fol. 1.

64. Ibid. fol. 5.

65. Ibid. fol. 25.

66. Ibid. fol. 27.

67. Lassell, 'Description of a machine ...', *MRAS* 1849 [n. 39] 3.

68. Lassell's abandonment of arsenic in the composition of the 48-inch speculum does not to my knowledge appear in his own published accounts of the instrument.   It was pointed out, however, by Margaret Huggins in her Lassell obituary in *The Observatory* 43 (Nov. 1880) 589.

69. Lord Oxmanton (Lord Rosse), 'An Account of experiments on the reflecting telescope', *Philosophical Transactions* 1840, 503-27.   See also Lassell's 'Description of a machine ...' [n. 39] 2.

70. Lassell, 'Description' [n. 39] 11.

71. A wooden polisher seems to have been placed beneath this sheet, for it formed the only feature of the machine's design which John Herschel believed could be improved because of its tendency to warp: Herschel, 'Address' [n. 2].

72. Lassell Mss. 3, fols. 87-88.

73. Lassell Mss. 3, fols. 79-81.   The damage which could be done by over-polishing is discussed by J. Norman Lockyer, *Stargazing: Past and Present* (Macmillan, London, 1878) 128.

74. John Mudge, 'Directions for making the best composition for the metals of reflecting telescopes', *Phil. Trans.* 67 (1777) 296-349, 336-8. Mudge's mirror tests, which were the ancestors of modern testing, consisted of looking at distant objects under a high power through diaphragms which exposed selected parts of the speculum.

75. Lassell Mss. 3, fols. 87-95.

76. Lassell Mss. 3, fol. 130.

77. Lassell Mss. 3, fol. 137.

78. Lassell Mss. 9:2, 7 Aug. 1837.

79. Lassell Mss. 7:1, fol. 98.

80. The use of the diaphragms and zones was based on the method described by Mudge in 1777 [n. 74]. See Lassell, 'Description of an Observatory ...', *MRAS* 1842 [n. 51] 268.

81. Lassell, 'Description of a machine ...' [n. 39] 13-14.

82. Derek Howse, *Greenwich Observatory, III: the Buildings and Instruments* (Taylor and Francis, London, 1975) 117. [367]

83. Lord Oxmanton (Rosse), 'An account ...' [n. 69].

84. Lassell, 'Description of a machine ...' [n. 39] 5. The paper was read before the R.A.S. on 8 Dec. 1848.

85. Ibid. 14.

86. William Simms, 'On self-acting circular dividing engines', *MRAS* xv (1846) 83-90.

87. Allan Chapman, *Dividing the Circle. The Development of Critical Angular Measurement in Astronomy 1500-1850* (2nd edn., Wiley-Praxis, Chichester and New York, 1995) 123-37 [originally cited as Oxford University D.Phil. Thesis, 1978]. Much of this book deals with the 'technical ceilings' placed upon research by instruments, and their resulting stimulus to fresh innovation. See also A. Chapman, 'The accuracy of angular measuring instruments used in astronomy between 1500 and 1850', *JHA* xiv (1983) 133-7.

88. John Herschel seized upon the great importance of Lassell's work in the design of balanced equatorial mounts, and described the 9-inch as representing a new 'epoch' in telescope design: 'Address' [n. 2] 193-4.

89. Lassell, 'Description of an Observatory' [n. 51] 269.

90. Herschel, in 'Address' [n. 2] 194, states that it was Lassell who developed the equatorial mounts of Fraunhofer and Reichenbach to accommodate the more complex requirements of reflectors.

91. Lassell, 'Description of an Observatory' [n. 51] 271.

92. In a letter to Struve, 8 Feb. 1848, Lassell Mss. 8:2, fol. 32, he states that the 24-inch was 'constructed entirely by myself with the exception of the prism [micrometer] which is the work of M. Merz of Munich'. Lassell's direct personal involvement in all aspects of the construction of his instruments is displayed in his 1842 and 1849 papers, not to mention his notebooks. Only his large castings and some very large pieces of industrial ironwork seem to have been contracted out. For instance, the mirror for the 24-inch was cast by Messrs. Preston and Ross as he did not have 'a crucible large enough to contain the requisite quantity of metal': see his 'Description of a machine ...' [n. 39] 2-3. Some castings for the heavy machinery for the 48-inch were done by professional founders; there is an account for

£246 3s. 6d. for various castings in Lassell Mss. 7, fol. 21, while the first 48-inch mirror was also cast commercially at Forester's Foundry in December 1857. It was still warm on 9 January 1858: Lassell Mss. 7, fols. 65-70. In his 'Observations of planets and nebulae at Malta', *MRAS* xxxvi (1867) 1-32; 1, he makes it clear that the 48-inch equatorial telescope was made 'at Bradstones', his observatory residence since 1855. Lassell published three consecutive major papers on his Malta work in the same volume of *MRAS*, xxxvi (1867). All of the papers were described as having been read on 9 Nov. 1866.

93. 'The 20-foot [24-inch] telescope has been mounted so exactly on the plan of my 9-foot [9-inch] equatoreal ...' that a description of one could suffice for the other: 'Description of a machine ...' [n. 39] 15. The 48-inch, on its fork mount, was illustrated as part of its description in 'Observations of planets ...' [n. 92] 1-32.

94. Lassell to Struve, 26 Nov. 1847, Lassell Mss. 8:2, fol. 28. Herschel, 'Address' [n. 2] 15, also pointed out Liverpool's 'bad climate (nothing I understand can be much worse)'. Yet Lassell's notebooks reveal him as a man who enjoyed looking at the heavens for their own sake, and showing its wonders to his family and friends. In Lassell Mss. 9:3, 19 Sept. 1840, for instance, he records how he and his friend the Reverend James Roberts succeeded in observing *all* the then known planets in one day at Starfield, in early morning and evening.

95. Lassell, 'Observations on the Orion Nebula made at Valletta [sic] with the twenty-foot Equatoreal', *MRAS* xxxiii (1854) 53-7, dated 'Starfield, 10th January 1854'. His paper 'Observations of the planet Saturn with the 20-foot Equatoreal at Valletta [sic]', *MRAS* xxii (1854) 151-65, records observations of the planet from 5 October 1852. This paper was signed and dated from Valetta on 25 February 1853 and read to the R.A.S. on 11 March. I have not checked whether Lassell read the paper himself, on a short visit to England, or had it read by a friend.

96. Lassell Mss. 7, fols. 65-68. The 48-inch equatorial had first been set up, probably for test purposes, in Sandfield Park, near Liverpool, where it stood in January 1860. An engraving of the complete instrument mentioning its current location, prior to its despatch to Malta, exists as a loose sheet in Lassell Mss. 13:2. A similar engraving was reproduced in *MRAS* 1867, op. cit. [n. 92].

97. Lassell left Liverpool for Malta on 6 Sept. 1861, travelling overland by train to Marseilles. The passage from Marseilles to Valetta took two days, arriving at the island on 14 September. The telescope was sent on by sea. On 26 September 'arrived the Damascus with 270 [368] packages, telescopic apparatus & household effects'. See 'Journal of Malta Expedition 1861-1864', Lassell Mss. 11:14.

98. Lassell, 'Observations of planets and nebulae ...', *MRAS* 1867 [n. 92] 1-4.

99. Lassell's engineering design abilities, not to mention his meticulous draughtsmanship, are apparent in the detailed plans and drawings, made under his own hand, for the 48-inch: see Lassell Mss. 13:1, 13:2, 13:3. The drawing folio 13:3 contains his original plans for the skeleton iron tube.

100. His first 48-inch mirror 'Speculum A' began rough grinding on his machine in Liverpool on 27 April 1858: Lassell Mss. 7, fol. 70. It is clear, however, that on his two visits to Malta he took his polishing machine and full workshop to re-figure specula on site. In a letter from Valetta, 24 Nov. 1852, in *MNRAS* xiii (1853) 43-4, it is clear that he was experimenting with mirror finishes suitable to local

viewing conditions with the 24-inch, while in an earlier letter to De la Rue of 8 Nov. 1852, he discusses adjustments to his polishing machine to obtain a 'perfect figure', *MNRAS* xiii (1853) 14. In his article 'On polishing the specula of reflecting telescopes', *Phil. Trans.* 1875, 303 he describes the polishing experiments of the 48-inch mirror. The machine was steam powered, ibid., 311. His 'Malta Journal' [n. 97] indicates that he changed the 'A' and 'B' speculae for the 48-inch about every two weeks, or seven times in the autumn of 1864, and re-polished them: Lassell Mss. 11:14, fol. 59. In the same 'Journal', from fol. 29 onwards, he lists the changes and re-figurings of the two mirrors. As Lassell claimed to have a laboratory as well as workshops at the Malta observatory in the 1860s, one presumes that his outfit was comprehensive.

101. Ray Lodge is the only one of Lassell's houses still standing, and is an elegant three-storey brick mansion of the Regency (?) period in its own grounds. I am indebted to Mr. Tim Haynes of the Maidenhead Astronomical Society and Mr. Gerard Gilligan of Liverpool for the information and photographs. No sign of his observatory remains in the grounds.

102. Lassell, letter, 'Ray Lodge August 27th 1877', in *The Observatory* 6 (1877) 178-9. It is unclear why Lassell kept the 48-inch in a dismantled condition for a dozen years after bringing it back from Malta. He had tried to give it away without success some years earlier when he offered it to the Melbourne Observatory. As it was broken up and sold for scrap in 1877, in the depths of the 1870s financial recession, and in which Lassell admitted in the 1878 Codicil to his Will [n. 10] that he had suffered heavy property losses, one suspects that he may have been in need of ready cash.

103. When Lassell received a copy of Nasmyth's book *The Moon* in 1875, he wrote to congratulate the author saying that it stimulated him (Lassell) to 'a deeper interest in the Moon than I ever felt before': Smiles, *James Nasmyth Engineer* [n. 37], 396. There was some suspicion a few years ago, however, that Lassell had presented a map which he had made of the moon to Liverpool Library in 1852, though nothing conclusive was ever shown; see Stephen A. Clarke's letter on 'Lassell's Lunar Work' to the *Journal of the British Astronomical Association* 82, 5 (1972) 382. I have found no substantiation for this gift in any of Lassell's papers, nor any interest in lunar observation.

104. Lassell Mss. 9:2, date order.

105. Lassell to Le Verrier, 6 March 1848: Lassell Mss. 8:2, fol. 36.

106. Lassell, 'Donati's Comet, Remarks by Mr. Lassell', *MRAS* xxx (1862) 58-9.

107. Lassell, 'Observations of the planet Saturn with the 20-foot Equatoreal at Valletta [sic]', *MRAS* xxii (1854) 151-65: 151.

108. *Proceedings of the Liverpool Literary and Philosophical Society* vi (1849-51, see 5th meeting, 16 December 1850) 195-6. Lassell learned of Bond's discovery of the third ring of Saturn, made at the Harvard College Observatory, on 15 November 1850, from the *Times* newspaper over breakfast on 4 December, the morning after having first seen the 'crape veil'. At first, Lassell had not taken the 'veil' to be an independent third ring. This incident of the double discovery was mentioned by the two obituarists in both *MNRAS* 1880, 190 and *The Observatory* 1880, 589 [n. 2]. Robert Ball, in *The Story of the Heavens* (Cassell, London, 1897) 240-1, attributes

the discovery of Saturn's third ring to Bond and Dawes, but not Lassell. Simon Newcomb in *Popular Astronomy* (Macmillan, London, 1898) 356, states that Lassell and Dawes had prior observation of the third ring in December 1850, before Bond's discovery was known in England.

109. Lassell, 'Observations of the planet Saturn ...', *MRAS* 1854 [n. 95] 152-3.

110. Lassell to De la Rue, Valetta, 8 November 1852, *MNRAS* xiii (1853) 14.

111. The discovery of Hyperion was the second which he shared virtually simultaneously with W.C. and G.P. Bond at Harvard. Lassell saw Hyperion on 18 September 1848, and confirmed it on 19: *MNRAS* viii (1848) 195-7. G.P. Bond had seen it at Harvard on 16 September, however, as communicated to the R.A.S. by W.C. Bond in the November [369] *MNRAS* ix (1848) 1. See also *MNRAS* ix, 4 (1849) 68.

112. Lassell saw satellites on 24 October 1851, and obtained subsequent confirmations: Lassell, letter, 3 November 1851, *MNRAS* xi Supplement 9 (1851) 248. The announcement of his discovery to the Liverpool Lit. and Phil. was made at the third, 17 November, meeting, 1851: *Proceedings* [n. 108] 20-1.

113. Allan Chapman, 'Private research and public duty; George Biddell Airy and the search for Neptune', *JHA* xix (1988) 121-39. There was, indeed, a 'legend', published by E. Holden in 1893, that Lassell had been 'begged' by his friend Dawes to search for 'planet X' prior to its official discovery, on coordinates given to Dawes by G.B. Airy. Lassell was unable to search, however, because of a sprained ankle and a housemaid who subsequently destroyed the figures giving the calculated position while tidying up. The legend was analysed and effectively laid to rest by Robert W. Smith in 'William Lassell and the discovery of Neptune', *JHA* xiv (1983) 30-3. Neither Smith nor I found any evidence to support the story in either Lassell's or Dawes's papers. From my own work on Airy, moreover, I have found no evidence to suggest that the Astronomer Royal gave sufficient credence to Adams's calculations to prompt any of his colleagues to launch an urgent search. In Lassell Mss. 9:7, moreover, when Lassell recorded his early observations of Neptune on 3 and 10 October 1846, he simply called it 'Le Verrier' and not 'Adams', and expressed no regret that the kudos for the discovery could have gone to him, and not to the Berlin astronomers, if only he had written to Dawes or Airy for a set of replacement figures for the planet's place.

114. The first sighting of Neptune in England was by John Russell Hind at the South Villa Observatory, on 30 September 1846: see Robert Main to Challis, Cambridge University Library, 'Neptune File', item 14. See also my 'Private research ...', *JHA* 1988 [n. 113] 131. On 2 October 1846, Lassell noted down the now publicly announced coordinates for 'Le Verrier's Planet' in his notebook, Lassell Mss. 9:7, but did not cite an observation of the satellite. Though 10 October 1846 is the generally accepted date for Lassell's discovery of Neptune's satellite Triton (Obituary *MNRAS* 1880 [n. 2] 190), a note for 3 October in Lassell Mss. 9:7 implies that he saw 'the planet' (Neptune) on the night of the 2nd. A very faint pencil note, probably jotted down at the telescope, for 10 October 1846 in Lassell Mss. 9:7, is the first official record for 'Le Verrier': 'I see a satellite or most suspicious looking star ...'

115. Lassell to Struve, 26 November 1847, Lassell Mss. 8:2, fol. 28.

XVII

116. I have not been able to find these observations from Bond amongst Lassell's papers, but they are referred to in the letter to Struve, 26 Nov. 1847, ibid.

117. Lassell to Augustus de Morgan, 8 February 1848, Lassell Mss. 8:2, fol. 32. The micrometer referred to here may, or may not have been, a newly-commissioned instrument from Merz, for on 10 October 1846, when he first saw Neptune's satellite for certain, he already had a Merz micrometer: 'On this night I first used Merz's prism of 2 inches aperture', Lassell Mss. 9:7.

118. The 'ring' of Neptune seems to have impressed Lassell right from his first observation of the planet. On 3 October 1846, which was *probably* the first time on which he recorded seeing the planet in his notebook: Lassell Mss. 9:7, 'I observed the planet last night the 2nd & suspected a ring & could not verify it. I showed the planet to all my family & certainly tonight have the impression of a ring thus [sketch] ♂ .'

119. *MNRAS* vii and viii (1847-8) contain several references to Neptune's ring. For a good modern study, see Robert W. Smith and Richard Baum, 'William Lassell and the ring of Neptune: a case study in instrument failure', *JHA* xv (1984) 1-17. Also Richard Baum, 'The phantom ring of Neptune', *JBAA* 99, 2 April (1989) 77.

120. Lassell Mss. 9:8, 29 September 1847. Lassell also got his family and friends to look at Neptune and make independent sketches of the supposed ring. As early as 3 October 1846, he was showing it to his family (n. 118, op. cit.), while on 10 November 1846, he showed it to Messrs. Hartnup, Lee, Alfred King, and either Mr. S. [or else Mrs.] Kearsley, and got them to make drawings: Lassell Mss. 9:7, 10 Nov. 1846. The following night, 11 November, he got his daughter Maria to do the same, as well as verifying the position of his newly discovered satellite. At N.P.D. 105°3', the planet, ring and satellite appeared thus ' •♂ '.

121. Quite dramatically, after I had already completed and submitted the text for this article, and was typing the last of the footnotes, the U.S. spacecraft Voyager 2 made its historic flypast of Neptune, approaching the planet within a few thousand miles. In the *Daily Telegraph* newspaper, 22-23 August 1989, recently released data from the Voyager 2 Control Station, the Jet Propulsion Laboratory in Pasadena, California, announced that Neptune does indeed possess a ring. Though at the time of writing this note the news had only been available for a few hours and little detail was available, it is interesting that the idea of a Neptunian ring, after 137 years of oblivion, has re-entered scientific currency. I should emphasise however that I am in no way claiming that the ring discovered by Voyager is in any way connected with the object reported by Lassell. Better telescopes than his [370] 24-inch have failed to detect anything over the intervening years and Lassell's ring may turn out to be a chance coincidence like Swift's prediction of the moons of Mars in *Gulliver's Travels*, or a genuine instrumental mistake such as Schiaparelli's Martian canals. On the other hand it will be interesting to see if parallels are found.

122. Lassell, 'Observations of the Nebula of Orion, made at Valletta [sic] with the twenty-foot Equatoreal', *MRAS* xxiii (1854) 53-7: 55.

123. J.R. Hind was taken on by Airy at Greenwich as an Assistant in the Magnetic Department in October 1841, and left the establishment to go to Mr. Bishop's South Villa Observatory in the summer of 1844: see Mss. RGO 6, 1/368; 1/407-8. I have

looked at the career of Hind and many other 'employed' astronomers of the period as part of my forthcoming biography of G.B. Airy. During the first 75 years of the nineteenth century there were some eleven public and university observatories in the British Isles, along with some twenty-five that were privately owned: see article 'Observatory' in *Encyclopaedia Britannica*, 9th ed. (1884) 711. When the Liverpool Observatory was equipped with a new 8½-inch Merz refractor in 1848, Lassell went to inspect it with Dawes: Lassell Mss. 9:9, 27 May 1848.

124. *D.N.B.* Huggins.
125. *D.N.B.* South.
126. See article 'Johannes Hevelius' by John North, *Dictionary of Scientific Biography* (Scribners, New York, 1981).
127. This was Flamsteed's confession to von Uffenbach in 1710 on the occasion of the latter's visit to the Royal Observatory: see Z.C. von Uffenbach, *Merkwürdige Reisen durch Niedersachsen* (Ulm 1753, translated by W.H. Quarrell and Margaret Mare as *London in 1710* (1934) 22).
128. See n. 92.
129. Dreyer's 'Index' to the Lassell papers is written on a double sheet of foolscap and currently kept in Lassell Mss. No. 1.

Reprinted from Vistas in Astronomy *32 (1988), pp. 341-70. Numbers in square brackets indicate original pagination.*

INDEX

Aberration of light: X 6, 8
Académie Royale des Sciences: IV 441;
   X 7-8; XIV 135; XV 78
Accuracy, Airy: XVI 323-8
  angular instruments: II 133-7
  astrolabes: I 473-88
  dividing engine: XI 424
  Gassendi: IX 103-6
  Graham: X 3-8
  Herschel, W.: XII 410*ff*
  Horrocks: V 138*ff*
  increase in, at Greenwich: VIII 141-56
  Lassell: XVII 367
  marine sextant: XI 419*ff*
  quadrants: VII 457-71
  Tycho Brahe: III 71*ff*
  Verbiest: IV 442
Acosta, Josephus: VI 13
Adams, J.C.: XIV 121-39; XVI 322;
   XVII 369
  character: XIV 133-4, 138
  mistaken by Airy: XIV 134
  radius vector: XIV 126
  visits to Airy: XIV 125-6
Aiken, J.: XI 420, 429
Air pump: IV 440; XII 400
Airy, George Biddell: XII 412; XIII 2-4,
   8; XIV 121-39; XV 70-8; XVI 321-8;
   XVII 342, 346, 361-2, 365, 369-70
  daily observation: XIV 127
  doubts about Adams: XIV 133-4, 138
  engineer: XIV 123-4
  fastidiousness: XIV 131-2
  health: XIV 129
  'Journal': XIV 123-6, 129*ff*, 135*ff*
  knighthood: XIV 136
  pendulum work: XIV 139
  personal circumstances: XVI 322
  professionalism: XVI 325-6
  recreations: XVI 326
  salary: XIV 122, 136
Airy, Richarda: XIV 125-6
Airy, Wilfred: XVII 365
Allen, Elias: V 340; VII 458, 460-3, 466,
   468, 470; IX 107

Almanack: VI 4-6, 12-13
'Amateur Astronomer' as noble title:
   XVII 347, 365
Antikythera instrument: I 474
Apian, Petrus: V 354
Aries, First Point of: III 72, 74
Arkwright, Richard: XI 419*ff*, 425
Armagh Observatory: XII 423, 426
Armillary sphere, Tycho Brahe's: III 72*ff*
  Verbiest's: IV 420-43
Arsenius, Regnerus: I 476, 482, 488
Ashmolean manuscripts: VI 1, 12
Aske, Henry: X 3
Aspinwall, Mary: V 334
Assistant astronomers: XIV 122-9;
   XV 76-8; XVI 321-8; XVII 342, 348,
   362, 369-70
Astrea, asteroid: XIV 123, 137
Astrolabes, analysis of: I 473-88; V 341
Astrology: VI 1-14
  Shakerley's scepticism of: VI 10
Astrometer: XIII 15*ff*
Astronomical radius: V 341*ff*, 354;
   IX 103, 109-16

Babbage, Charles: XIII 3-4, 19
Bailey, John E.: V 350-7
Baily, Francis: V 351*ff*; VIII 152-3, 155
Banks, Joseph: XII 403
Bath Philosophical Society: XII 405, 416
Baum, Richard: XVII 369
Bennett, J.A.: XI 420, 426; XII 416;
   XVI 328
Berge, Matthew: XI 429
Berlin Observatory: XIV 129-31
Bessel, W.F.: VIII 155; XII 412; XIII 5
Big Ben clock: XIV 124
Bion, Nicholas: VIII 154
Bird, John: II 134-6; VII 468, 470;
   VIII 150, 152, 154-6; X 5-8; XI 418-
   19, 421, 423-4, 428; XII 403, 407;
   XVI 323
Birr Castle: XVII 346
Bishop, George: XVII 361, 370
Bode's Law: XIV 126-8, 138